AF179577

Library of Congress Cataloging-in-Publishing Data

Módis, László.
Organization of the extracellular matrix : polarization
microscopic approach / author. László Módis.
p. cm.
Includes bibliographical references.
Includes index.
ISBN 0-8493-5786-1
1. Extracellular matrix — Microscopy. 2. Polarization microscope.
I. Title.
[DNLM: 1. Extracellular Matrix — ultra structure. 2. Microscopy,
Polarization. QU 105 M692o]
QM563.M64 1990
611'.018—dc20
DNLM/DLC
for Library of Congress 90-2588
CIP

This book represents information obtained from authentic and highly regarded sources. Reprinted material is quoted with permission, and sources are indicated. A wide variety of references are listed. Every reasonable effort has been made to give reliable data and information, but the author and the publisher cannot assume responsibility for the validity of all materials or for the consequences of their use.

All rights reserved. This book, or any parts thereof, may not be reproduced in any form without written consent from the publisher.

Direct all inquiries to CRC Press, Inc., 2000 Corporate Blvd., N.W., Boca Raton, Florida 33431.

© 1991 by CRC Press, Inc.

International Standard Book Number 0-8493-5786-1

Library of Congress Card Number 90-2588
Printed in the United States

PREFACE

Growing appreciation of the role played by connective tissues in various normal and pathological processes has led to the rapid development of research in this field. The extracellular matrix dominates the connective tissues, and is responsible for the basic functions of these tissues.

Furthermore, it is known that the extracellular matrix is also present in all tissues, and that it influences the differentiation, morphology, and functions of the cells. In light of this fact, matrix research has been attracting the intense interests of cell biologists, too.

Owing to the efforts of biochemists, much progress has been made in the exploration of the molecular anatomy of the matrix constituents. The contribution of morphologists is more modest due to the limitations of their methodology. Their role in contemporary matrix research may be compared to that played by Sancho Panza in Cervantes' classic tale, but we know that his thoughtful and dedicated service was invaluable to his master, and he proved to be a faithful and, at the same time, a helpful companion.

This volume is devoted to a peculiar morphological approach to matrix research. Polarization microscopy had been the leading technique of utrastructure research in the pre-electron microscopic era until electron microscopy provided the means of direct visualization of the ultrastructure, and the indirect techniques like polarization microscopy became more and more neglected.

The potentialities of polarization microscopy, however, were greatly enhanced when it became possible to combine them with relevant histochemical methods. These histochemical reactions, today called topo-optical reactions after the custom of G. Romhányi, a pioneer of this field, can selectively increase the optical anisotropy of a tissue constituent if it is spatially oriented. This modern polarization microscopy proved to be a powerful tool, especially in the exploration of oriented submicroscopic structures of different glycoconjugates like glycosaminoglycans in the matrix of different embryonic and differentiated animal tissues.

Despite the fact that these methods are being used more and more, we cannot call them well known. Therefore, when I was asked to write a volume, I decided that this contribution would be an attempt to display the state of the art of this research technique. Its advantages and limitations will be illustrated predominantly by the results achieved in our laboratory. Our work shows that the old assumption — believed even today by many researchers — that the matrix is an amorphous system, is no longer tenable.

It is hoped that the merits of this volume will overshadow its weaknesses. The author welcomes any constructive criticism and comments. *"Artes discuntur peccando."*

THE AUTHOR

László Módis, M.D., Ph.D., is Associate Professor of Anatomy, Histology, and Embryology at the University School of Medicine, Debrecen, Hungary.

Dr. Módis obtained his M.D. degree *(summa cum laude)* in 1963 and his Ph.D. in 1975, and served as an Assistant Professor from 1965 at the same University. He became Associate Professor in 1980. As a post-doctoral fellow, he spent 2 years in the Department of Histology and General Embryology, University of Fribourg, Switzerland (1969—1971) thanks to a grant from Swiss National Foundation for Scientific Research. More recently, as a visiting scientist, he spent 1 year (1989—1990) in the Department of Pathology and Cell Biology, Thomas Jefferson University, Philadelphia, supported by a NIH grant.

He is currently the elected Secretary of the Histochemical Section of the Hungarian Society of Anatomists, Histologists, and Embryologists, and member of the boards of the Hungarian Society of Anatomists, Histologists, and Embryologists, the Hungarian Connective Research Club, and the Osteoarthrological Society of Hungarian Orthopaedic Surgeons. He was the managing chairman of two international scientific meetings held in Debrecen (12th Symposium of the European Society of Osteoarthrology: Biology of the Cartilage, 1983; International Symposium on Histochemistry, 1985).

Among other awards, he has received the Lenhossék Medal from the Hungarian Society of Anatomists, Histologists, and Embryologists, and the title "Professor of the Year, 1981" from the medical students of his University.

Dr. Módis has published 90 papers and 1 book. He has presented over 30 invited lectures at International Meetings and over 20 invited seminars at different foreign Universities. His current major research interests include the differentiation of the structure of the extracellular matrix and cell-matrix interactions, predominantly in cartilage tissue.

TABLE OF CONTENTS

Dedicated to GEORGE ROMHÁNYI,
Emeritus Professor of Pathology, Pécs, Hungary,
one of the pioneers of polarization microscopy

Chapter 1

A SHORT OVERVIEW OF MAJOR CONSTITUENTS OF THE EXTRACELLULAR MATRIX

TABLE OF CONTENTS

I. INTRODUCTION

The extracellular matrix (ECM) constitutes a significant part of the multicellular animal organism. Except blood, there are no tissues without an ECM. In connective tissue the matrix is abundant, while its bulk is much less in other tissues where sometimes special techniques are needed for its detection. The matrix molecules are produced and degraded by cells.

After deposition the ECM, or some of its constituents, influence a number of functions of contacting cells. The cell-matrix interactions are generally mediated by specific receptors in the cell membrane. The ECM can control the metabolism, growth, differentiation, migration, and adhesion of cells, the cytoskeletal organization and cell shape, and the organization of the tissues[1-5] (Figure 1). There are experimental data suggesting that ECM exerts an influence on gene expression via transmembrane proteins and cytoskeletal compartments which are associated with polyribosomes as well as the nuclear matrix.[6] According to this theory, the cytoskeletal association with polyribosomes may affect the rate of protein sythesis, while the matrix-cell membrane-cytoskeleton-nuclear matrix interaction could affect mRNA processing and rates of transcription.

Thus, the ECM is not a simple and passive structural scaffold. Multicellular animals can function only because single cells learned to synthesize, and maintain macromolecular connections between them, and they have become dependent upon the matrix they made.[7] This has been realized in the past decade due to the large amounts of new data accumulated about the biosynthesis, molecular anatomy, functions, and interactions of the ECM constituents. It is now, therefore, generally accepted that the interest of cell biology extends beyond the cell membrane.

There are different major trends in the contemporary ECM research. One of them is — the most classical — the biochemical approach with the aim of a detailed description of the matrix constituents and their interactions. The second and tremendously increasing field is the molecular biology of the ECM.[8] The third trend — which is rapidly developing — is the study of the role of the ECM in diseases.[9-11] A fourth direction is to get new data about the ECM structure in different normal and pathological tissues. In this respect, immunohistochemistry has contributed much to our present knowledge.[12] Other morphological techniques like electron microscopy and polarization microscopy, however, are also indispensable tools of ECM research. As this volume is dedicated to a morphological approach of the ECM organization, it is beyond the scope of the present work to give a detailed description of the ECM constituents. This chapter shortly summarizes some basic informations about the major macromolecular components of the ECM, other details are found in the references given as well as in the relevant chapters of this volume.

II. MAJOR CONSTITUENTS OF THE MATRIX

A. COLLAGEN

Collagen is the most ubiquitous protein in animal kingdom. It is a heterogenous family of proteins, but the members share common features both in primary structure and conformation, e.g., the repeating Gly-X-Y triplet in the polypeptide chain, and the presence of hydroxyproline and hydroxylysine.

Today, 14 collagens are distinguished. They differ genetically, chemically, and immunologically.[13-16] There are 5 of the collagen types which are known to be involved in the formation of crossbanded fibrils (types I, II, III, V, XI), but other collagens (e.g., type IX) may also be associated with the fibrillar surface. The collagen fibrils and fibers represent systems where the degree of the orientation of the macromolecular constituents is the highest in the ECM. Some collagen types are tissue-specific: types II, IX, X and XI are round in cartilage (and also in vitreous body and notochord) (see Chapter 4). Collagen type I molecules

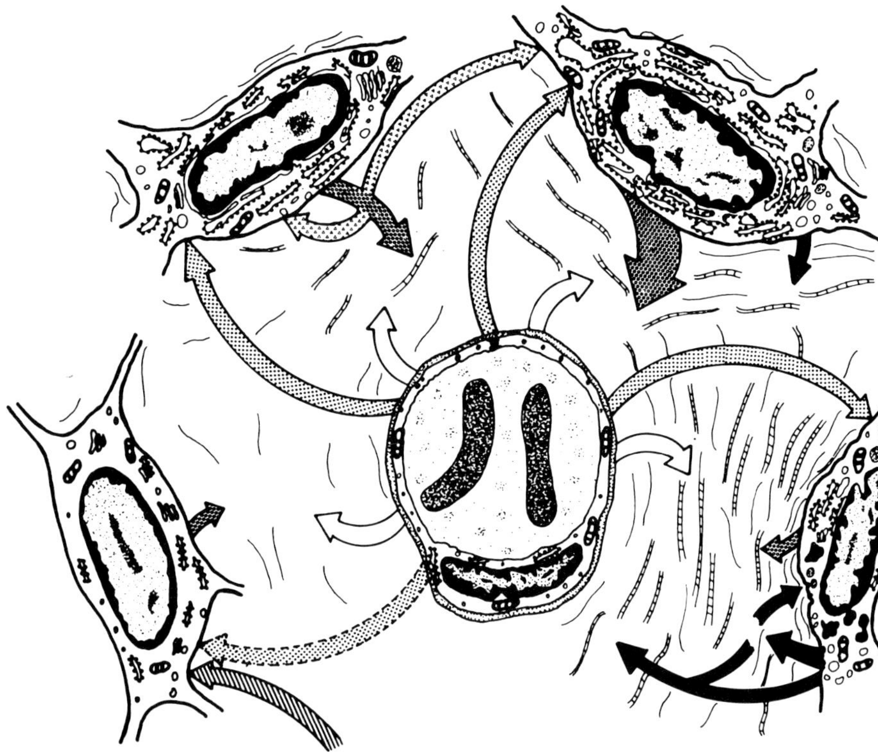

FIGURE 1. A scheme illustrating some selected aspects of cell-matrix interactions. As an example, a portion of a connective tissue is depicted. In the center of the diagram, a blood capillary can be seen. Around the capillary, cells of varying degree of differentiation are shown (from left to right: undifferentiated, differentiating, mature and aging cell). The space between the cells and the capillary is occupied by the extracellular matrix (the fibrillar components are masked by a few fine lines). Empty arrows: anabolic transport; dotted arrows: humoral factors regulating cell functions; arrows filled with circles: secretion of matrix constituents; arrows filled with triangles: effects of matrix on cell functions; dense arrows: matrix degradation and catabolic transport. (From Módis, L., *Acta Biol. Acad. Sci. Hung.*, 29, 197, 1978. With permission.)[4]

form the thick, interstitial collagen fibers (in skin, bone, tendon, etc.), while type III is present in a wide variety of tissues, generally associated with type I collagen, forming thinner fibrils (''reticulin'' fibers of the classical histology). Type IV collagen is specific to basement membranes, they form an orientad micellar network but not fibrils (see Chapter 7). Type V collagen shows a wide distribution. It is generally found in the pericellular matrix in form of small fibrils, and, therefore, it is frequently named as pericellular collagen and considered as a part of the exocytoskeleton.[13,17] Collagen type VI is present in a wide variety of tissues as interstitial collagen. Type VIII collagen is secreted by endothelial cells. Type XII collagen is expressed in embryonic tendon, but its role is still unknown.

Collagen may be involved in different interactions. Collagen molecules of the same type can be associated with each other forming fibrils or micelles (e.g., type I fibrils and type IV micelles). Different collagen types can also be bound with each other (e.g., in collagen fibrils of the hyaline cartilage[18]). Type VII — as a major part of anchoring fibrils — interconnects basement lamina to the underlying connective tissue matrix.[19] Different collagen types may be attached to the cell membrane via collagen receptors or by nectin molecules (fibronectin, chondronectin, laminin, see later). Proteoglycans also interact with collagen, predominantly by electrostatic forces but more specific interactions may also occur.[20] Considering these multisided interactions, collagens can be regarded as key molecules of the extracellular matrix organization.

B. ELASTIN

The elastic behavior of some tissues (lung, artery wall, skin) is mainly due to the presence of elastic fibers in the ECM. The morphological appearance of elastic fibers is simple, the mature fibers appear as homogeneous structures under electron microscope. They have been, therefore, considered to possess an amorphous structure. Polarization microscopy revealed that elastic fibers have submicroscopically oriented constituents.[21] These may correspond to the microfibrils of 10 to 12 nm in diameter recognized on the surface of developing and mature elastic fibers electron microscopically. The composition of the microfibrils is not yet well established because of the difficulties in solubilizing them.[22-24] It seems that microfibrils appear first in the extracellular matrix during the differentiation of the elastic fibers, and subsequently they are embedded in an amorphous component called elastin. The elastin composes about 50% of the mature elastic fibers. It is a rubber-like protein, rich in glycine, proline, and hydrophobic amino acids, and its conformation is like a random coil. Hydrophobic regions of the mobile spiral chains of elastin are held together by desmosine crosslinks. Between these crosslinks, the elastin molecules align when the fiber is stretched and recoil when the force is released.

Due to the technical difficulties encountered in elastic fiber research (e.g., the insolubility of the components), this component of the ECM is still the poor relative compared to collagen, proteoglycan, and glycoprotein biochemistry and morphology.

C. PROTEOGLYCANS

Proteoglycans (PGs) contain a protein core to which sulfated glycosaminoglycans (GAGs) (chondroitin sulfate, dermatan sulfate, keratan sulfate, heparan sulfate, heparin) are bound covalently. The only GAG which is not bound to a protein core is the hyaluronic acid. PGs represent a diverse group of glycoconjugates, with the only common basic feature that all of them contain at least one GAG side chain.[25-27] The core protein is generally linear in shape — except some globular domains — and the extended GAGs are perpendicularly oriented to the protein backbone. Due to these, larger PGs have a shape similar to a bottlebrush (for illustration, see Chapter 4). The GAGs have a helical conformation, and they — due to its high density of anionic charges (sulfates and carboxyls) can easily interact with microcations (Na^+, K^+, Ca^{2+}) and polycations (e.g., collagen).[4,20]

PGs can be classified according to (1) their size; (2) their GAG type; (3) their GAG content; (4) their core protein; (5) their abilities for interactions; (6) their localization (Figure 2). Based on these criteria, about 20 PGs can be distinguished. This structural diversity certainly implies considerable functional heterogeneity in the PG family.

PG molecules are located in different compartments of the ECM. One group is found at the cell surface. The best known example is the heparan sulfate/chondroitin sulfate PG group. These PGs are either integral membrane proteins or they are attached to the cell membrane. They are involved in cell attachment by binding with their GAG chains to the GAG binding sites of matrix proteins such as fibronectin.[27] Another class of PGs is bound to collagen. Besides the electrostatic interactions of GAG side-chains of PGs and collagen, more specific protein-protein interactions also occur.[20,27] The most typical example is the dermatan sulfate/chondroitin sulfate PG (decorin) which has binding sites for collagen on its protein core.[27] In addition to ion and water binding as well as mechanical (space filling and shock-absorber) functions, PGs are shown to be involved in a number of biological processes such as cell differentiation and morphogenesis. The elucidation of their role in pathological processes like neoplastic phenotype expression and tumor progression also attracts more and more attention.[28]

D. NONCOLLAGENOUS PROTEINS

A large number of matrix glycoproteins have been isolated and characterized. Contrary to the PGs, their protein constituents bear only oligosaccharides and not GAGs. This family

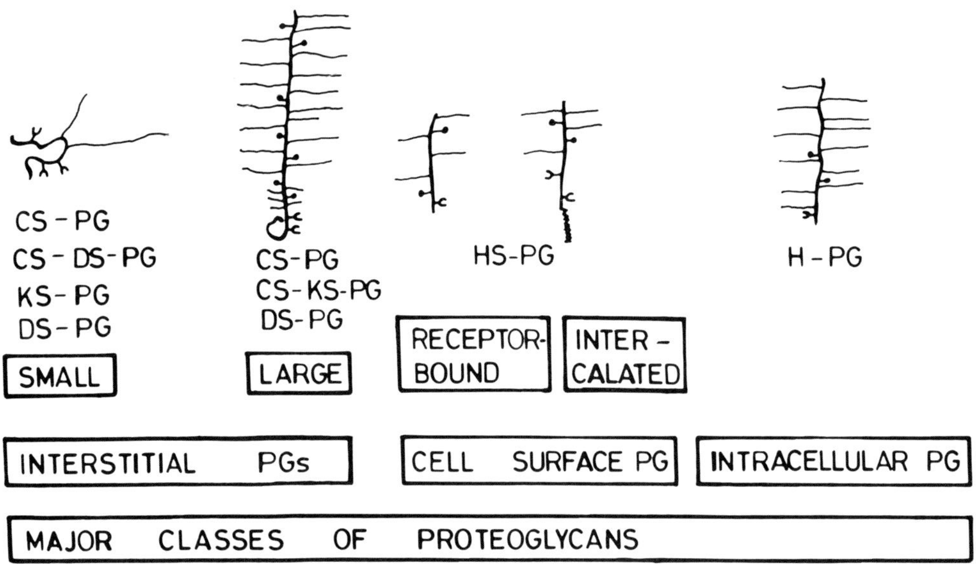

FIGURE 2. Schematic representation of the main PGs. CS: chondroitin sulfate; DS: dermatan sulfate; KS: keratan sulfate; HS: heparan sulfate; H: heparin.

comprises structurally and functionally very heterogenous macromolecules. They are involved in cell attachment and matrix organization.

The major cell-substratum adhesion molecules (fibronectin, laminin) are multifunctional and well-characterized glycoproteins.[1,7,29-33] They interconnect cell surface receptors (integrins) and collagen of the pericellular space. The cell binding domain of these molecules is a tripeptide recognition sequence Arg-Gly-Asp. Fibronectin is primarily synthesized by fibroblasts and endothelial cells and is involved — due to its multiple functional domains — in cell attachment, spreading, migration, differentiation, and matrix organization. Laminin is synthesized by epithelial cells and found in basement membranes. Like fibronectin, it also binds to other ECM components including collagen type IV, heparan sulfate proteoglycan, and nidogen/entactin (see Chapter 7). Chondronectin is supposed to bind collagen type II to the chondrocyte membrane.[34]

These molecules form a bridge between the pericellular matrix constituents (collagen and PG) and the cytoskeleton, because the microfilament compartment of the cytoskeleton is associated via a vinculin + talin complex with the cellular receptors (integrins) of these cell adhesive proteins (Figure 3, see also Chapter 9).

III. RECEPTORS INVOLVED IN THE CELL-MATRIX INTERACTIONS

Different cell-surface receptors have been identified which mediate cell-matrix interactions. The most important member of this receptor group is the integrin family.[35,36] Integrins consist of two subunits, and both of them have a large extracellular domain, a transmembrane, and a short cytoplasmic domain. The extracellular domain recognizes and binds the arg-gly-asp sequence of different matrix proteins including fibronectin, laminin, vitronectin, etc. The cytoplasmic domain interacts with the cytoskeleton via a talin-vinculin complex. Supposedly, this interaction mediates signal transfer from the ECM to the cytoplasm (and through a cytoskeleton-nuclear matric interaction to the nucleus).

Other receptors have also been identified, e.g., for hyaluronic acid[37,38] and collagen.[39,40]

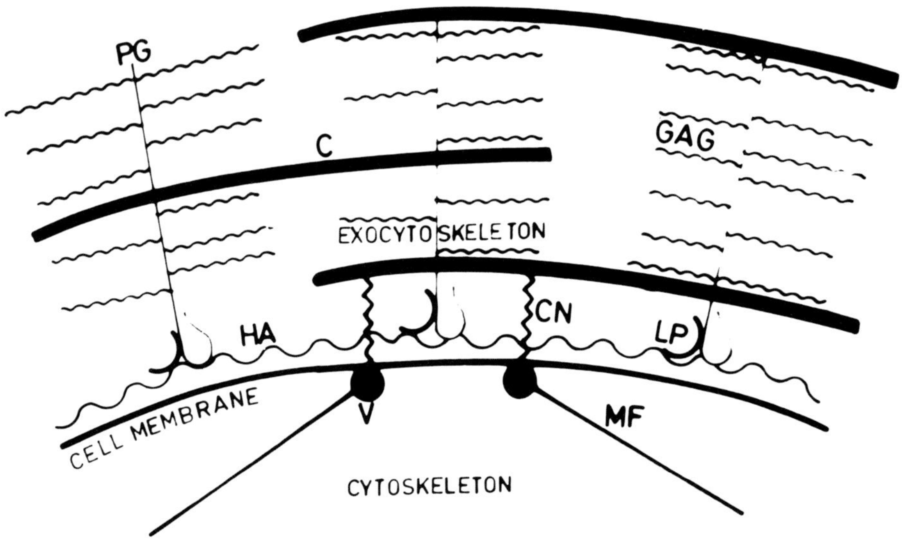

FIGURE 3. An oversimplified scheme showing the interconnection of cytoskeleton with the pericellular matrix (exocytoskeleton) in cartilage. PG: proteoglycan; GAG: glycosaminoglycan; C: collagen; HA: hyaluronic acid; CN: chondronectin; LP: link protein; V: vinculin; MF: microfilament. Receptors for collagen and hyaluronic acid are not shown. Instead of chondronectin, fibronectin binds collagen to integrin receptor of the cell membrane in other tissues.

REFERENCES

1. **Hay, E. D.,** *Cell Biology of Extracellular Matrix,* Plenum, New York, 1981.
2. **Hay, E. D.,** Cell and extracellular matrix: their organization and mutual dependence, *Mod. Cell Biol.,* 2, 509, 1983.
3. **Hay, E. D.,** Extracellular matrix, cell skeletons, and embryonic development, *Am. J. Med. Genet.,* 34, 14, 1989.
4. **Módis, L.,** The molecular structure of the interfibrillar matrix in connective tissue, *Acta Biol. Acad. Sci. Hung.,* 29, 197, 1978.
5. **Trelstad, R. E., Ed.,** *The Role of Extracellular Matrix in Development,* Alan R. Liss, New York, 1984.
6. **Bissell, M. J. and Barcellos-Hoff, M. H.,** The influence of extracellular matrix on gene expression: is structure the message, *J. Cell. Sci. Suppl.,* 8, 327, 1987.
7. **Piez, K. A. and Reddi, A. H., Eds.,** *Extracellular Matrix Biochemistry,* Elsevier, New York, 1984.
8. **Olsen, B. R.,** The next frontier: molecular biology of extracellular matrix, *Conn. Tissue Res.,* 23, 115, 1989.
9. **Prockop, D. J. and Kivirikko, K. I.,** Heritable diseases of collagen, *N. Engl. J. Med.,* 311, 376, 1984.
10. **Uitto, J. and Perejda, A. J., Eds.,** *Connective Tissue Disease: Molecular Pathology of the Extracellular Matrix,* Marcel Dekker, New York, 1987.
11. **Rojkind, M., Ed.,** *Connective Tissue in Health and Disease,* CRC Press, Boca Raton, 1989.
12. **Furthmeyer, H., Ed.,** *Immunohistochemistry of the Extracellular Matrix,* Vols. I and II, CRC Press, Boca Raton, FL, 1982.
13. **Gay, S. and Miller, E. J.,** What is collagen, what is not, *Ultrastructural Pathol.,* 4, 365, 1983.
14. **Martin, G. R., Timpl, R., Muller, P. K., and Kuhn, K.,** The genetically distinct collagens, *Trends in Biochem. Sci.,* 10, 285, 1985.
15. **Piez, K. A.,** Collagen types, *J. Biol. Chem.,* S10B, 101, 1986.
16. **Nimni, M., Ed.,** *Collagen: Biochemistry, Biomechanics, Biotechnology,* Vols. I, II, III, IV, CRC Press, Boca Raton, 1988.
17. **Gay, S., Martinez-Hernandez, A., Rhodes, R. K., and Miller, E. J.,** The collagenous exocytoskeleton of smooth muscle cells, *Coll. Relat. Res.,* 1, 377, 1981.

18. **Mayne, R.,** Cartilage collagens. What is their function, and are they involved in articular disease, *Arthritis Rheum.*, 32, 241, 1989.
19. **Sakai, L. Y., Keene, D. R., Morris, N. P., and Burgeson, R. E.,** Type VII collagen is a major structural component of anchoring fibrils, *J. Cell. Biol.*, 103, 1577, 1986.
20. **Scott, J. E.,** Proteoglycan-fibrillar collagen interactions, *Biochem. J.*, 252, 313, 1988.
21. **Romhányi, G.,** Submicroscopic structure of elastic fibers as observed in the polarization microscope, *Nature,* 182, 930, 1958.
22. **Sandberg, L. B., Gray, W. R., and Franzblau, C., Eds.,** *Elastin and Elastic Tissue,* Plenum Press, New York, 1976.
23. **Sandberg, L. B., Soskel, N. T., and Leslie, J. G.,** Elastin structure, biosynthesis, and relation to disease states, *N. Engl. J. Med.*, 304, 566, 1981.
24. **Robert, L. and Hornebeck, W., Eds.,** *Elastin and Elastases,* Vols. I and II, CRC Press, Boca Raton, FL, 1989.
25. **Lennarz, W. J., Ed.,** *The Biochemistry of Glycoproteins and Proteoglycans,* Plenum Press, New York, 1980.
26. **Poole, A. R.,** Proteoglycans in health and disease: structures and functions, *Biochem. J.*, 236, 1, 1986.
27. **Ruoslahti, E.,** Structure and biology of proteoglycans, *Ann. Rev. Cell Biol.*, 4, 229, 1988.
28. **Iozzo, R. V.,** Cell surface heparan sulfate proteoglycan and the neoplastic phenotype, *J. Cell. Biochem.*, 37, 61, 1988.
29. **Timpl, R., Rohde, H., Robey, P. G., Rennard, S. I., Foidart, J., and Martin, G. R.,** Laminin — a glycoprotein from basement membranes, *J. Biol. Chem.*, 254, 9933, 1979.
30. **Hynes, R. O. and Yamada, K. M.,** Fibronectins: multifunctional modular glycoproteins, *J. Cell Biol.*, 95, 369, 1982.
31. **Yamada, K. M., Akiyama, S. K., Hasegawa, T., Hasegawa, E., Humphries, M. J., Kennedy, D. W., Nagata, K., Urushihara, H., Olden, K., and Chen, W. T.,** Recent advances in research on fibronectin and other cell attachment proteins, *J. Cell. Biochem.*, 28, 79, 1985.
32. **Von der Mark, K. and Kuhl, U.,** Laminin and its receptor, *Biochim. Biophys. Acta,* 823, 147, 1985.
33. **Ruoslahti, E. and Pierschbacher, M. D.,** Arg-Gly-Asp: a versatile cell recognition signal, *Cell,* 4, 517, 1986.
34. **Hewitt, A. T., Kleinman, H. K., Pennypacker, J. P., and Martin, G. R.,** Identification of an attachment factor for chondrocytes, *Proc. Natl. Acad. Sci. U.S.A.*, 77, 385, 1980.
35. **Buck, C. A. and Horwitz, A. F.,** Cell surface receptors for extracellular matrix molecules, *Ann. Rev. Cell Biol.*, 3, 179, 1987.
36. **Hynes, R. O.,** Integrins: a family of cell surface receptors, *Cell,* 48, 549, 1987.
37. **Lacy, B. E. and Underhill, C. B.,** The hyaluronate receptor is associated with actin filaments, *J. Cell Biol.*, 105, 1395, 1987.
38. **Perides, G., Lane, W. S., Andrews, D., Dahl, D., and Bignami, A.,** Isolation and partial characterization of a glial hyaluronate-binding protein, *J. Biol. Chem.*, 264, 598, 1989.
39. **Möllenhauer, J. and von der Mark, K.,** Isolation and characterization of a collagen-binding glycoprotein from chondrocyte membranes, *EMBO J.*, 2, 45, 1983.
40. **Pfaffle, M., Ruggiero, F., Hofmann, H., Fernandez, M. P., Selmin, O., Yamada, Y., Garrone, R., and von der Mark, K.,** Biosynthesis, secretion and extracellular localization of anchorin CII, a collagen-binding protein of the calpactin family, *EMBO J.*, 7, 2335, 1988.

Chapter 2

PHYSICAL BACKGROUNDS OF POLARIZATION MICROSCOPY

TABLE OF CONTENTS

I. INTRODUCTION

In this Chapter, an attempt is made to summarize the theoretical basis and the practical aspects of polarization microscopy. For more details, other works should also be consulted.[1-31]

First, we describe the most important optical principles. This short summary is supplemented with a glossary of selected terms used in polarization microscopy of biological specimens. Finally, some operations with the polarization microscope are also given for those who want to try this technique in ultrastructure research of biological specimens.

II. A SHORT SUMMARY OF OPTICAL PRINCIPLES

The oscillating electric field of linearly polarized light — generated by a polarizer — interacts with the valence electrons of anisotropic, birefringent, double refracting) material. The anisotropic material — due to its submicroscopically oriented structure — is characterized by two (in some crystals more than two) refractive indices (ordinary and extraordinary) perpendicular to each other. The state of polarization of light is influenced by this material. In the birefringent structure, the linearly polarized light splits into two components, the ordinary and extraordinary rays, which oscillate in two directions at right angles to each other and each at 45° with respect to the vibration plane of the incident polarized light and pass through the material at different velocities. Both beams will be slowed down inside the structure, but one (oscillating in the direction of the extraordinary refractive index) will be slowed down more than the other. There will be, therefore, a phase difference between the two beams. From the superposition of the two linearly polarized vibrations with a certain phase difference a resultant vibration is constructed after emerging from the birefringent specimen. The resultant vibration is generally an elliptical polarized light (may also be linear and circular), and its component parallel to the vibration plane of the analyzer — which is perpendicular to that of the polarizer — is transmitted by the analyzer. This light is observed in the eyepiece and used for qualitative and quantitative analyses of the birefringence.

The analysis of the ultrastructure at a biological specimen in polarized light is not limited to examination only between crossed polaroids. Some structures exhibit a direction dependent absorption of polarized light called dichroism. It can be quantified by cytophotometric measurements.

From qualitative and quantitative analysis of birefringence and dichroism, information can be indirectly obtained about the spatial orientation of asymmetrical submicroscopic constituents (molecules, micelles) of the specimen. Birefringent systems are extremely common in biology but some of them can only be recognized after special histochemical reactions which enhance the optical anisotropy of a structure. Classical 65polarization microscopy comprises techniques used mainly for unstained structures. Modern polarization microscopy is supplemented by a number of histochemical reactions which make this technique more powerful. The primary aim of this volume is to show the usefulness of the combination of histochemistry and polarization microscopy in connective tissue research.

III. GLOSSARY OF SELECTED TERMS FOR POLARIZATION MICROSCOPY

Analyzer — A polaroid containing highly oriented dichroic particles which transmit light with only one plane of vibration. It is inserted between the specimen and eyepiece in a polarization microscope. The plane of vibration of the analyzer is crossed to the plane of vibration of the polarizer when studying birefringence of a specimen. In this position, the light coming from the polarizer is extinguished. The microscopic field appears completely dark, while birefringent specimens appear bright. The analyzer is removed when dichroism is observed. There are fixed and rotatable analyzers in use. Rotating analyzers are indispensable when measuring the retardation value of a specimen using a fixed quarter-wave plate.

Anisotropic material — A material whose properties are different in different directions. *Optically anisotropic materials* influence the state of polarization of the illuminating light because their structure is characterized by anisometric (asymmetric) submicroscopic particles oriented in two or three preferential directions. As the electronic resonators of this material are oriented in preferred directions, the material transmits light with different velocities in these directions, i.e., it shows two or three refractive indices. In biological specimens, there are generally two — ordinary (n_o) and extraordinary (n_e) — refractive indices and they are perpendicular to each other. Therefore, such materials are called birefringent (double refracting). The geometric symbol is the indicatrix or its cross section called index ellipse. Its long axis represents the n_e and the short axis the n_o.

Anomalous dispersion of birefringence — Variation of birefringence ($n_e - n_o$) with wavelength of the illuminating light. It is due to variation of the refractive index with wavelength (dispersion of the refractive index). The dispersion is normal if the two refractive indices (n_e and n_o) parallelly change, consequently the birefringence remains constant at different wavelengths. The dispersion is anomalous (anormal) if the two refractive indices change differently as a function of the wavelength of illuminating light. The ordinary refractive index (n_o) slightly changes (generally decreases towards the longer wavelengths), while the extraordinary refractive index (n_e) varies considerably with wavelength because more light is absorbed in the direction of n_e than in that of n_o. In this case, the birefringence changes as a function of the wavelength. If the dispersion curves of the two refractive indices cross each other at a certain wavelength (generally near to the absorption maximum of colored structures), vivid anomalous polarization colors appear in the polarization microscope and the sign of birefringence is different before and after the crossing point.

Anomalous dispersion of birefringence appears in highly oriented light-absorbing structures. It can be investigated by plotting curves of anomalous dispersion of retardation in the following way. Retardation values of a birefringent (and generally stained) material are measured with compensator plates at different wavelengths. Retardation values (Γ) are plotted against the wavelength (λ). As for

$$\Gamma = d(n_e - n_o)$$

where d = geometric thickness of the structure measured which was constant during the measurements, the curve obtained shows the anomalous dispersion of birefringence. The shape of the curve reflects the degree of orientation of the structure: the steeper the curve, the greater the degree of orientation. Examples are shown in Chapter 3.

The anomalous optical rotatory dispersion (ORD) is the dependence of rotary birefringence on wavelength. If the sign of birefringence is also changed (near to the absorption maximum) this phenomenon is called the Cotton effect. If the sign of birefringence is positive at longer wavelengths, the Cotton effect is positive, and in a reversed situation it is negative. Intrinsic Cotton effect is a property of chiral molecules, while extrinsic Cotton effect is induced by reagents (e.g., dye molecules) bound to the molecule under investigation. Curves of anomalous dispersion of birefringence and ORD curves may be similar in the same material.

ORD is a stereochemical method to elucidate molecular conformations. Using this technique, the helical conformation of different glycosaminoglycans has been described.[32-38] X-ray diffraction studies have also revealed the helical conformation of these macromolecules.[39-43]

ORD of cells and tissues can also be studied by a microspectropolarimeter and it is suggested that conformational changes of intracellular biopolymers may be detected with this method.[44,45]

Anomalous polarization color — This is produced by the orientation of light — absorbing bonds (e.g., dye molecules), and it is detected by examining the specimen between crossed polaroids and varying the compensation by rotating the compensator plate. A given amount of compensation will only accomodate certain wavelengths. Light of other wavelengths is uncompensated and the specimen appears colored. For example: it is known that congo red stains amyloid and enhances its optical anisotropy very strongly. Congo red molecules absorb green light, therefore, the dye molecules deposited on amyloid fibrils in an oriented manner are strongly birefringent in green but weak in blue and red light. Thus the polarization color of the amyloid stained with congo red is green. After additive compensation in white illuminating light, yellow anomalous polarization color appears, while after subtractive compensation greenish blue anomalous color can be observed. Another example concerning toluidine blue will be shown in Chapter 3. This dye stains polyanionic structures in a metachromatic red color. Due to the ordered (generally axiperpendicular) deposition of the dye molecules, toluidine blue enhances the birefringence of the structure stained. The metachromatic toluidine blue aggregate absorbs green light, therefore it is strongly birefringent in green. Its polarization color is, therefore, green. After additive compensation, light of longer wavelength (orange), and after subtractive compensation, light of shorter wavelength (blue) appears in the structure. Knowledge of anomalous polarization colors helps greatly if we want to describe the orientation patterns of dyes bound to oriented biological structures. If the geometry of dye-binding to the structure is known, the ultrastructure of the biological specimen can be inferred from the dye orientation pattern.

Artifacts in polarization microscopy — Two kinds of artifacts can be distinguished. First, there are foreign materials exhibiting optical anisotropy (paraffin residues, dust, talc, textile and paper fibers, dye precipitates, other birefringent dirts) whose presence in the microscopic specimen disturb the polarization microscopic analysis or microphotography.

For histologists, the presence of paraffin is especially important. Due to the thermic orientation of the molecules, cooled (solid) paraffin is highly birefringent and melted paraffin is isotropic. Insufficiently deparaffinized sections contain small birefringent particles — especially in cell nuclei[46-48] — which may cause an incorrect interpretation. Paraffin residues can be recognized by a simple probe: the specimen is warmed up to 55 to 60°C and immediately checked in polarization microscope. If the birefringence disappears after warming and gradually reappears during cooling on the microscope stage, the birefringence is caused by paraffin. The complete removal of paraffin from the embedded material is, therefore, essential before polarization microscopic investigation. Generally, a prolonged deparaffinization in xylene at 37° or 56°C (4 to 24 h, depending on the type of paraffin) or in an 1:1 mixture of methanol-chloroform at room temperature (2 to 12 h) must be carried out.

It is of the same importance to keep any kind of microscopic preparations as well as the optics of the microscope free of dust.

The second group of polarization microscopic artifacts comprises the cases of pseudoisotropy (see under pseudoisotropic materials).

Biaxial crystals — These have three different refractive indices: n_α (lowest), n_β (middle) and n_γ (highest). Its geometric symbol is a triaxial (general) ellipsoid with three unequal principal axes. Such crystals have two optical axes (i.e., there are two positions where the material seems to be isotropic). Both optical axes are in the plane of the smallest and the largest refractive index. Biaxial materials are rarely met in biological structures.

Birefringence — Birefringence or double refraction is an intrinsic property of a material, and is a modification of polarized light by a structurally and optically anisotropic material.

It can be observed between crossed polars if the plane of vibration of the polarizer and the reference axis of anisotropic material are at 45° to each other. Birefringence (Δn) can be described by the following equation:

$$\Delta n = n_e - n_o$$

where n_e and n_o are the extraordinary and ordinary refractive indices, respectively. The linearly polarized illuminating light splits into two components which oscillate in the two directions of the refractive indices.

Major categories of birefringence are as follows: intrinsic (crystalline), form (textural), strain (tension or accidental birefringence, accompanied by a photoelastic effect), rotary (optical rotation), flow, and electric birefringence. In many cases, these categories may be combined. Filamentous molecules (e.g., hyaluronic acid)[49,50] exhibit flow birefringence. Electric birefringence (Kerr effect) is due to the orientation of the solute molecules as the solution is subjected to a pulsed electric field.[51] This is a sensitive indicator of molecular conformation changes and associations. A prolate ellipsoid shape and a 1.8 ± 0.1 μm length of the proteoglycan aggregate, characteristic for the cartilage matrix, were inferred from electric birefringence studies.[52]

Circular polarized light — If the electric vector of a circularly polarized light beam is projected on a plane normal to the axis of propagation, the projection of the tip is a circle. It is produced by birefringent materials showing $\lambda/4$ and $3/4\lambda$ retardation (Figure 1). Circular polarized light is sometimes used as illuminating light in the polarization microscope. Its advantage over the use of linear polarized light is that birefringence of the structure can be observed in any direction including the vibration planes of the polarizer and analyzer where no birefringence can be seen using linear polarized illuminating light. This technique is useful in investigating both spherical (e.g., cell membranes) and fibrous structures [53,54] (Figures 2,3). In practice, two $\lambda/4$ compensator plates are necessary for microscopic investigations in circular polarized light. One plate is inserted between the polarizer and specimen; the other plate is placed between the specimen and eyepiece. The n_γ directions of the plates should be crossed with each other, and the n_γ direction of the lower plate be at 45° with respect to the plane of vibration of the polarizer.

Compensation — Operating with a compensator plate of the polarization microscope, we can establish the sign and retardation value of birefringence a structure. It is carried out by rotating the compensator (or in the case of fixed compensator plate, by rotating the analyzer) until the birefringence of the spencimen is extinguished (compensated). When the higher refractive indices of the compensator plate and specimen are crossed, they will subtract (subtractive position). This is the generally used position for measurements. When the direction of higher refractive index of the compensator plate runs parallel to the direction of higher refractive index in the specimen, their effects will be added (additive position). Unstained or colorless specimens are generally compensated in white or green light, while colored (stained) biological specimens are compensated in appropriate monochromatic light.

Compensators — Birefringent plates of known retardation are used to determine the sign and retardation of birefringence. The direction of their slow ray (n_γ) is usually marked with γ. Compensator plates are inserted between the object and analyzer.

There are different types of compensator plates. The wedge compensators are used in cystallography. The tilting compensators (Berek and Ehringhaus plates) of great measuring range (retardation of several thousand nm) are not accurate enough for the measurement of biological structures. Compensators of fixed azimuth generally belong to the standard equipment of a polarization microscope. One of them is the "red I order plate" (its retardation value is one full wavelength), and it is used to estimate the relative sign of birefringence of colorless structures of greater retardation.

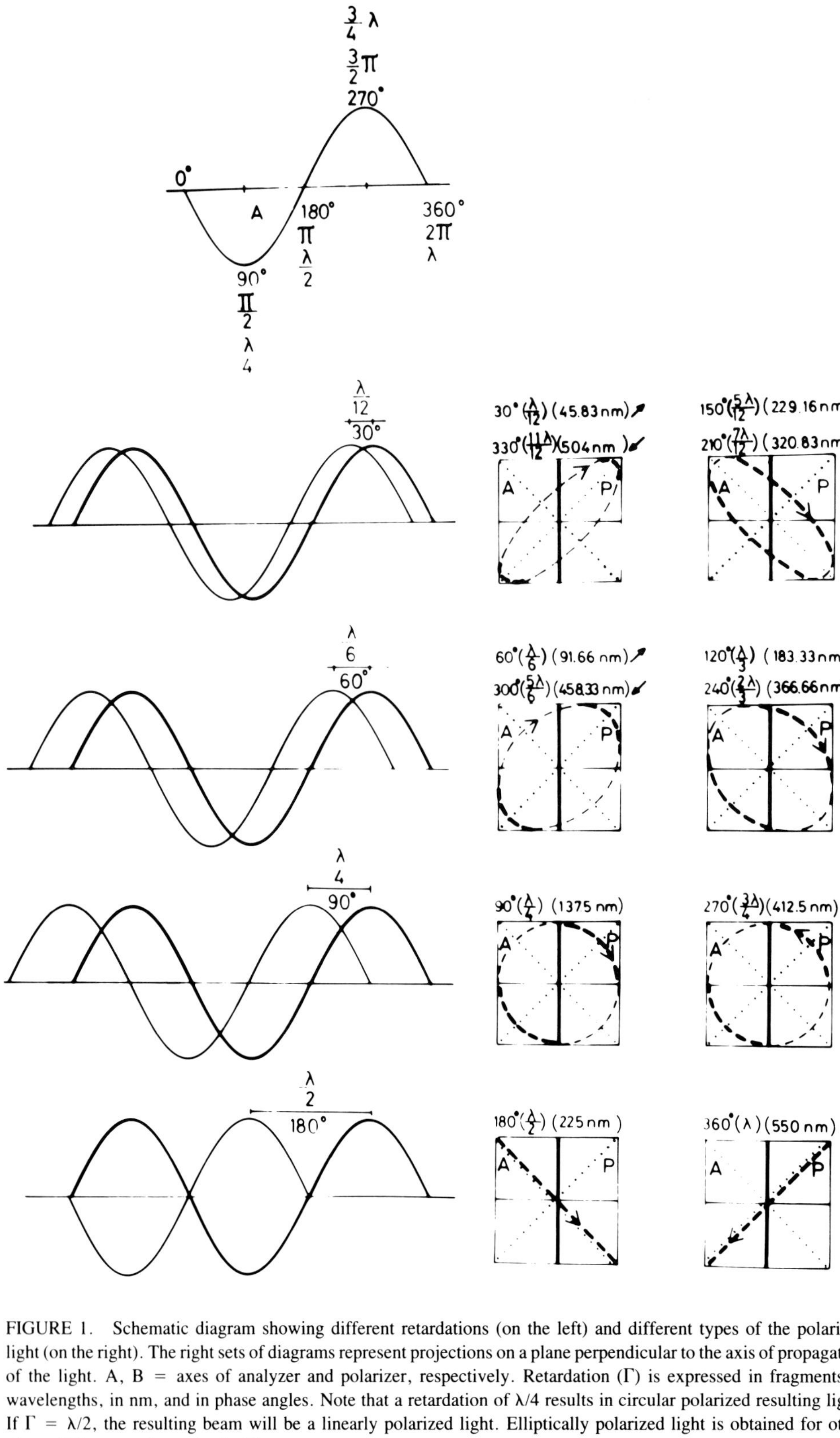

FIGURE 1. Schematic diagram showing different retardations (on the left) and different types of the polarized light (on the right). The right sets of diagrams represent projections on a plane perpendicular to the axis of propagation of the light. A, B = axes of analyzer and polarizer, respectively. Retardation (Γ) is expressed in fragments of wavelengths, in nm, and in phase angles. Note that a retardation of λ/4 results in circular polarized resulting light. If Γ = λ/2, the resulting beam will be a linearly polarized light. Elliptically polarized light is obtained for other phase differences shown in the Figure.

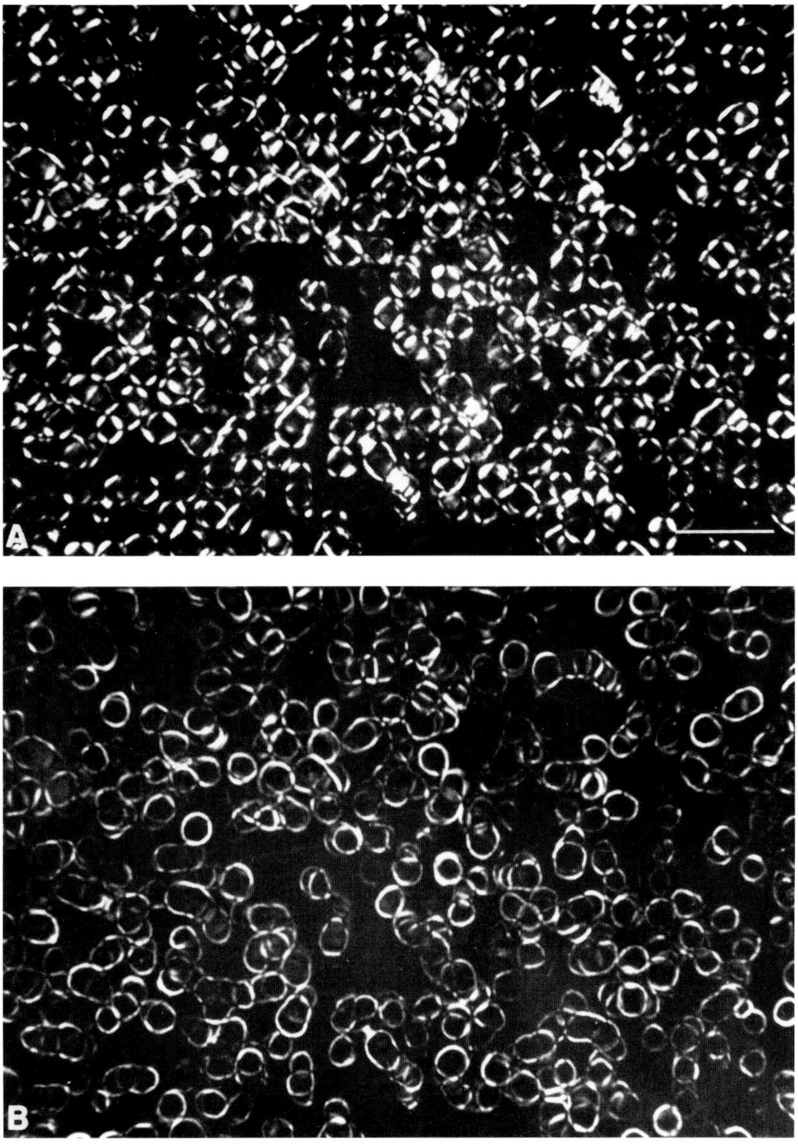

FIGURE 2. Human peripheral blood smear stained with toluidine blue at pH 7.0 and taken in linear (A) and circular (B) polarized light. Note the birefringence of the cell membranes of the erythrocytes. Using linear polarized light, the cell membranes appear as Maltese crosses. Birefringent membranes appear as continous bright circles when illuminated with circular polarized light. Scale: 30 μm.

The other one is the quarter-wave plate (λ/4) which is widely used for measurement of retardation values according to the Sénarmont method[20] and estimation of the relative sign of birefringence. Using this method, the λ/4 plate is inserted with its n_γ in subparallel position below the analyzer, and the analyzer is rotated until the birefringence is extinguished (the specimen becomes dark). The retardation value of the birefringent structure (Γ) is calculated as follows:

$$\Gamma = \frac{\alpha \times \lambda}{180}$$

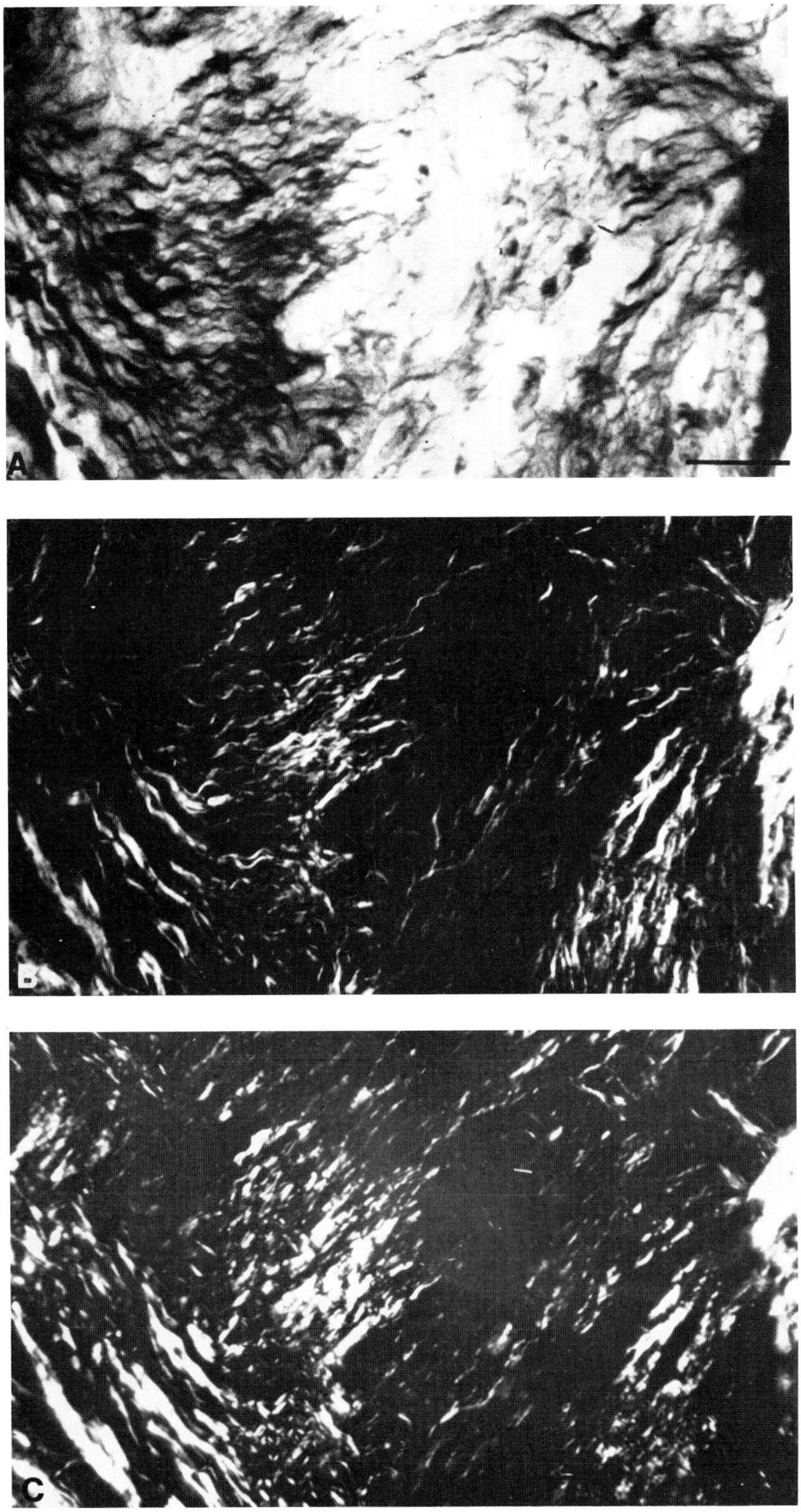

FIGURE 3. Collagen fibers in pericapsular region of knee joint of a 4.5-month-old human fetus stained with picrosirius red F3B and taken in normal (A), in linear (B) and circular polarized light (C). Scale: 100 μm.

where α = an angle reading of the rotating analyzer in extinction position, λ = wavelength of the illuminating light.

Two quarter-wave plates are needed if polarization microscopic investigations are carried out in circular polarized light (see under this key-word).

Goldstein[55] proposed a new fine analytical procedure for transparent, birefringent, and dichroic specimens to measure separately the retardation and dichroism. In this elegant technique, which is a further development of the classical Sénarmont method, two quarter-wave plates (one fixed and one rotatable) are inserted between the specimen and the rotatable analyzer.

For accurate measurement of small retardation values (0.5 to 1.0 nm), Brace-Koehler rotatable compensators are used. Their retardations are generally $\lambda/10$, $\lambda/20$, $\lambda/30$. They are used for measurement as follows: The polarizer and analyzer are crossed. The reference axis of the specimen is oriented at ± 45° to the polarizer. The slow axis (n_γ) of compensator is oriented parallel to the vibration plane of the polarizer. The compensator is rotated until the specimen is extinguished and the angle (α) is read from the drum vernier linked to the rotating mica plate. The retardation value of the specimen (Γ) is then calculated as follows:

$$\Gamma = \Gamma_\kappa \times \sin 2\alpha$$

where Γ_κ is the maximal retardation value of the compensator plate at the wavelength of the illuminating light (this value is provided by the manufacturer).

Combination of the polarization microscope with a cytophotometer has been suggested to increase the accuracy of measurement of the retardation with compensator.[56]

Although, measurement with compensator plates is relatively simple, it is a time consuming and tiring technique and requires well-trained eyes of the microscopist. There is also a serious operational problem: the microscopist is required to make a rapid change in both visual accommodation and dark adaptation in the transition from viewing through the microscope to reading the drum vernier. To overcome these difficulties, an attempt has been made to construct a computer-compatible system to measure and record retardation.[57,58] The system consists of a high precision potentiometer linked directly to the drum vernier of a compensator plate or analyzer. A highly stabilized fixed voltage is applied to the potentiometer and the voltage difference between the potentiometer wiper and one end of the resistance element is directly proportional to the compensator angle. This voltage is converted to a digital time base and stored for subsequent computer analysis. Recently, a computer-compatible digital compensator called Retarmet has been developed by Carl Zeiss (Jena, GDR) for the Jenapol polarization microscope of the same factory. It is hoped that these modern compensators will be widespread and will greatly facilitate quantitative polarization microscopic investigations with their simplicity and great accuracy.

Composite (mixed) bodies — Idealized textures containing oriented asymmetrical units (rods, layers, platelets or discs) which can be regarded as models of biological structures exhibiting optical anisotropy.[4,12,28,56,59] The units are small in diameter with respect to the wavelength of light. If their units are isotropic, the composite bodies exhibit form (textural) birefringence. According to the theory of Wiener,[59] the rods with an index of refraction n_1 are dispersed in a homogeneous medium with refractive index n_2, and all of them are oriented parallel to the optic axis. In this case, the sign of form birefringence is positive with respect to the optic axis. From biological structures, the skeletal muscle fiber and the sperm tail approximate this type of composite bodies. In other types of composite bodies, layers or discs are perpendicularly oriented with respect to the optical axis. In this case, the sign of birefringence is negative with respect to the optic axis. Form birefringence is frequently combined with intrinsic birefringence in biological structures. In this case, the units of the composite body possess an oriented molecular structure. For example, the collagen fibers show positive form and positive intrinsic birefringence, and cell membranes exhibit positive form and negative intrinsic birefringence (with respect to the tangent of the cell). Form and intrinsic birefringence can be distinguished by analysis of the imbibition curves.

Dichroic dyes — These are deposited on a structure in an oriented manner thereby modifying the optical properties of the structure (birefrigence, dichroism). Generally, dyes with asymmetrical structure and planar configuration comprise this group. The absorption of polarized light by such dyes is direction-dependent. If the plane of the light vibration is parallel with the long axis of the oriented dye molecules, the absorption is stronger when compared with a perpendicular position. The material stained with a dichroic dye appears, therefore, to vary in color when observed in polarized light of different vibration directions. There are some naturally occuring structures containing dichroic dyes (e.g., feathers of some birds). The great majority of these dyes, however, are used for staining of tissues (e.g., congo red for amyloid, toluidine blue for glycosaminoglycans, sirius red F3B for collagen; for more details see Schmidt,[60] Ruch,[61] Missmahl).[62]

Dichroic staining — An oriented array of dye molecules are deposited on a structure during this procedure which results in the modification of the optical properties of the structure stained. Its synonyms are the topochemical reactions[6] and topo-optical reactions.[63] In this volume, the term topo-optical reaction will be used. Dichroic stainings have been used for a long time.[2,60,64] The recent progress of the polarization microscopical techniques used in biology can be attributed to the development of new dichroic stainings among which a number of histochemical reactions are also found by means of which dichroism can be induced.

Dichroism (pleochroism) — Anisotropic absorption of polarized light. In linear dichroism, the absorption varies changing the azimuth of transmitted linear polarized light. Certain crystals and molecules absorb different wavelengths of polarized light to different degrees depending on the relation of the plane of polarization to the optic axis of the dichroic material. Due to this, different colors may be produced when transilluminating the material in different directions with polarized light. In circular dichroism (CD), the absorption of circular polarized light changes upon reversing the direction of rotation of the transmitted light. CD is a valuable tool in stereochemistry to describe molecular configuration.

Form and intrinsic dichroism can be distinguished (see the theory of composite bodies). Form dichroism is shown by structures in which submicroscopic elongated units are oriented in a preferential direction. Intrinsic dichroism is a property of the molecules themselves. Depending on the absorption of the material, dichroism can be observed in the visible spectral range and in infrared or ultraviolet light. For example, oriented filaments of DNA exhibit a negative dichroism at 260 nm which is due to the perpendicular orientation of purine and pyrimidine bases to the long axis of the molecules.

The sign of dichroism may be positive and negative. A material exhibits positive dichroism related to its preferential axis if it shows maximum absorption of polarized light vibrating parallel to the direction of the preferential axis. Amyloid fibers stained with congo red or collagen fibers stained with sirius red F3B show positive dichroism. The sign of dichroism is negative with respect to some distinguished axis of the material if the absorption is maximal in a direction perpendicular to that axis. For example, collagen fibers or cell membranes stained with toluidine blue show negative dichroism (in the case of cell membranes the reference axis is the tangent of the cell). In both of the latter examples, the negative dichroism is due to axiperpendicular orientation of the dye molecules to the structures.

Dichroism occurs naturally in some tissues containing colored pigments, or it can be induced by appropriate staining techniques. Dichroism provides information on the submicroscopic orientation of the structure, and it is a powerful tool to demonstrate dye orientation in the tissue. For microscopic demonstration of dichroism, the specimen is placed above the polarizer and observed without an analyzer. To detect linear dichroism, the azimuth of linear polarized light is varied and color changes are observed. It can be measured by a cytophotometer: absorption curves are taken from different planes of the illuminating polarized light (perpendicular and parallel with respect to the direction of a preferential axis of the structure). The intensity of dichroism can be calculated by subtracting the extinction values obtained at perpendicular position from that of those measured at parallel position:

$$\Delta E = E_{\parallel} - E_{\perp}$$

Circular dichroism is examined by illuminating the specimen by circular polarized light produced by a quarter-wave plate placed just above the polarizer in a 45° position. The plate is rotated through an angle of 90° and the color changes of the object are observed. For more details concerning dichroism see other references.[18,20,28,61,62,65-67]

Dispersion of birefringence — This is the variation of birefringence with wavelength (λ), and it is a consequence of the dispersion of the refractive index (n). The dispersion is normal if n decreases with λ, and abnormal if n increases with λ. The formula of the dipersion of birefringence (D):

$$D = \frac{n_{e1} - n_{o1}}{n_{e2} - n_{o2}}$$

where n_{e1} and n_{o1} are the refractive indices at short wavelength, and n_{e2} and n_{o2} are the refractive indices at long wavelength.

Elliptically polarized light — The terminal points of the electric vector in a propagated ray rotates in a path which can be projected as an ellipse on a plane normal to the axis of propagation (Figure 1).

Extraordinary ray — When a light beam enters birefringent material it is split into two plane polarized beams. Their vibration planes are perpendicular to each other. One beam obeys the ordinary laws of refraction, and this is called ordinary ray. The velocity of this light beam is maximum. The other ray is designated as the extraordinary ray, and its velocity is minimum. It does not obey the ordinary laws of refraction. The refractive index for the extraordinary ray is n_e, and it is parallel with the major axes of the molecules forming the anisotropic structure (=slow axis). The extraordinary ray is retarted because this light interacts with chemical bonds.

Fluorescence polarization — Fluorescence emitted by dye molecules bound in an oriented manner to a structure is highly polarized because of the dichroic nature of the molecules (fluorescence polarization). If the molecules rotate in the period between absorption (excitation) and emission (it takes about 10^{-8} s), the fluorescent light is depolarized. The depolarization can be measured.[68,69] The depolarization is related to the rotational diffusion constant of the dye molecule. If such dyes are bound to a macromolecule, from the depolarization of the fluorescent dye the rotational diffusion constant of the macromolecule can be obtained. This technique is mostly used for analysis of particles in solution.[69,70] The application of fluorescence polarization for the investigation of submicroscopic orientation in cell and tissue structure is also possible although not widely used.[71-81]

Form (textural) birefringence — This is found in mixed bodies wherein asymmetrical particles (rodlets, platelets) of a given refractive index are dispersed in an oriented manner in a medium of different refractive index. Form birefringence can be abolished by replacing the medium with one having a refractive index equal to that of the dispersed bodies. This is the basis of the imbibition analysis.

Imbibition analysis — This is a classical and well established method of polarization microscopic ultrastructure research. Unstained structures are mounted in media of different refractive indices, the retardation values of the birefringence are determined in each medium, and they are plotted against refractive indices. The curves obtained are called form birefringence curves. If there is a distance between the site of minimum retardation value of the curve and the abscissa, the structure also has intrinsic birefringence (Figures 4, 5) (see also Chapter 12). It is a useful and sensitive technique to show structural alterations of the collagen fibers.[82,83] Using this method, Vidal[84] was the first to show the axiparallel orientation of sulfated GAGs to the collagen fibers in unstained tendons. A similar orientation pattern of sulfated GAGs has been revealed by imbibition analysis in cartilage matrix[85] (Figures 4, 5).

Intrinsic (or crystalline) birefringence — This is due to asymmetrical alignment of chemical bonds or molecules, and it is an inherent property of the structure. The intrinsic birefringence does not vary in the function of the refractive index of the mounting medium. For this reason the two contributing factors of birefringence, form and intrinsic birefringence, can be separated from each other by imbibition analysis.

Isotropy — This occurs in isotropic materials where a large number of electronic resonators are distributed throughout a medium in a perfectly random manner. Isotropic materials do not influence the state of polarization of light, they are optically homogeneous, and can be characterized by one refractive index.

Linear (plane) polarized light — The electric vector of the light wave oscillates in a single plane. It is produced by structures where the $\Gamma = n\lambda$, where $n = 1,2,3$, etc.; or $\Gamma = n\lambda/2$ where $n = 1,3$, etc. If $\Gamma = n\lambda$, the resultant beam oscillates parallel to the polarizer axis and it is not passed by a crossed analyzer (Figure 6). The birefringence of such structures cannot be seen. Contrary to this, the resultant beam produced by a retarding object of $\Gamma = n\lambda/2$ oscillates parallel to the analyzer axis and it passes the analyzer with maximum brightness (Figure 6). Linear polarized light is produced by the polarizer in a polarization microscope.

Micelle (micella) — This term was introduced by Nageli in 1858 (see Reference 12) to describe long, submicroscopic colloid particles of crystalline structure. Micelles are supramolecular units, more often packets of chain molecules in parallel arrangement. Micella is a diminutive of the latin mica, a crumb.

Optic axis — This is a direction parallel to which a linear polarized light can be propagated with equal transmission velocities for all azimuths of polarization of the electric vector of the beam (e.g., the long axis of a collagen fiber). If light arrives in this direction, no optical anisotropy can be observed. The optic axis can be determined by finding an orientation in which the birefringent structure appears dark at all angles of rotation of the stage between crossed polaroids. In this position the optic axis of the specimen is parallel to the axis of the optical system of the microscope. The orientation of the optic axis of uniaxial materials may be longitudinal or transverse with respect to the long axis of the structure. The former is a positive uniaxial body (e.g., a collagen fiber), and the latter is negative (e.g., a collagen fiber after phenol reaction). In addition, the optic axis may be radial, circumferential, or tangential, and helical or spiral.

Ordinary ray — One of the split polarized light beams in a birefringent material. It oscillates in a plane perpendicular to the extraordinary ray (fast axis of transmission). The velocity of this light beam is maximum, and not retarded.

Polarization colors — They may appear in unstained, colorless or stained birefringent structures. Polarization color appears in an unstained specimen if one spectral region of the transilluminating white light is missing. In this case the complementary color to the missing wavelength is seen in the polarization microscope. If we see, for example, a purple color in an unstained structure, we know that the retardation value of this structure is about 550 nm. In this structure the green light ($\lambda = 550$ nm) is linearly polarized and it will not be transmitted by the analyzer, and we see the complementary purple color (= "red I").

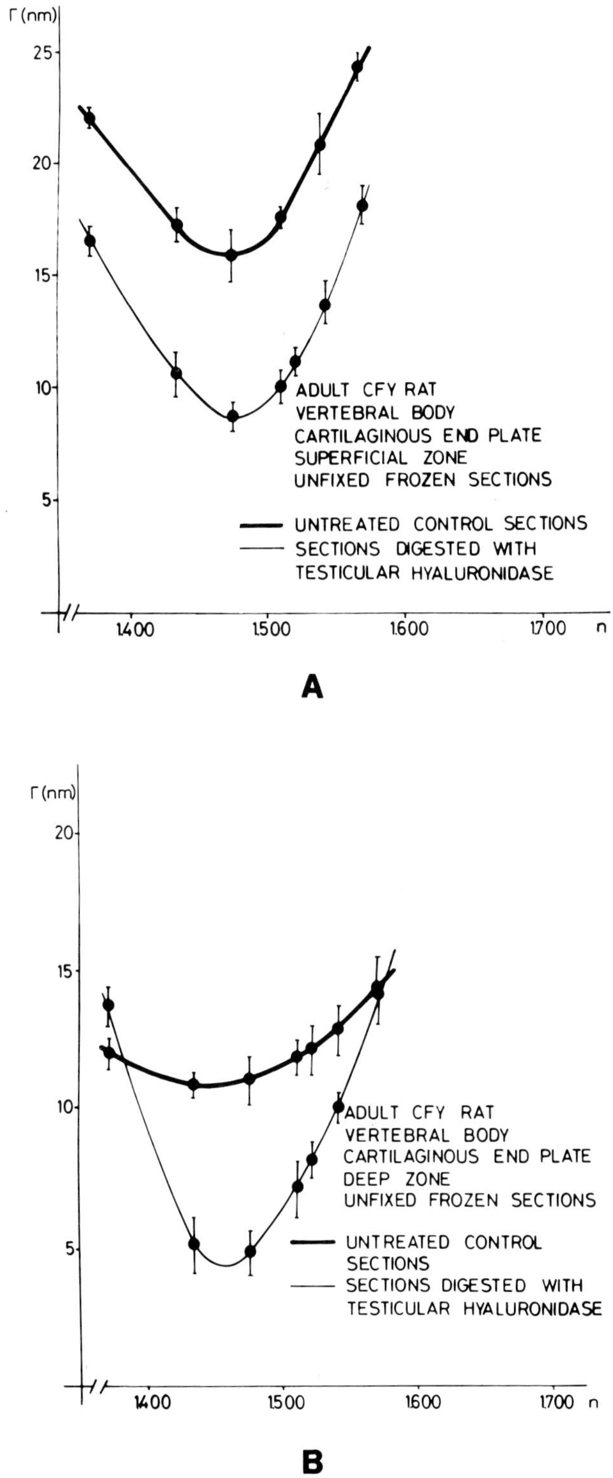

FIGURE 4. Form birefringence curves of unfixed cartilage matrix taken after an imbibition analysis carried out without and after digestion with testicular hyaluronidase. The decreased retardation values obtained after the digestion clearly indicate that sulfated GAGs are also oriented in the cartilage matrix. These molecules, or at least a population of them, are arranged parallel to the collagen fibers.

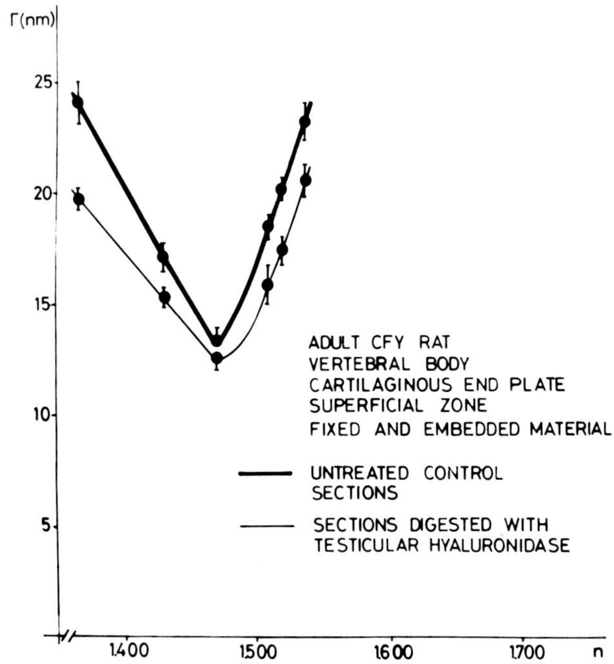

A

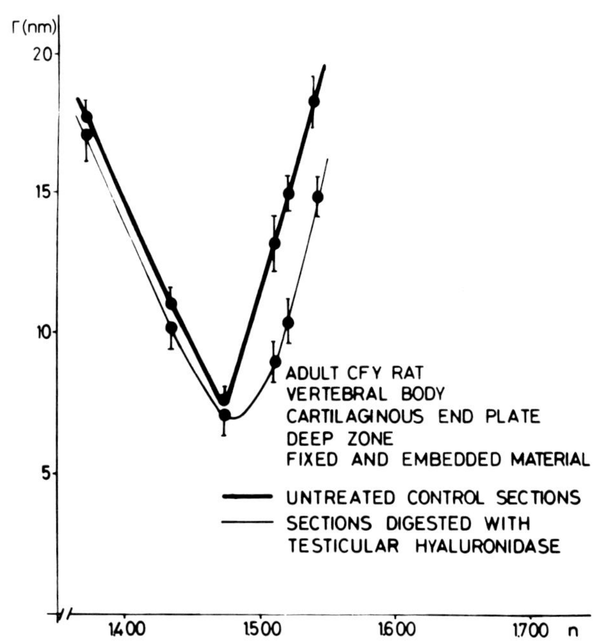

B

FIGURE 5. Form birefringence curves recorded from the extracellular matrix of cartilage fixed and embedded without and after digestion with testicular hyaluronidase. The digestion did not significantly alter the form birefringence curves indicating that unfixed specimens are better for imbibition analysis of the GAG molecules (compare with Figure 4).

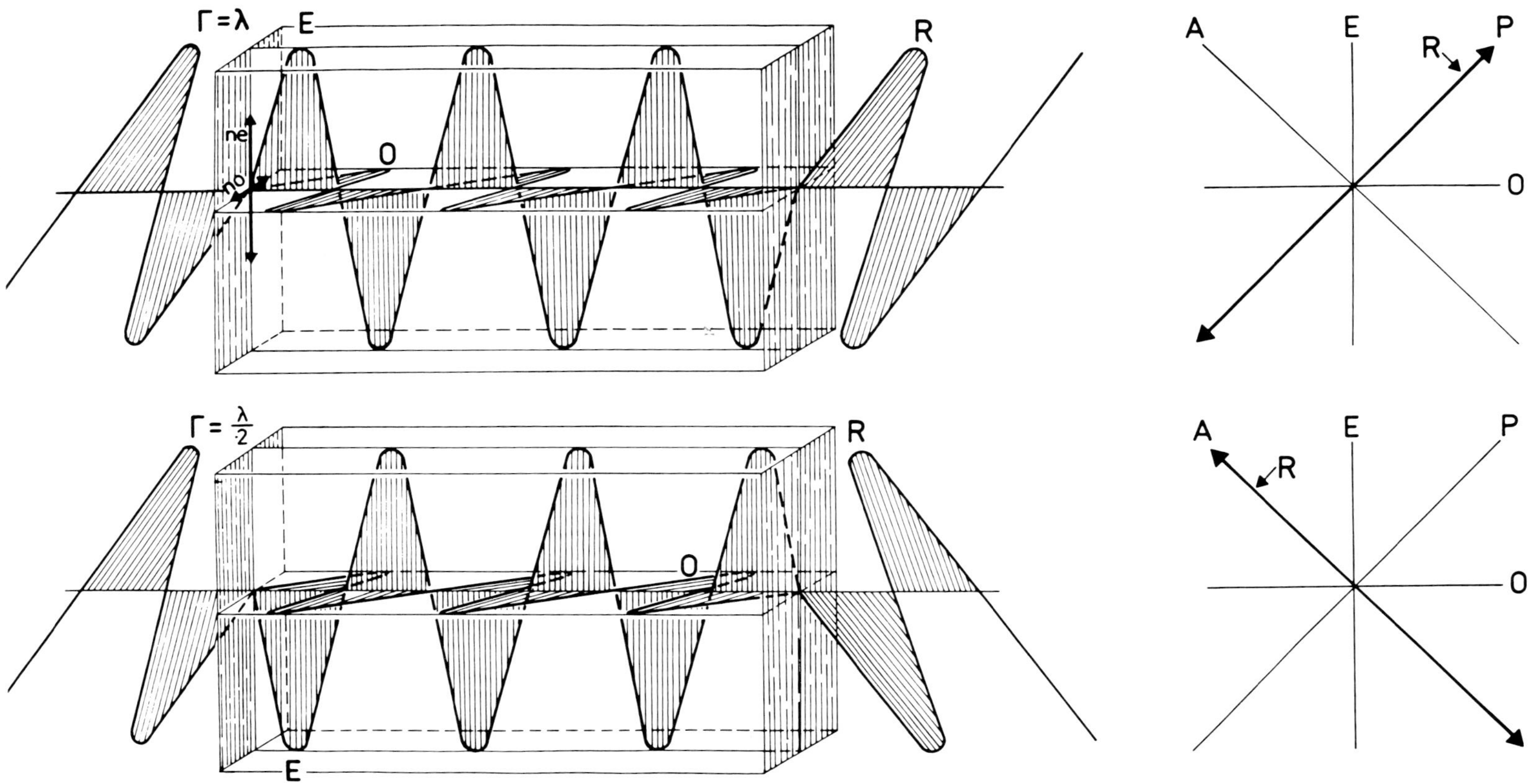

FIGURE 6. A scheme showing two birefringent materials. The upper one has a retardation value $\Gamma = \lambda$ with $\Gamma = \lambda/2$ in the lower one. A linear polarized light enters the structure (on the left) and splits into two components, the extraordinary (E) and the ordinary (O) rays, which oscillate in two directions at right angles to each other. The resulting beam (R) is linear polarized light in both cases, but it cannot pass the analyzer if $\Gamma = \lambda$. On the right side of the figure, the vibration directions are shown on a plane perpendicular to the axis of propagation of the light. A, P = axes of analyzer and polarizer, respectively; n_e, n_o = extraordinary and ordinary refractive indices of the birefringent structure.

Polarization colors appearing in unstained structures are divided into "orders" according the units of the green wavelengths. The first order appears in structures of retardation values between 0 and 550 nm, the second order between $\Gamma = 550$ and 1100 nm, etc. The polarization colors of the higher orders become fainter. Polarization colors are summarized in the Michel-Lévy interference color chart.[31] Polarization colors of the stained structures are generally the complementary colors. For example, an extracellular matrix area rich in sulfated GAGs is red after a metachromatic staining with toluidine blue when viewed in normal white light. If the dye molecules are highly ordered (i.e., the birefringence is maximum), the polarization color will be green, because the maximum retardation occurs in the spectral range where the absorption maximum of the dye is found.

Polarized light — This is a light beam in which the planes of vibration of all light waves are parallel to each other and where the state of polarization is uniform for all light waves. Its forms: linear, elliptical, and circular polarized light. It can be obtained from nonpolarized (normal) light using a polarizer.

Polarizer (polaroid, polar) — This is a material that transmits only beams with one plane of vibration. Generally, filter polars are used in the polarization microscopes. They consist of thin sheets of plastic material containing oriented dichroic particles (e.g., polyvinyl alcohol complexed with iodine). These anisometric particles absorb very little in one direction and the absorption is very high in a perpendicular direction (which corresponds to the long axes of the particles). The polars are used in the polarization microscope: the polarizer and the analyzer. The polarizer is mounted below the condenser, and it produces linearly polarized light by which the specimen is illuminated. The analyzer is mounted above the objective. Its plane of vibration is crossed to the plane of vibration of the polarizer. The background is dark when viewed between crossed polaroids, but a birefringent specimen appears bright, because it changes the state of polarization.

Pseudoisotropic materials — Occasionally, oriented structures are seen to be isotropic. In these cases the optical isotropy does not correlate to the real ultrastructure. Isotropy is observed in a material with oriented microstructure where the $\Gamma = n\lambda$, or if the light arrives in the direction of optic axis. Furthermore, when the slow axes of two anisotropic structures having identical retardation values cross each other at a right angle, they extinguish the birefringence of each other and isotropy appears in the crossing. If an anisotropic structure exhibiting form birefringence is imbibed by a medium with a refractive index identical to that of the structure, isotropy will appear. Isotropy may also occur after inversive topo-optical reactions of oriented structures. Such possibilities should be kept in mind when polarization optical analyses are carried out.

Reference axis — This is a direction to which the larger refractive index (n_γ or n_e) is referred. It is used for determination of sign (or character) of the birefringence. In biological systems generally the long axis of the elongated structures is the reference axis. In spherulites either the tangent or the radius may serve as reference axes.

Retardation (phase difference, optical path difference) — This develops between the beams travelling with different velocities in two perpendicular planes through the birefringent structure. One of the beams is slowed down more (retarded) than the other. The retardation is dependent on the geometric thickness of the specimen (d) and the refractive index difference in the two directions (= birefringence):

$$\Gamma = d(n_e - n_o)$$

It is usually expressed in linear distance (nm). If the d is small, birefringence may be unrecognizable, or it is recognized when using an improved optical system.[86-91]

Sign of birefringence — It denotes the relation of slow axis of transmission to a specified direction of the structure (reference axis). It may be positive or negative. The sign of birefringence is positive if the larger refractive index (n_γ or n_e) is parallel to the reference axis. If they are perpendicular to each other, the sign is negative. The sign of birefringence is generally indicated by an index ellipsoid. Its long axis is the larger refractive index, and the short axis represents the smaller refractive index (n_α or n_o). The large refractive index is in a direction parallel to the long axes of the oriented particles. In determining the relative sign of birefringence, we find out which is the larger and which is the smaller refractive index thereby determining the orientation of the molecules and micelles of the structure under examination. To determine the sign of birefringence, a compensator plate is used.

Structural anisometry — This means that structural elements (molecules, micelles) are not randomly distributed, but they are oriented in a preferential direction. It causes differences in optical properties (optical density, refractive index). It is accompanied by optical anisotropy. Analysis of optical anisotropy permits conclusions as to the structural anisometry (ultrastructure).

Topo-optical reactions — These are histochemical reactions which modify, and generally enhance the optical anisotropy of a tissue or cell constituent.[63,92-95] Alternative names are dichroic stainings[60] and topo-chemical reactions.[6] For details, see Chapter 3.

Uniaxial crystals — They have one preferred optic axis. The optic axis coincides with one of the principal axes of the index ellipsoid (with minimum or maximum refractive index). If the large refractive index is parallel with the optic axis, the material is called positive birefringent (collagen, muscle). The index ellipsoid has the shape of a cigar. If the small refractive index is parallel to the long axis of the structure, the sign of birefringence is negative. In this case, the indicatrix has a lenticular shape.

The major sources of this glossary are the references.[13,17,18,20-23,25,31,61]

IV. OPERATIONS WITH THE POLARIZATION MICROSCOPE

The section is placed on the stage of a polarization microscope equipped with a strong light source. Stained specimens are analyzed first in normal light. Then polaroids are crossed, and a compensator plate is inserted. The stage is slowly turned in one direction and presence or absence of the birefringence is observed. In birefringent structures, a reference axis is chosen, and the sign of birefringence is determined. After these steps, we are informed about the following features of a birefringent structure: (1) it contains spatially oriented constituents (2) what is the preferential axis of the orientation.

If an unstained specimen is the subject of investigation, imbibition analysis is carried out by measuring the retardation values after using different mounting media. Form birefringence curves are plotted, and the presence or absence of intrinsic birefringence is observed.

If working with stained sections, the orientation of the dye or other reagent to the chromotrope should be known prior to the analysis, otherwise the orientation of the stained constituent cannot be established from the sign of birefringence. The sign of birefringence is determined, and from this the orientation pattern is inferred. For quantification, retardation values of birefringence are measured at different wavelengths of the transilluminating light, and anomalous dispersion curves are constructed. Alternative possibility of quantification is the measurement of the retardation value at a certain wavelength which is generally in the spectral range of the complementary color of the polarization color of the structure stained. Data are plotted in histograms, and histograms obtained in different structures are compared mathematically.

Analyzer and compensator are removed, and the dichroism of the stained structure is investigated by turning the polarizer or the stage in one direction. A reference axis is chosen, and the sign of dichroism is determined. The intensity of dichroism is measured using a cytophotometer. Extinction values are measured at different wavelengths in two different positions of the polarizer which are perpendicular to each other: 1. where the structure exhibits maximum color intensity, 2. where the structure shows minimum color intensity. Two absorption curves are obtained, and they are compared mathematically.

ACKNOWLEDGMENT

Thanks are due to Drs. M. Antal and Mária Kern by whom the imbibition analyses are carried out in our laboratory.

REFERENCES

1. **Ambronn, H.,** *Anleitung zur Benützung des Polarisationsmikroskops bei histologischen Untersuchungen,* Leipzig, 1892.
2. **Ambronn, H. and Frey, A.,** *Das Polarisationsmikroskop. Seine Anwendung in der Kolloidforschung und in der Farberei,* Akademische Verlagsgesellschaft mbH, Leipzig, 1926.
3. **Schmidt, W. J.,** Polarisationsoptische Analyse des submikroskopischen Baues von Zellen und Geweben, in *Handbuch der biologischen Arbeitsmethoden,* Abt. V, Teil 10, Heft 1, Abderhalden, E., Ed., Urban and Schwarzenberg, Berlin, 1933, chap. 10.
4. **Schmidt, W. J.,** Polarisationsoptische Analyse des submikroskopischen Baues von Zellen und Geweben, in *Handbuch der biologischen Arbeitsmethoden,* Abt. V, Teil 10, Abderhalden, E., Ed., Urban und Schwarzenberg, Berlin, 1934, 435.
5. **Schmidt, W. J.,** *Die Doppelbrechung von Karyoplasma, Zytoplasma und Metaplasma,* Protoplasma-Monographien, 11, Verlag von Gebrüder Borntraeger, Berlin, 1937.
6. **Schmidt, W. J.,** Der Wandel der optischen Anisotropie bei topochemischen Reaktionen histologischer Strukturen, *Ber. d. Oberhess. Ges. f. Natur-u. Heilkunde N. F., Naturw. Abtg.,* 23, 56, 1948.
7. **Schmidt, W.,** Wie erforscht man den submikroskopischen Bau von Zellen und Geweben, *Ber. d. Oberhess. Ges. f. Natur-u. Heilkunde N. F., Naturw. Abtg.,* 24, 5, 1949.
8. **Schmidt, W. J.,** Einiges über die Polarisationsoptik von Proteinstrukturen, *Acta Histochem. (Jena),* 4, 175, 1957.
9. **Schmidt, W. J.,** Anwendung der Polarisationsmikroskopie in der Histochemie, in *Handbuch der Histochemie, Allgemeine Methodik,* Vol. 1, Part 1, Graumann, W. and Neumann, K., Eds., G. Fischer Verlag, Stuttgart, 1958, 171.
10. **Burri, C.,** *Das Polarisationsmikroskop,* Birkhäuser, Basel, 1950.
11. **Hallimond, A. F.,** *Manual of the Polarizing Microscope,* 2nd ed., Cooke, Troughton and Simms Ltd., York, 1953.
12. **Frey-Wyssling, A.,** *Submicroscopic Morphology of Protoplasm,* Elsevier, Amsterdam, 1953.
13. **Frey-Wyssling, A.,** *Die pfanzliche Zellwand,* Springer-Verlag, Berlin, 1959.
14. **Frey-Wyssling, A.,** The scientific work of W. J. Schmidt, *Microscopica Acta,* 77, 105, 1975.
15. **Frey-Wyssling, A.,** Submikroskopische Morphologie mit Hilfe indirekter und direkter Methoden, *Ber. Bot. Ges.,* 89, 525, 1976.
16. **Rinne, F. and Berek, M.,** *Anleitung zu optischen Untersuchungen mit dem Polarisationsmikroskop,* 2nd ed., Claussen, S. H., Driesen, A., and Rösch, S., Eds., Schweizerbarth, Stuttgart, 1953.
17. **Barer, R.,** Phase-contrast, interference-contrast, and polarizing microscopy, in *Analytical Cytology,* Mellors, R. C., Ed., Blakiston Division, McGraw Hill, New York, 1955, chap. 3.
18. **Oster, G.,** Birefringence and dichroism, in *Physical Techniques in Biological Research,* Vol. 1, Oster, G. and Pollister, A. W., Eds., Academic Press, New York, 1955, chap. 8.
19. **Vickers, A. E. J.,** The polarizing microscope in organic chemistry and biology, in *Modern Methods of Microscopy,* Vickers, A. E. J., Ed., Butterworths, London, 1956, 103.
20. **Bennett, H. S.,** The microscopical investigation of biological materials with polarized light, in *McClung's Handbook of Microscopical Technique,* 3th ed., McClung Jones, R., Ed., Hafner Inc., New York, 1967, 597.
21. **Brewer, D. B.,** Differences in the fine structure of collagen and reticulin as revealed by the polarising microscope, *J.Pathol.Bact.,* 74, 371, 1957.
22. **Bartels, P. H.,** Principles of polarized light, in *Introduction to Quantitative Cytochemistry,* Wied, G. L., Ed., Academic Press, New York, 1966, 519.
23. **Wied, G. L., Ed.,** *Introduction to Quantitative Cytochemistry,* Academic Press, New York, 1966, 559.

24. **Chayen, J. and Denby, E. F.**, *Biophysical Technique as Applied to Cell Biology*, Methuen, London, 1968.
25. **Wolman, M.**, On the use of polarized light in pathology, *Pathol. Annu.*, 5, 381, 1970.
26. **Wolman, M.**, Polarized light microscopy as a tool of diagnostic pathology, *J. Histochem. Cytochem.*, 23, 21, 1975.
27. **Schmidt, W. J. and Keil, A.**, *Polarizing Microscopy in Dental Tissues. Theory, Methods and Results*, Pergamon Press, Oxford, 1971.
28. **Scheuner, G. and Hutschenreiter, J.**, *Polarisationsmikroskopie in der Histophysik*, VEB G. Thieme, Leipzig, 1972.
29. **Czaja, A. T.**, *Einführung in die praktische Polarisations-Mikroskopie*, G. Fischer Verlag, Stuttgart, 1974.
30. **Módis, L.**, *Topo-Optical Investigation of Mucopolysaccharides (Acid Glycosaminoglycans)*, Handbuch der Histochemie, Vol. 2, Part 4, Graumann, W. and Neumann, K., Eds., G. Fischer Verlag, Stuttgart, 1974.
31. **McCrone, W. C., McCrone, L. B., and Delly, J. G.**, *Polarized Light Microscopy*, Ann Arbor Science Publishers, Inc., Ann Arbor, 1979.
32. **Stone, A. L.**, Optical rotatory dispersion of mucopolysaccharides and mucopolysaccharide-dye complexes, *Biopolymers*, 3, 617, 1965.
33. **Stone, A. L.**, Asymmetric modes of metachromatic binding of dyes to glycosaminoglycans and the influence of polymer conformation on the induced Cotton effects *in vitro*, in *Chemistry and Molecular Biology of the Intercellular Matrix*. Vol. 2, Balazs, E. A., Ed., Academic Press, London, 1970, 1067.
34. **Stone, A. L.**, Helical conformations in acidic polysaccharides in solution, *Biopolymers*, 11, 2625, 1972.
35. **Stone, A. L.**, Optical conformation of heparin and heparin complexes with cationic dyes, amines, and protein models, *Fed.Proc.*, 36, 101, 1977.
36. **Németh-Csóka, M., Kajtár, J., and Kajtár, M.**, Biological significance of helical conformation of acid polysaccharides, *Connect. Tissue Res.*, 3, 207, 1975.
37. **Németh-Csóka, M., Kajtár, M., and Kajtár, J.**, Secondary structure of glycosaminoglycans, *Acta Morphol. Acad. Sci. Hung.*, 24, 261, 1976.
38. **Chakrabarti, B. and Balazs, E.A.**, Induced Cotton effect of hyaluronic acid—acridine orange complex and conformation of the polymer, *Biochem. Biophys. Res. Commun.*, 52, 1170. 1973.
39. **Atkins, E. D. T. and Laurent, T. C.**, X-ray diffraction pattern from chondroitin-4-sulphate, dermatan sulphate and heparan sulphate, *Biochem. J.*, 133, 605, 1973.
40. **Nieduszynski, I. A. and Atkins, E. D. T.**, Conformation of the mucopolysaccharides. X-ray fiber diffraction of heparin, *Biochem. J.*, 135, 729, 1973.
41. **Phelps, C. F. and Atkins, E. D. T.**, Implications of the tertiary structure of glycosaminoglycans and proteoglycans, in *Biopolymere und Biomechanik von Bindegewebssystemen*, Hartmann, F., Hartung, Ch., and Zeidler, H., Eds., Springer-Verlag, Berlin, 1974, 317.
42. **Arnott, S. and Winter, W. T.**, Details of glycosaminoglycan conformations and intermolecular interactions, *Fed. Proc.*, 36, 73, 1977.
43. **Sheehan, J. K., Gardner, K. H., and Atkins, E. D. T.**, Hyaluronic acid: a double helical structure in the presence of potassium at low pH and found also with the cations ammonium, rubidium, and caesium, *J. Mol. Biol.*, 117, 113, 1977.
44. **West, S. S.**, Optical rotatory dispersion and the microscope, in *Introduction to Quantitative Cytochemistry*, II, Wied, G. L. and Bahr, G. F., Eds., Academic Press, New York, 1970, 451.
45. **West, S. S. and Lorincz, A. E.**, Fluorescent molecular probes in fluorescence microspectrophotometry and microspectropolarimetry, in *Fluorescence Techniques in Cell Biology*, Thaer, A. A. and Sernetz, M., Eds., Springer-Verlag, Berlin, 1973, 395.
46. **Nedzel, G. A.**, Intranuclear birefringent inclusions, an artefact occuring in paraffin sections, *Q. J. Micr. Sci.*, 92, 343, 1951.
47. **Hamperl, H.**, Remnants of paraffin (''paraffin-inclusions'') in histological sections, *Virchows Arch.(A)*, 334, 79, 1961.
48. **Vlachos, J. D.**, Birefringence and paraffinophilia of cell nuclei, *Stain Technol.*, 43, 89, 1968.
49. **Blix, G. and Snellman, O.**, On chondroitin sulfonic acid and hyaluronic acid, *Ark. F. Kemi. Mineral. O. Geol.*, 19, 1, 1945.
50. **Balazs, E. A.**, Physical chemistry of hyaluronic acid, *Fed. Proc.*, 17, 1086, 1958.
51. **Hanss, M. and Roux, B.**, Biréfringence électrique. Application á l'étude des macromolecules, *Ann. Phys. Biol. Med.*, 3, 133, 1972.
52. **Foweraker, A. R., Isles, M., Jennings, B. R., Hardingham, T. E., and Muir, H.**, Electric birefringence studies of cartilage proteoglycan aggregation, *Biopolymers*, 16, 1367, 1977.
53. **Schnabel, R.**, Zur mikroskopischen Untersuchung optisch anisotroper Strukturen des Nervensystems mit circular polarisiertem Licht,. *Acta Neuropathol. (Berl.)*, 7, 180, 1966.
54. **Frohlich, M. W.**, Birefringent objects visualized by circular polarization microscopy, *Stain Technol.*, 61, 139, 1986.

55. **Goldstein, D. J.**, Analysis of polarized light with two quarter-wave plates, *J. Microsc,*. 91, 19, 1970.
56. **Scheuner, G. and Hutschenreiter, J.**, Verwendungsmöglichkeiten eines Sekunderelektronenvervielfachers bei quantitativen polarisationsoptische Untersuchungen, *Acta Histochem. (Jena)*, 35, 153, 1970.
57. **Dvorak, J. A. and Stotler, W. F.**, A controlled environment culture system for high resolution light microscopy, *Exp. Cell Res.*, 68, 144, 1971.
58. **Dvorak, J. A., Clem, T. R., and Stotler, W. F.**, The design and construction of a computer-compatible system to measure and record optical retardation with a polarizing or interference microscope, *J. Microsc.*, 96, 109, 1972.
59. **Wiener, O.**, Die Theorie des Mischkörpers für das Feld der stationaren Strömung, *Abh. Sachs. Ges. Wiss.*, 32, 507, 1912.
60. **Schmidt, W. J.**, Dichroitische Farbung tierischer und pflanzlicher Gewebe, in *Handbuch der Biologischen Arbeitsmethoden*, Abt. V. Teil 2, Abderhalden, E., Ed., Urban und Schwarzenberg, Berlin, 1932, 1835.
61. **Ruch, F.**, Birefringence and dichroism of cells and tissues, in *Physical Techniques in Biological Research*, Vol. 3, Oster, G. and Pollister, A. W., Eds., Academic Press, New York, 1956, chap. 4.
62. **Missmahl, H. P.**, Birefringence and dichroism of dyes and their significance in the detection of oriented structures, in *Introduction to Quantitative Cytochemistry*, Wied, G. L., Ed., Academic Press, New York, 1966, 539.
63. **Romhányi, G.**, *On the Ultrastructure of Intercellular Substance of Connective Tissue on the Basis of Polarization Optical Investigations by Means of Topo-Optical Reactions*, D. Sc. Dissertation (in Hungarian), Pécs, 1966.
64. **Frey, A.**, Die Technik der dichroitischen Metallfarbungen, *Z. Wiss. Mikr.*, 42, 421, 1925.
65. **Romhányi, G.**, Uber die submikroskopische strukturelle Grundlage der metachromatischen Reaktion, *Acta Histochem. (Jena)*, 15, 201, 1963.
66. **Frey-Wyssling, A.**, Quantitative Bestimmung des Formdichroismus, *Z. Wiss. Mikr.*, 66, 45, 1964.
67. **Goldstein, D. J.**, Detection of dichroism with the microscope, *J. Microsc.*, 89, 19, 1969.
68. **Förster, Th.**, *Fluoreszenz organischer Verbindungen*, Vandenhoeck and Ruprecht, Göttingen, W. Germany, 1951.
69. **Weber, G.**, Polarized fluorescence, in *Fluorescence Techniques in Cell Biology*, Thaer, A. A. and Sernetz, M., Eds., Springer-Verlag, Berlin, 1973, 5.
70. **Witholt, B. and Brand, L.**, The effect of the excitation wavelength on the fluorescence depolarization of protein-dye complexes or conjugates, in *Fluorescence Techniques in Cell Biology*, Thaer, A. A. and Sernetz, M., Eds., Springer-Verlag, Berlin, 1973, 283.
71. **Ziegenspeck, H.**, Langspolarisierte Fluoreszenz von Faserfarbungen, *Kolloid. Z.*, 106, 62, 1944.
72. **Ziegenspeck, H.**, Emission polarisierten Fluoreszenz (Difluoreszenz) durch gefarbte Zellulose und Kutinmembranen von Pflanzen. *Mikroskopie Suppl.*, 1, 71, 1949.
73. **Scharf, J. H.**, Fluoreszenz und Fluoreszenzpolarisation der Nervenfaser nach Farbung mit Phenyloxyfluoronen. Versuch einer Interpretation, I, *Mikroskopie*, 11, 261, 1956.
74. **Scharf, J. H.**, Fluoreszenz und Fluoreszenzpolarisation der Nervenfaser nach Farbung mit Phenyloxyfluoronen. Versuch einer Interpretation, II, *Mikroskopie*, 11, 349, 1956.
75. **McInnes, J. W. and Uretz, R. B.**, Organization of DNA in dipteran polytene chromosomes as indicated by polarized fluorescence microscopy, *Science*, 151, 689, 1966.
76. **Aronson, J. F. and Morales, M. F.**, Polarization of tryptophan fluorescence in muscle, *Biochemistry*, 8, 4517, 1969.
77. **Dos Remedios, C. G., Millikan, R. G. C., and Morales, M. F.**, Polarization of tryptophan fluorescence from single striated muscle fibers, *J. Gen. Physiol.*, 59, 103, 1972.
78. **Dos Remedios, C. G., Yount, R. G., and Morales, M. F.**, Individual states in the cycle of muscle contraction, *Proc. Natl. Acad. Sci. U.S.A.*, 69, 2542, 1972.
79. **Tasaki, I., Watanabe, A., and Hallett, M.**, Properties of squid axon membrane as revealed by a hydrophobic probe, 2-*p*-toluidinylnaphthalene-6-sulfonate, *Proc. Natl. Acad. Sci. U.S.A.*, 68, 938, 1971.
80. **Nihei, T., Mendelson, R. A., and Botts, J.**, Use of fluorescence polarization to observe changes in attitude of S-1 moieties in muscle fibers, *Biophys. J.*, 14, 236, 1974.
81. **Schmidt, Th.**, Physikalische Grundlagen der Fluoreszenz und ihre Anwendung in der Histochemie, *Acta Histochem. (Jena) Suppl.* 19, 11, 1977.
82. **Whittaker, P., Boughner, D. R., Perkins, D. G., and Canham, P. B.**, Quantitative structural analysis of collagen in chordae tendineae and its relation to floppy mitral valves and proteoglycan infiltration, *Br. Heart J.*, 57, 264, 1987.
83. **Whittaker, P., Schwab, M. E., and Canham, P. B.**, The molecular organization of collagen in saccular aneurysms assessed by polarized light microscopy, *Connect. Tissue Res.*, 17, 43, 1988.
84. **Vidal, B. C.**, The part played by the mucopolysaccharides in the form birefringence of collagen, *Protoplasma*, 59, 472, 1964.

85. **Vidal, B. C. and Vilarta, R.,** Articular cartilage: collagen type II-proteoglycan interactions. Availability of reactive groups. Variation in birefringence and differences as compared to collagen I. *Acta Histochem. (Jena),* 83, 189, 1988.
86. **Inoué, S.,** A method for studying small retardations in structures in living cells, *Exp. Cell Res.*, 2, 513, 1951.
87. **Inoué, S.,** Studies on depolarization of light at microscope lens surfaces. I. The origin of stray light by rotation at the lens surface, *Exp. Cell Res.*, 3, 199, 1952.
88. **Inoué, S.,** Improvements of the polarization microscope for biological purposes, *Trans. Am. Microsc. Soc.*, 71, 311, 1952.
89. **Inoué, S.,** Polarization optical studies of the mitotic spindle. I. The demonstration of spindle fibers in living cells. *Chromosoma,* 5, 487, 1953.
90. **Inoué, S. and Dan, K.,** Birefringence of the dividing cell. *J. Morphol.*, 89, 423, 1951.
91. **Inoué, S. and Koester, C. J.,** Optimum half-shade angle in polarizing instruments, *J. Opt. Soc. Am.*, 49, 556, 1959.
92. **Romhányi, G.,** Submicroscopic structure of interphase nuclei as revealed by topo-optical staining reaction with cationic dyes, *Acta Morphol. Acad. Sci. Hung.*, 15, 131, 1967.
93. **Romhányi, G., Deák, G., and Bukovinszky, A.,** Collagen-specific topo-optical staining reaction with congo red and its ultrastructural interpretation, *Acta Morphol. Acad. Sci. Hung.*, 18, 261, 1970.
94. **Romhányi, G.,** Selective differentiation between amyloid and connective tissue structures based on the collagen specific topo-optical staining reaction with congo red, *Virchows Arch.(A),* 354, 209, 1971.
95. **Romhányi, G.,** Ultrastructure of biomembranes as shown by topo-optical reactions, *Acta Biol. Acad. Sci. Hung.*, 29, 311, 1978.

Chapter 3

TOPO-OPTICAL REACTIONS USED IN POLARIZATION MICROSCOPIC ULTRASTRUCTURE RESEARCH

TABLE OF CONTENTS

I. INTRODUCTION

Electron microscopy, X-ray diffraction, and polarization microscopy are three powerful and complementary tools in ultrastructure research. The potentialities of both electron microscopy and polarization microscopy are considerably increased when combined with histochemistry. While electron-microscopic histochemical methods are widely used, the application of relevant histochemical reactions for polarization microscopic investigations is more scarce. This may be due to at least three things: (1) they are not sufficiently known; (2) the ultrastructural analysis based on these techniques is an indirect — and not always simple — approach which has, therefore, less popularity and credit among some scientists; (3) occasional abuse may cause distrust of this technique. The aim of this chapter is to sum up some theoretical aspects of histochemical reactions used in polarization microscopic ultrastructural analysis. The methods are described in Chapter 12.

II. DEFINITION AND TAXONOMY

It has been known for a long time that the optical anisotropy of different biological structures can be modified by different reagents. These methods are called dichroic stainings,[1-3] topo-chemical reactions,[4] or topo-optical reactions.[5-7] Organic or inorganic reagents are used in these methods. The most classical organic reagent is the phenol molecule which selectively inverts the positive sign of birefringence of collagen, and this almost 100-year-old method is one of the most selective histochemical reactions of collagen.[8] Among the inorganic reagents, silver and gold salts, which are used in different impregnation methods, should be mentioned. Tissue constituents impregnated by silver or gold exhibit birefringence indicating the oriented deposition of the metal particles.[3,4,9-11]

We use the term "topo-optical" reactions introduced by Romhányi;[5-7] to define histochemical reactions resulting in optical anisotropy and to distinguish these methods from other histochemical reactions.

In a broader sense, topo-optical reactions can be defined as all possible methods which modify the native birefringence of tissue constituents. For example, if demineralized compact bone matrix is stained by some metachromatic cationic dyes at low pH values (below pH 3), the orginal birefringence of the bone collagen with positive sign — depending on the age of the specimen — is frequently diminished or disappears although the staining reaction cannot be observed in normal light or at best only slightly (see in Chapter 5). This phenomenon is due to the axioperpendicular binding of the cationic dyes to the sulfated GAGs associated with the bone collagen. The local concentration of GAGs is low in compact bone matrix, therefore, the quantity of dye bound to these molecules is not always sufficient to cause light absorption visible in normal light. But the axiperpendicular orientation of the planar dye molecules to the collagen fibers can extinguish the birefringence of the collagen.

Generally, topo-optical reactions considerably enhance the birefringence of a constituent in cells and tissues. This results from the oriented binding of the reagent molecule to the tissue constituents (Figures 1, 2). The enhancement does not modify the sign of native birefringence in some cases. These are the so-called additive topo-optical reactions, for example picrosirius red F3B staining for collagen and congo red staining for amyloid. In additive topo-optical reactions, the reagents are bound parallel to the chromotropes (Figure 3). Other reagents bind perpendicularly to the long axis of the chromotrope, and this orientation results in an inversion of the sign of native birefringence. These are the so-called inversive topo-optical reactions (Figure 3), for example the metachromatic stainings using thiazine dyes for polarization microscopic investigation of the GAG molecules (for more details see Romhányi[7,12] and Módis.[13]

It is essential to bear in mind that the development of any type of topo-optical reaction

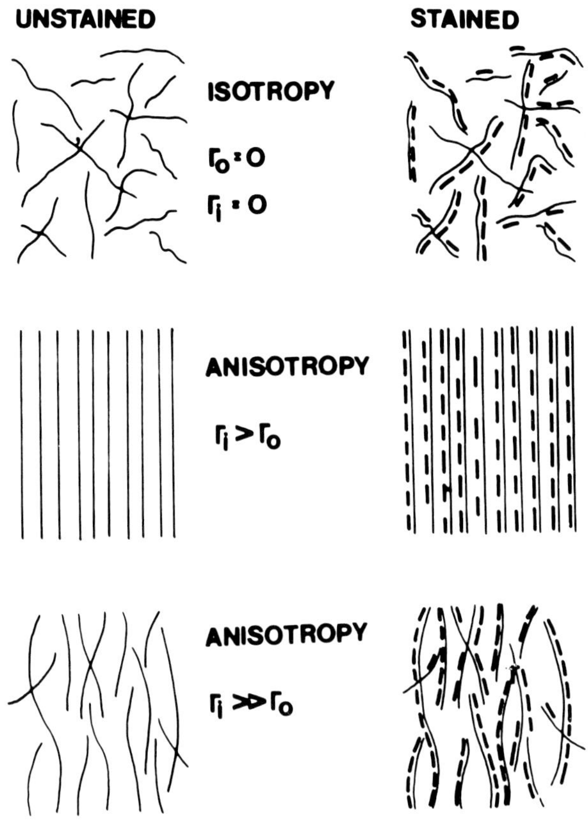

FIGURE 1. Diagrams showing the effects of oriented dye-binding on the birefringence of different biological structures. On the left side of the figure, unstained structures containing randomly distributed (top), highly oriented (middle) and loosely oriented (bottom) constituents are depicted. Their retardation values are designated by Γ_0. On the right side, the same structures are seen after topo-optical reactions in which the reagent molecules are longitudinally arranged along the constituents. The retardation values of birefringence induced by the topo-optical reactions are designated by Γ_i. Note that the birefringence of amorphous structures cannot be intensified by topo-optical reactions.

has two equally important prerequisites: (1) the specific and oriented binding of the reagent to the chromotrope; (2) the spatial orientation of the chromotropes, which are generally macromolecular constituents of the tissues (Figure 1). If one of these two requirements is not met, induced birefringence will never be observed under a polarization microscope. These are the essential premises of any kind of polarization microscopic ultrastructural analyses using topo-optical reactions. The importance of the first point is obvious. To accept the validity of the second point, however, further considerations are needed. They will be discussed in this chapter. Just a simple anticipated example to this point: if hyaline cartilage is stained with a metachromatic dye at pH 2 to 2.5 to show sulfated GAGs, both the extracellular matrix and the cytoplasm of the differentiated chondrocytes are stained metachromatically indicating the presence of GAGs and their PGs in both of these structures. If this specimen is observed under a polarization microscope, the matrix exhibits birefringence indicating the spatial orientation of the sulfated GAGs which is due to the interactions of these molecules with other structures, mainly with collagen. In contrast to this, the metachromatic cytoplasm is never birefringent because PGs and their GAG side chains are

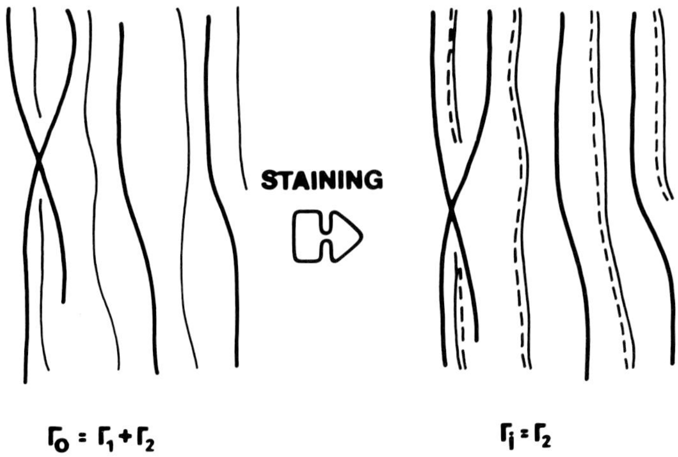

FIGURE 2. Scheme of a structure containing two different and loosely oriented constituents represented by thick and thin lines. Without any staining (on the left), the retardation value of the structure (Γ_0) is the mathematical sum of the retardation values of the two constituents Γ_1 and Γ_2. After a topo-optical reaction specific for constituent 2, the retardation value of birefringence induced by the topo-optical reaction Γ_i of the structure will be equal to the retardation value of the constituent 2 (Γ_2).

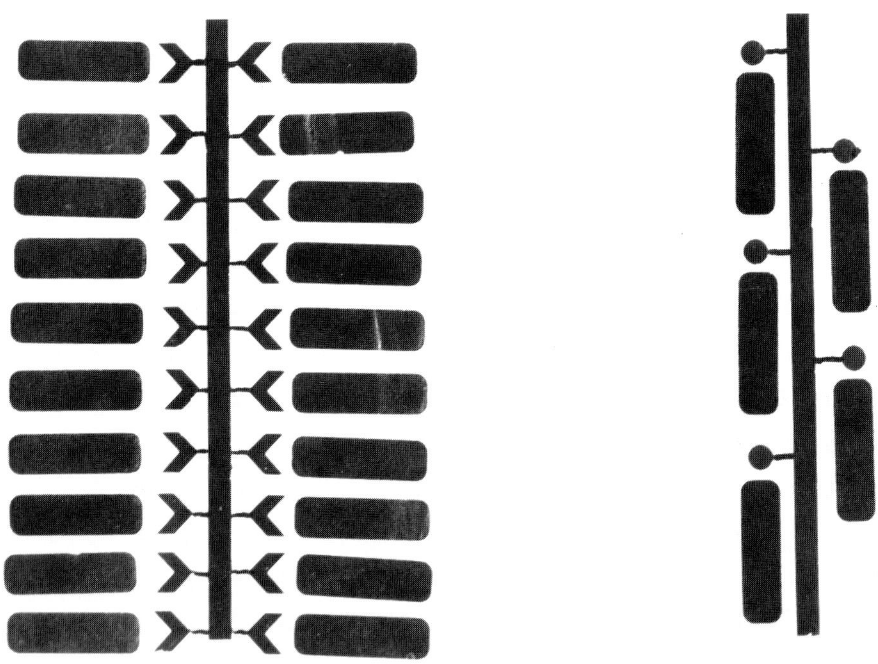

FIGURE 3. Scheme of inversive (on the left) and additive (on the right) topo-optical reactions. Reagent molecules of planar configuration are perpendicularly oriented to the chromotrope in the inversive reaction, while they are longitudinally arranged on the chromotrope surface in the additive reaction.

randomly distributed in the cell organelles where they are synthesized, transported, and prepared for secretion.

The topo-optical reactions of different ECM constituents will be discussed in this Chapter.

I have selected reactions whose staining mechanisms and specificity are known. The topo-optical reactions of amyloid will be separately discussed (see Chapter 10).

Polarization microscopy, mainly epipolarization microscopy, can also be used for qualitative and quantitative assessment of autoradiographs,[14] and for sensitive detection of immunohistochemical reactions when using immunogold-silver labelings.[15] The silver intensification of immunohistochemical reactions and silver impregnation of different tissue constituents using physical development may also result in optical anisotropy.[16,17] These methods are not discussed here, because it is quite uncertain whether the birefringence of silver and gold particles corresponds to any ultrastructural pattern of the tissue constituent labeled.

III. REACTIONS WITH CATIONIC DYES

A. SPECIFICITY

Cationic reagents including dyes bind to the anionic groups of different macromolecules with electrostatic bond. The major anionic groups are the sulfates, phosphates, and carboxyls after dissociation. Sulfate groups are present in high number in sulfated GAGs, phosphates are in different nucleic acids, phospholipids and phosphoproteins (e.g., in dentin), while carboxyl groups are found in hexuronic acid — and neuraminic acid — containing carbohydrates as well as in different proteins. A cationic probe can never be regarded as a specific reagent for one type of anion for one macromolecule bearing these groups because ionic interactions are highly aspecific. The specificity of cation-binding reactions depends on the staining conditions.

If the pH of a staining solution is 2.5 or less, only the sulfate groups are ionized and bind cations (Figure 4). It is known that sulfated GAGs are the only polysulfated macromolecules in the normal ECM of vertebrates. Using a cationic dye at low pH, they can be selectively stained. The drawback of the application of low pH is that $-NH_3^+$ groups are also generated in tissue which may interact with the dissociated sulfates thereby decreasing the dye-binding capacity of the polysulfates, i.e., the sensitivity of the staining reaction.[18] Increasing the pH of the reagent solution results in the dissociation of other types of anionic groups making the staining totally aspecific (Figure 4).

Carboxyl groups are ubiquitous anionic groups. Their blocking guarantees a selective uptake of cationic dyes by polysulfates in normal ECM except in tissues where phosphoproteins are present in higher local concentrations (e.g., in dentin). The dye-uptake of carboxyls can be abolished by two methods. One of them is the use of critical electrolyte concentration (CEC) technique.[19-21] In this procedure, the dye, which is generally a copper-containing phthalocyanin like alcian blue 8GX, is mixed with a salt, commonly $MgCl_2$. A series of staining solutions is prepared where the dye concentration is constant but the salt concentration is variable, and a staining of parallel sections of the same specimen is carried out. Due to the competition of the dye cations and Mg^{2+} ions the dye-uptake of different anionic groups (carboxyls and sulfates) and different GAGs will be gradually blocked as a function of the concentration of the $MgCl_2$ (Figure 5). The presence of 0.2 to 0.3 mol/l $MgCl_2$ in the staining solution, inhibits the binding of phthalocyanin dyes to the carboxyl groups. The CEC technique is widely used with phthalocyanins for light and electron microscopic staining of GAGs (see Chapter 11). Since these dyes are not the optimal topo-optical reagents of the GAGs (see later in this chapter), it has a limited value in polarization microscopy. Despite some critical remarks,[22,23] the CEC principle is a useful tool for selective electron microscopic visualization of PG molecules with phthalocyanins (see Chapter 11). It has been found in our laboratory that toluidine blue staining can also be combined with CEC.[13] We showed that the presence of 0.1 mol/l $MgCl_2$ in a 0.1% toluidine blue solution prevents staining of the carboxyl groups.

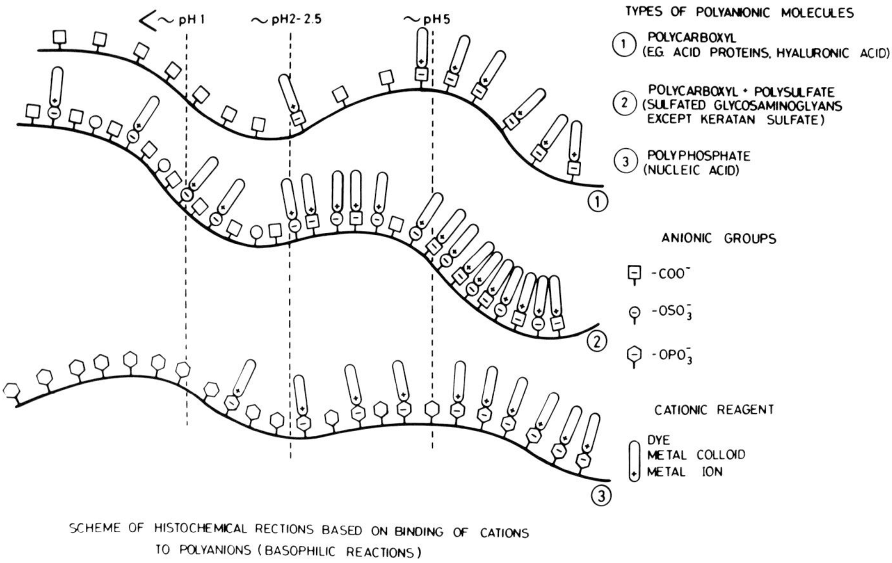

FIGURE 4. The scheme of the pH-dependence of cation-binding topo-optical reactions. Under pH 2 to 2.5 sulfate groups are stained but carboxyls are not ionized. Above pH 4.5 to 5 all of the anionic groups can interact with cationic reagents.

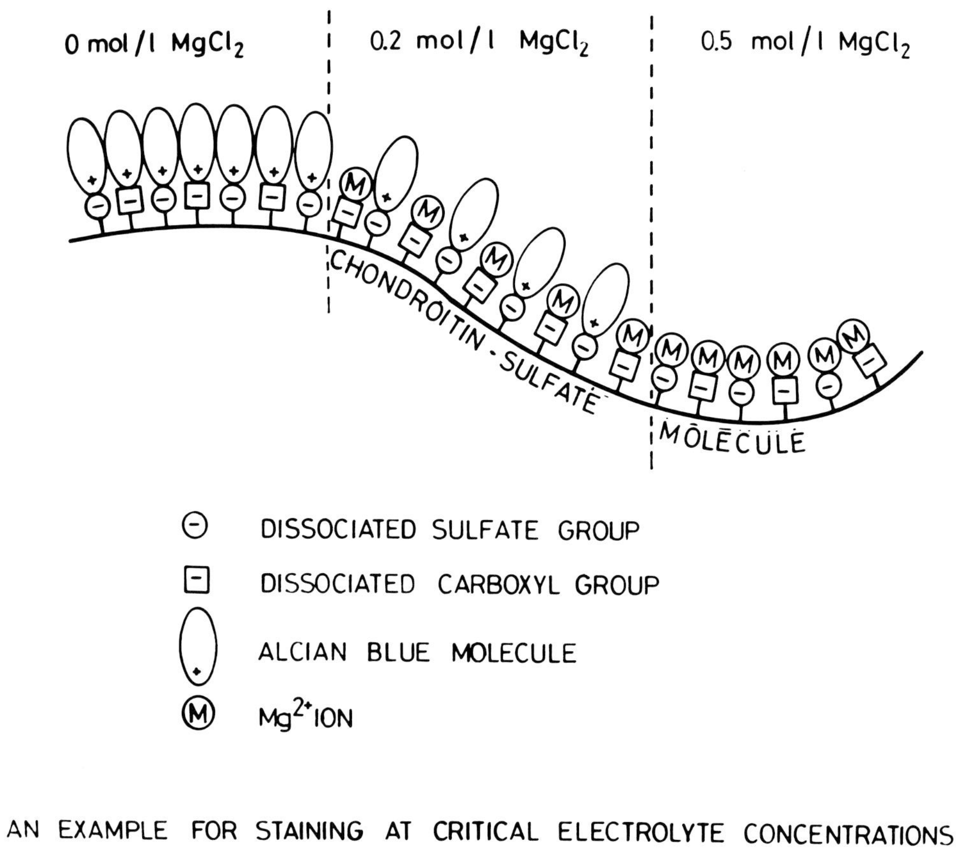

FIGURE 5. The diagram schematically illustrates the principle of the critical electrolyte concentration. Cationic dye molecules compete with Mg^{2+} ions for the dye-binding groups.

Carboxyl groups can be selectively alkylated by CH_3I.[24,25] If such alkylated sections are stained by any kind of cationic dyes at higher pH (pH 5 to 7), the sulfated GAGs can be selectively stained in the ECM. We prefer this combination as a selective topo-optical reaction of sulfated GAGs.

B. SENSITIVITY

The sensitivity of cation-binding reaction when evaluated in normal light is less than that of immunohistochemical reactions.[26] It has been shown that the stainability of hyaline cartilage with toluidine blue is completely abolished if 42% of its hexuronic acid content is lost.[27] Although the presence of cationic dyes bound to tissue polyanions in an oriented manner can be observed under the polarization microscope even if they are invisible in normal light, the amplification of the cation-binding staining reactions would be necessary. Two attempts have been made in this direction in our laboratory. One of them is the combination of an acetylation with a basophilic staining.[28] Acetylation blocks the $-NH_2$ and $-OH$ groups in the tissues. Blocking of $-NH_2$ groups has two favorable effects on the subsequent staining with a cationic dye: (1) the anionic dye-binding sites marked by interacting $-NH_2$ groups become accessible for the dye; (2) the repulsion of the dye cations by the dissociated $-NH_2$ groups ($-NH_3^+$) is diminished.

The other attempt was directed to increase the intensity of a cation-binding reaction of GAGs containing hexuronic acid. This reaction is called chemically intensified basophilic reaction (CIBR), and it is a combination of several histochemical reactions. It is known that sulfated GAGs do not react in the PAS reaction because their constituents are not oxidized by the periodic acid used. However, if a prolonged periodic acid oxidation is carried out, the hexuronic acid residues are oxidized yielding two aldehydes in one hexuronic acid.[29-31] It is also known that anionic bisulfite groups can be bound to the aldehydes,[32] which can be detected with toluidine blue or other cationic dyes at low pH.[33] Thus, a prolonged oxidation with periodic acid + bisulfite addition + staining with cationic dye at pH 2 results in an increased dye-binding of sulfate GAGs containing hexuronic acid residues. For example, a dissaccharide unit of a chondroitin sulfate molecule binds normally one cationic dye molecule at pH 2 to 2.5 while three dye molecules are bound to the same disaccharide after the CIBR (Figure 6). To avoid the aspecific bisulfite — and dye — binding of different monosaccharides of the glycoproteins, the first step of the CIBR is a conventional oxidation with periodic acid used in the PAS reaction. The aldehydes formed are blocked by a treatment with sodium borohydride, and then the sections are subjected to the prolonged oxidation with periodic acid (see Chapter 12). The sensitivities of the normal basophilic staining reaction and CIBR were compared using bovine nasal cartilage (BNC). Two BNC pieces isolated from the same specimen were embedded in the same block near to each other and cut. The twin sections mounted on the glass were cut to have identical thicknesses. After dewaxing, the two sections were separated by a nail varnish line. A simple staining was carried out on one section with a cationic dye at pH 2.5 and CIBR was performed on the other. The CIBR was developed by the same dye and at the same pH. CIBR resulted in a stronger staining than a simple staining (Figure 7). If toluidine blue or dimethylmethylene blue were used for staining, the birefringence of the cartilage matrix was also considerably increased after the CIBR. Scanning densitometry of the section-pairs revealed — depending on the dye used for development — a 1.4 to 2.7-fold increase in dye-uptake during the CIBR as compared to the simple staining. An example is shown in Figure 8.

C. SHAPE AND BIREFRINGENCE OF CATIONIC DYES

Birefringence can be generally induced by cationic dyes of planar configuration if they are bound to tissue polyanions. The great majority of the metachromatic dyes are planar molecules. Such molecules can easily be arrayed if they are bound to a polyanionic chromotrope where the dye-binding groups are close to each other.

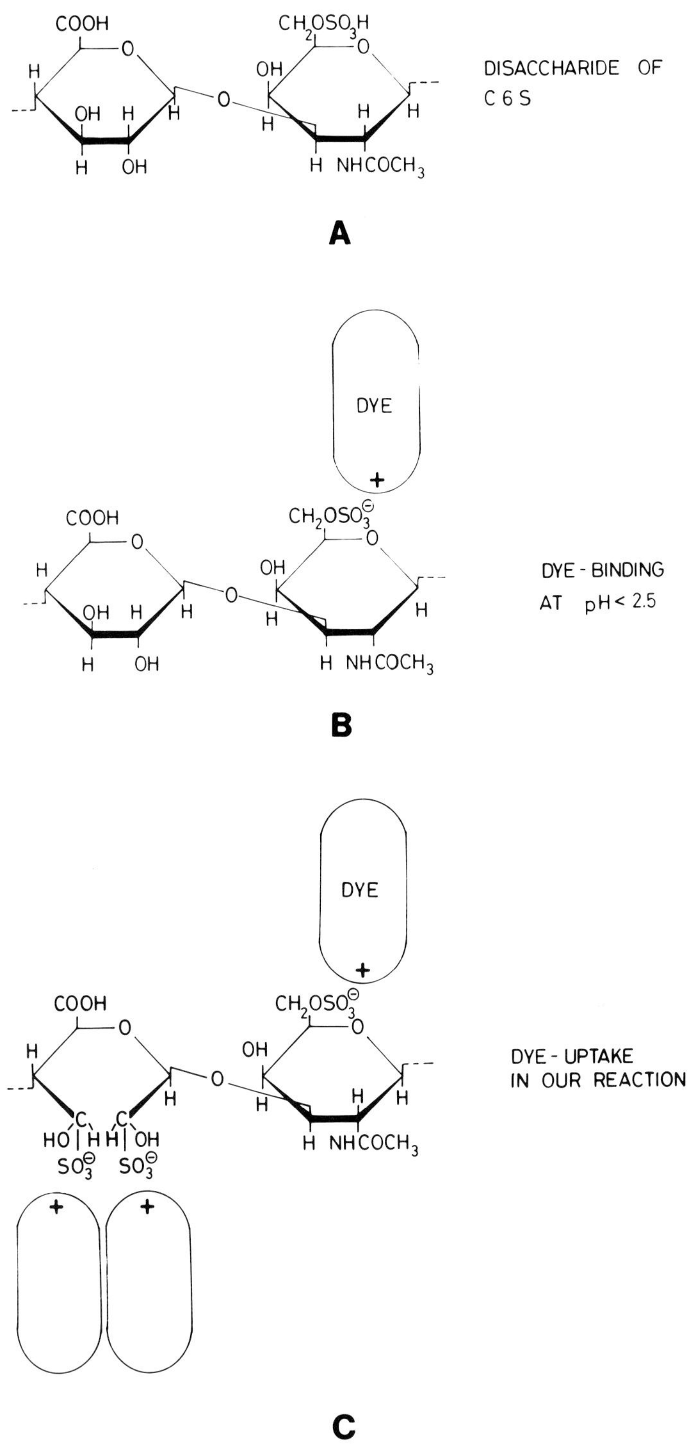

FIGURE 6. The mechanism of CIBR (chemically intensified basophilic reaction) illustrated schematically in a disaccharide unit of a chondroitin 6-sulfate (C6S). In a common basophilic reaction, one disaccharide (A) binds one cationic dye at about pH 2.5 (B). In CIBR (C), three dye molecules are bound to one disaccharide. Two of them detect the bisulfite groups bound to the aldehydes generated by a prolonged oxidation by periodic acid in the hexuronic acid.

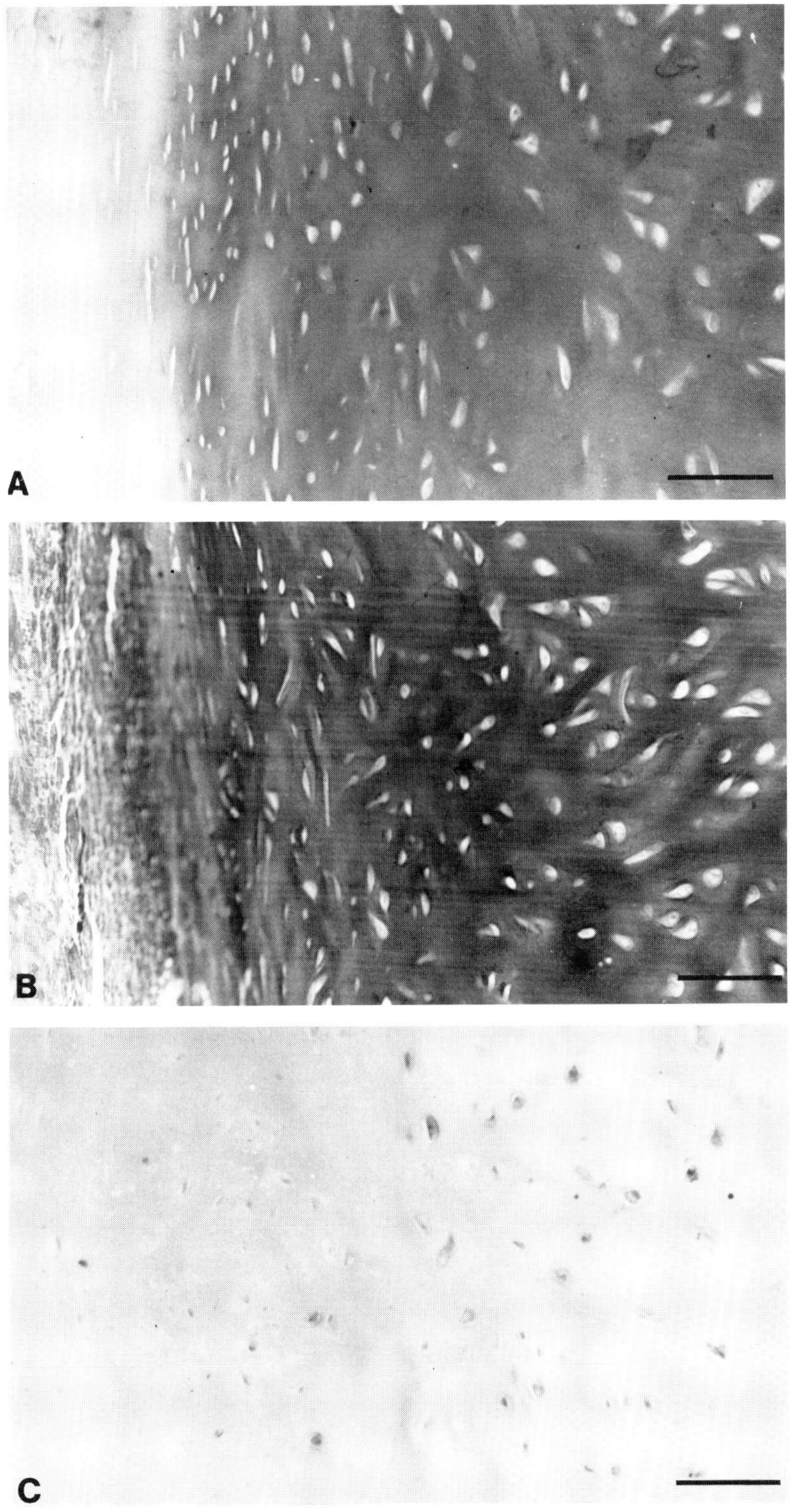

FIGURE 7. Bovine nasal cartilage stained with cuprolinic blue at pH 2.5 and taken in normal light. (A) A common basophilic reaction. Note that the perichondrium and superficial zone of cartilage are unstained. (B) CIBR. the staining of cartilage matrix is stronger than in A, and the perichondrium and superficial zone of cartilage are also stained. (C) CIBR after digestion with chondriotinase ABC. The staining is almost completely abolished. Scales: 100 μm.

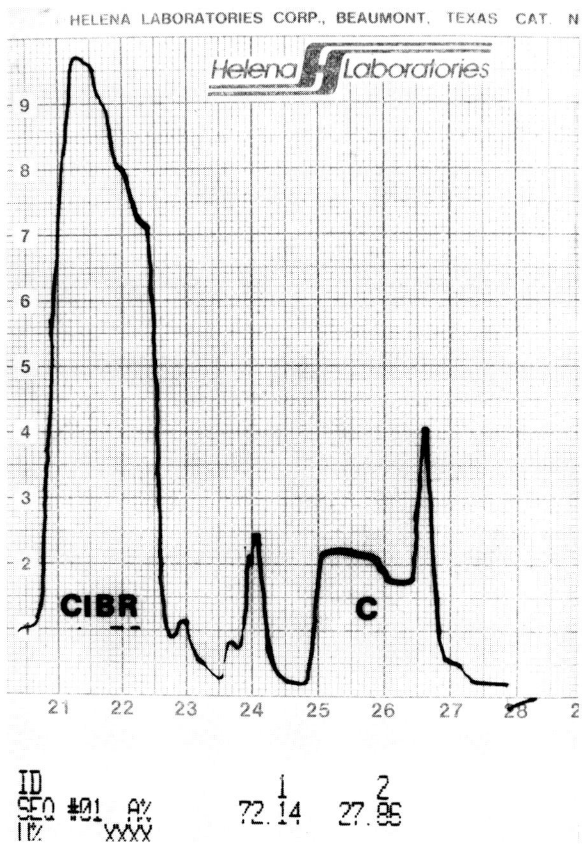

FIGURE 8. Scanning densitograms recorded from bovine nasal cartilage sections stained with cuprolinic blue at pH 2.5. C: control staining without any pretreatment. Note that the intensity of staining is considerably higher after CIBR than after the control staining.

The significance of the molecular shape of the dyes in induction of the birefringence of polyanions is illustrated with topo-optical reactions of the Janus dyes. Janus red, yellow, green, black, and blue are cationic azo dyes (Figure 9). All of these exhibit a moderate metachromatic effect with sulfated GAGs. Janus red and Janus yellow induce strong birefringence with negative sign in cartilage ECM (Figure 10). In contrast to this, Janus green, Janus black, and Janus blue molecules, which contain a benzene ring as a substituent on the quaternary nitrogen, either fail to induce birefringence (Janus blue) or exhibit a very moderate anisotropy (Figure 10). The perpendicular position of the benzene ring substituent (which can be rotated) to the major axis of the dye molecules (Janus green, Janus black, Janus blue) may explain the failure of dye orientation and birefringence. The presence of a β-naphtol component at the end of the Janus red molecule seemingly has no effect on the development of an oriented dye-aggregate on the chromotrope surface, but if it occurs together with a similarly positioned benzene ring, as in the Janus blue molecule, the oriented dye-binding is disturbed. It will be shown that the best cationic topo-optical reagents belonging to the thiazin, acridine, and quinoline dye groups also have a planar configuration. All these data suggest that the shape of the reagent molecule strongly influences the optical anisotropy of the dye-aggregate developed on the surface of the polyanionic chromotrope molecule. The more planar the configuration of the dye, the more intensive the birefringence induced.

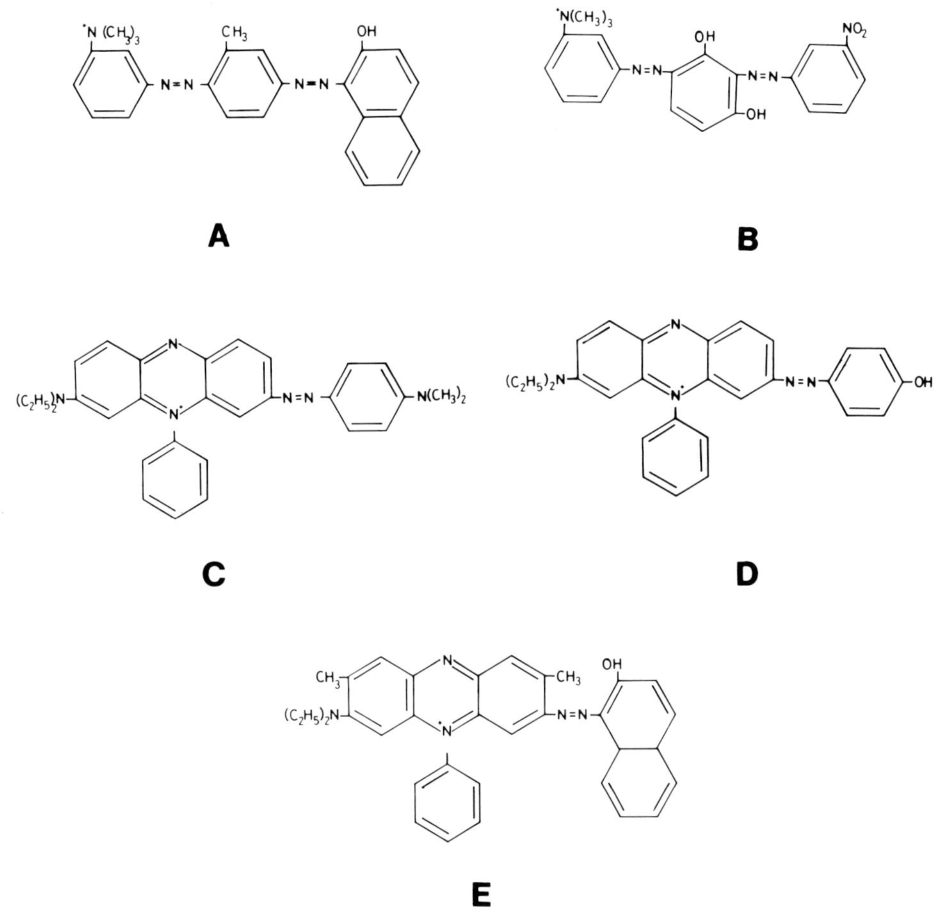

FIGURE 9. Structural formulae of different Janus dyes according to Lillie.[34] (A) Janus red; (B) Janus yellow; (C) Janus green; (D) Janus black; (E) Janus blue.

D. ORIENTATION OF DYE MOLECULES IN TOPO-OPTICAL REACTIONS

The spatial orientation of cationic dye molecules bound to the chromotrope can be approximated sometimes from the structure of the dye, i.e., from its charge distribution. *N,N′*-diethylpseudoisocyanine, for example, has its cationic charges along its long edge, and, due to this, it binds to chromotropes in an axiparallel manner.[13,35-37] Model experiments, however, provide more exact information about the arrangement of dye bound to a specific chromotrope.

Thiazins assume a predominant axiperpendicular orientation to the long axis of GAGs as revealed by the investigation of circular dichroism and optical rotatory dispersion of methylene blue-GAG complexes in solutions.[38,39] Acridine orange behaves similarly after binding to polyanions.[38,40] The cationic and metachromatic thiazin dye molecule, azur B, has been shown to bind almost perpendicularly with its longitudinal axis to the long axis of the chondroitin sulfate molecule.[41,42] An axiperpendicular orientation of toluidine blue bound to polyanionic chromotropes has also been suggested.[12,43,44] The cationic phthalocyanin dye, alcian blue, is also supposed to bind orthogonally to polyanions.[45]

The orientation pattern of dye molecules bound to tissue constituents can be estimated by determining the sign of birefringence and dichroism under a polarization microscope. If the spatial relationship of the dye to the chromotrope is known, the spatial orientation pattern of the chromotrope can be inferred. For example, if a collagen fiber bundle stained with

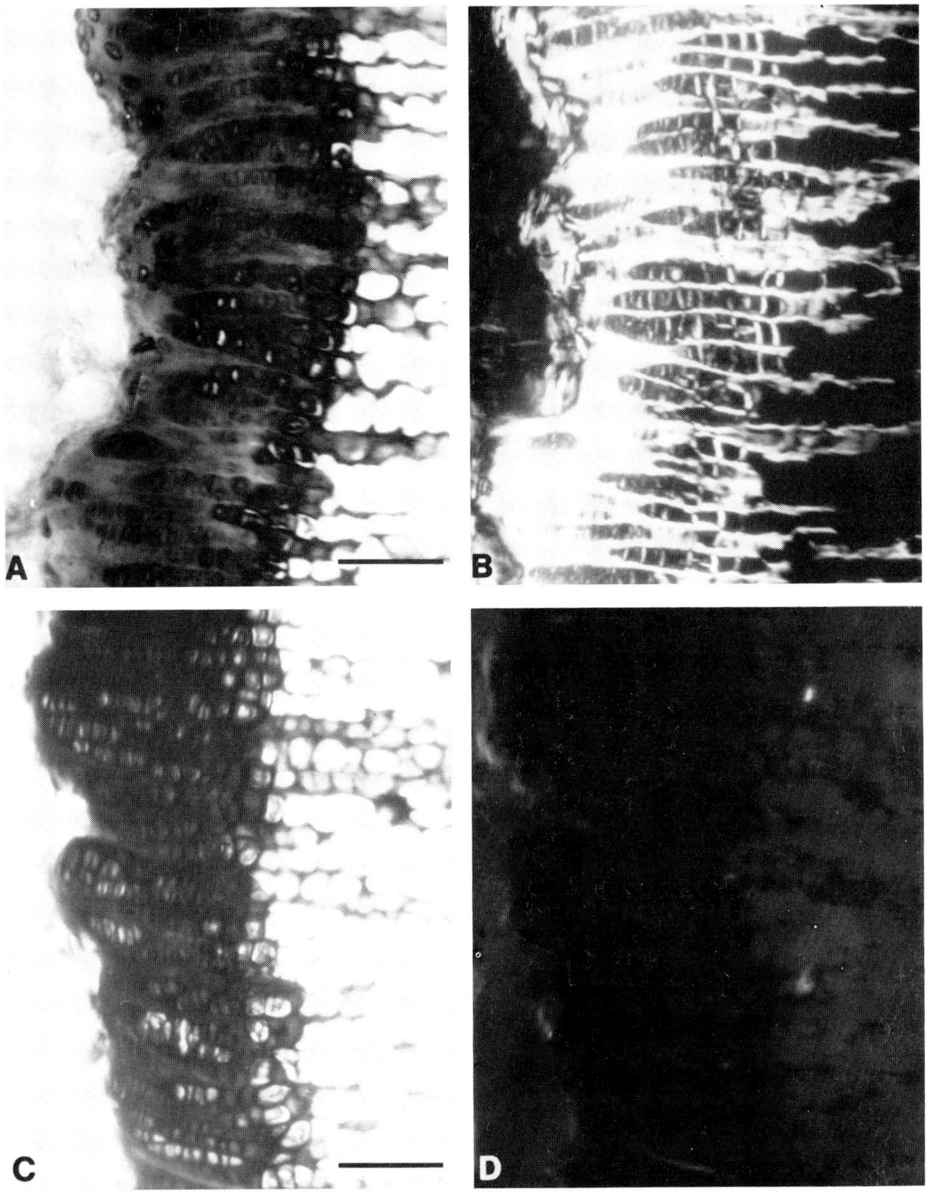

FIGURE 10. Epiphyseal cartilage of CFY-rat after staining with Janus red (A, B) and Janus blue at pH 2.5 (C, D) taken in normal (A, C) and in polarized light (B, D). Note that both of the dyes intensively stain the cartilage matrix, but only Janus red induces birefringence. Scales: 100 μm.

toluidine blue at pH 2.5 exhibits metachromasia in normal light and birefringence in polarized light, and if the relative sign of birefringence is negative relating to the long axis of the fibers, we can say that the dye molecules, or at least their great majority, are oriented perpendicularly to the fibers. Keeping in mind the aforementioned spatial relationships between toluidine blue and chromotrope molecules, we can conclude that the chromotropes (in our example sulfated GAGs) are oriented to the collagen fibers in an axiparallel manner.

E. METACHROMATIC TOPO-OPTICAL REACTIONS

1. Definition

Metachromasia is one of the oldest and most widely used histochemical reactions.[12,46-50]

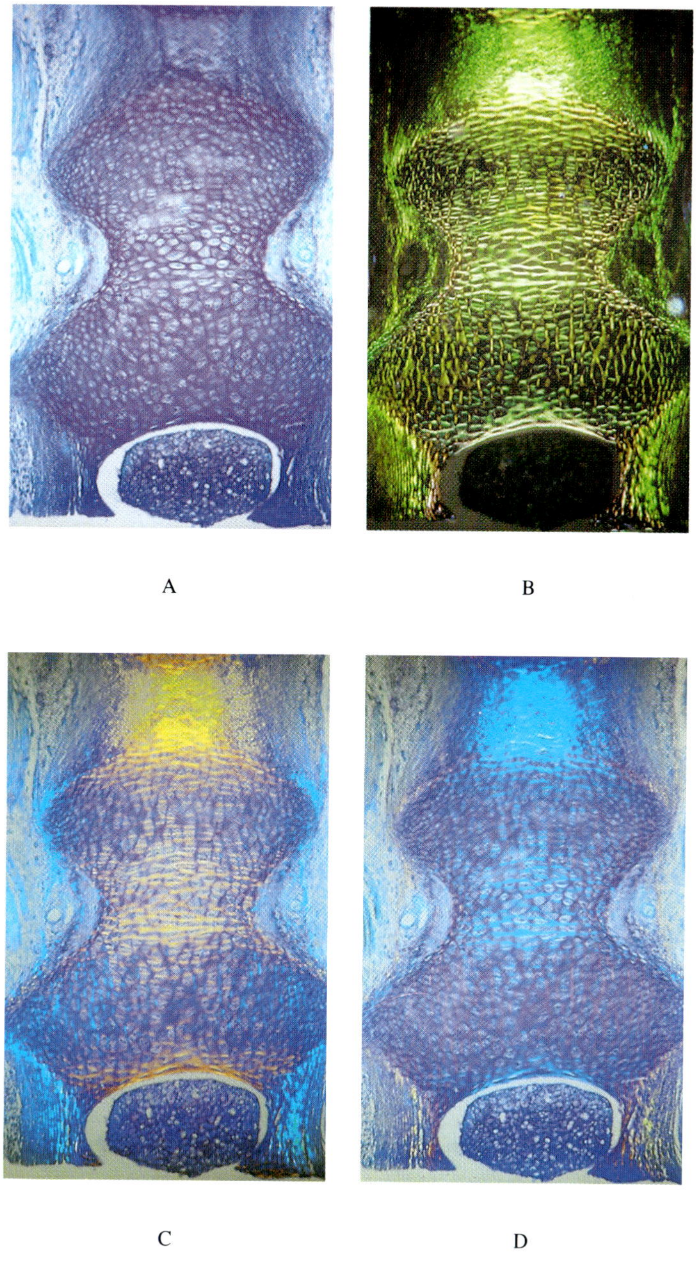

PLATE 1. Vertebral body and intervertebral disc of the tail of a newborn CFY-rat after staining with toluidine blue at pH 2.5 for sulfated GAGs to demonstrate the basic optical phenomena of a metachromatic topo-optical reaction. (A) Normal light microphotograph showing the violet metachromasia of the cartilage matrix and nucleus pulposus. The collagen fibers of the anulus fibrosus and the perichondrium are slightly metachromatic. Muscle fibers are orthochromatically stained (in blue). (B) The same visual field taken in polarized light between crossed polaroids. The metachromatic cartilage matrix and collagen fibers exhibit birefringence appearing as a yellowish-green polarization color. Nucleus pulposus is isotropic indicating the random distribution of its GAGs. (C and D). The same visual field taken in polarized light between crossed polaroids after compensation in different directions. Note the appearance of the anomalous polarization colors, the blue after subtractive and the orange after additive compensation. These colors appear in structures which run perpendicularly to each other and which contain oriented GAGs.

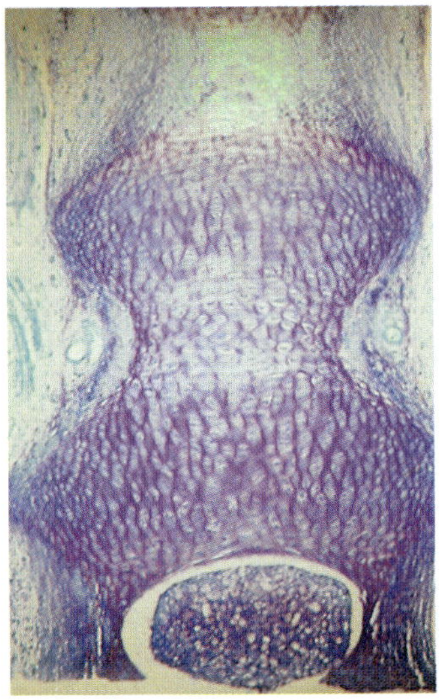

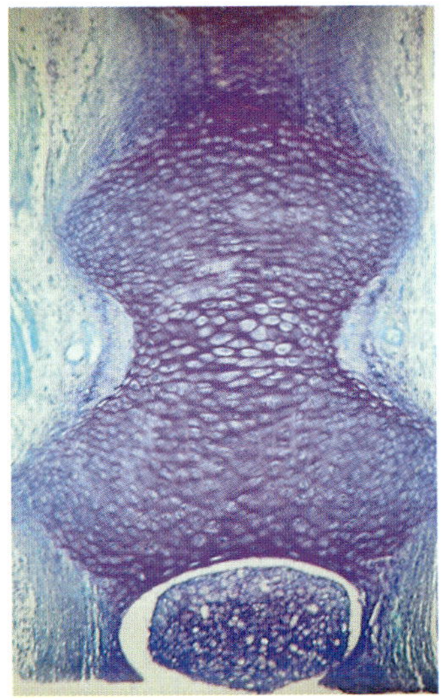

E F

PLATE 1 (E and F). The same visual field taken in linearly polarized light without analyzer to demonstrate the dichroism. Dark and pale red cartilage matrix trabecules are in perpendicular orientation to each other. Dark red signifies where the axis of the polarizer (i.e., the vibration plane of the polarized light) is parallel with the long axes of the dye molecules, and pale red is visible in the perpendicular position. The axis of the polarizer is roted 90° in (E) and (F).

The term metachromasia is used to describe a change in color of a dye after interaction with a chromotrope being either in solution or in tissue. In most cases, metachromasia develops if a cationic dye of planar configuration interacts with polyanions. These are two events which simultaneously occur during a metachromatic reaction: (1) a dye-chromotrope interaction which is ionic in nature; and (2) a dye-dye interaction. According to model experiments, the prerequisite of a metachromatic color shift is the high charge density on the surface of the chromotrope molecule. The distance between neighboring dye-binding groups must be within 0.5 to 0.7 nm, otherwise the dye-dye interaction cannot occur.[46,47] As a consequence of these interactions, coplanar dye-aggregates are formed on the chromotrope surface (Figure 11). Their length varies as a function of the charge density and length of the polyanionic chromotrope resulting in different color shifts which can be measured spectrophotometrically. It is obvious that — due to their polyanionic character — sulfated GAGs and nucleic acids are the ideal chromotropes to obtain a metachromatic reaction.

A metachromatic staining reaction exhibits optical anisotropy.[12,43,51-61] It is due to an oriented dye-binding to the chromotrope. A frequently neglected and salient precondition of induction of optical anisotropy by a metachromatic reaction is that the chromotrope molecules — or at least their great majority — must be spatially oriented along a preferential axis. If this is not the case, a metachromatic color shift is visible in normal white light, but it is isotropic when viewed in polarized light between crossed polaroids. This is the basis of any polarization microscopical ultrastructural analysis using metachromatic staining reactions.

During a staining carried out by a metachromatic reaction, some tissue constituents are stained in the original color of the dye. This staining effect is called orthochromasia and generally it is developed where the anionic charge density is low (Figure 11). Orthochromasia is a random dye-binding. Therefore, this staining phenomenon is not accompanied by optical anisotropy. There is an exception to this rule. Amyloid and corpora amylacea are stained orthochromatically using the metachromatic dye, toluidine blue at low pH, and the orthochromatic areas exhibit a strong birefringence (see Chapter 10). In this case, the staining mechanism is not electrostatic, and the special oriented ultrastructure of amyloid fibrils containing proteins of β-pleated sheet conformation is responsible for this unique phenomenon.

2. Application

According to Romhányi and his group[12,50,53] a metachromatic staining reaction — provided that the chromotrope is an oriented system — has three optical manifestations: 1 metachromatic color shift; 2 birefringence; and 3 dichroism (Plate 1). All of these phenomena can be quantified; the metachromasia and dichroism in a cytophotometer and the birefringence in a polarization microscope. Due to the fact that dyes being in metachromatic aggregate do not obey the Lambert-Beer's law, the measurement of optical density of a metachromatic structure is not recommended for quantification of the dye molecules bound to the chromotrope. The shift of the absorption maximum — which can be easily estimated by recording absorption curves — however, provides valuable information about the degree of metachromasia, i.e., the anionic charge density of the chromotrope.

A metachromatic topo-optical reaction is a useful tool in ultrastructure research if its specificity is known. As has been discussed earlier, there are several possibilities to selectively stain sulfated GAGs with a cationic reagent. Therefore, the metachromatic topo-optical reactions are frequently used in structural analysis of the sulfated GAG molecules and their proteoglycans. Using such staining conditions, the following method of microscopic analysis can be suggested.

The stained section is put on the stage of a polarization microscope, and the polaroid filters are removed. The staining properties are analysed, and the appearance of metachro-

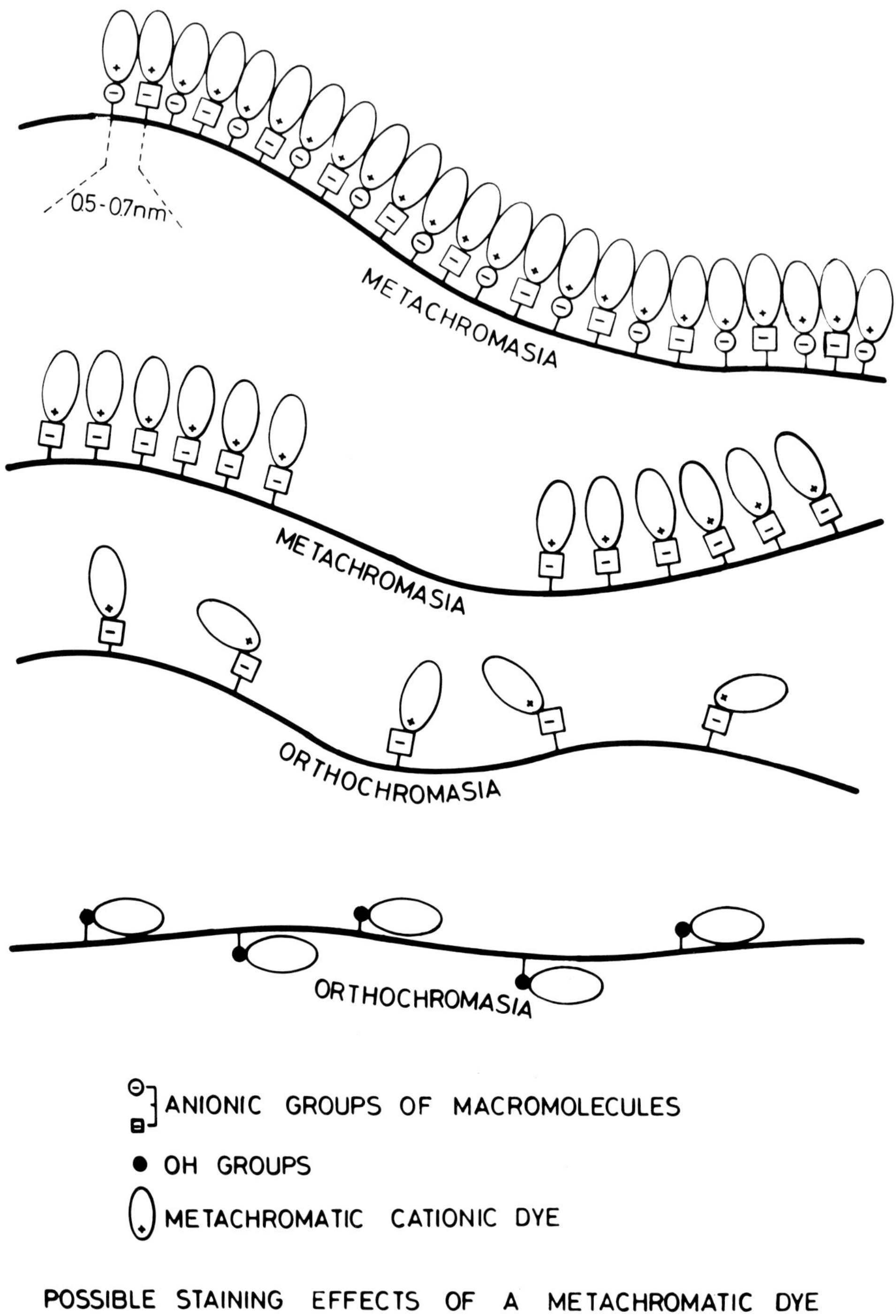

FIGURE 11. Diagram showing the dye-binding in metachromatic and orthochromatic staining. Chromotropes are illustrated by thick lines. Long dye aggregates can be developed if the anionic charge density is high on the chromotrope surface, e.g., in sulfated GAGs (top). In this case, sulfate (circles) and carboxyl groups (squares) alternate. If the anionic dye-binding groups are scarcer, smaller metachromatic dye aggregates are formed or an orthromatic staining effect develops with random dye-binding. In some cases (for example in amyloid) dye molecules are bound in an oriented manner in orthochromatic staining. Birefringence is induced by both metachromatic and orthromatic stainings if the dyes and chromotropes are spatially oriented.

masia is registered. The presence of metachromasia means that polyanions (e.g., sulfated GAGs) are present in appreciable amounts. The absence of metachromasia, however, does not mean the absence of the chromotrope (GAGs). The only correct statement is that in this case our method failed to detect the substance looked for. If our microscope is equipped with a cytophotometer, absorption curves can be recorded from the metachromatic areas. In the next step, the microscope is adjusted by polarizer and analyzer filters and a compensator plate. The polaroids are crossed, and the slow axis of the compensator plate is in parallel position with the axis either of the polarizer or analyzer. The use of an intense transilluminating light is advisable. The stage is slowly turned and the briefringence of the metachromatic areas is observed. Using toluidine blue or other cationic thiazin dyes for staining, the induced birefringence appears in green, greenish-yellow, or red polarization colors. Green polarization color appears if the dye molecules are perfectly oriented, because the metachromatic dye aggregate is red (in normal light), i.e., it absorbs green light, and the birefringence is maximal in the spectral region where the dye has its absorption maximum. Red polarization color appears either in loosely oriented structures or in overstained specimens or in thicker sections. If birefringence is seen, we can conclude that the chromotrope is spatially oriented.

The next step is the establishment of the relative sign of the birefringence. The major axis of the structure is put between the axes of the polaroids. Turning the compensator plate in different directions (or the analyzer if a $\lambda/4$ plate is used), the birefringence is extinguished in one direction and enhanced in the other direction. Anomalous polarization color appears after compensation. Using metachromatic thiazin dyes, blue color is visible after subtractive and orange appears after additive compensation in white transilluminating light. Using standard preparations (birefringent structures of known sign), the sign of birefringence can be determined. In the case of linear structures (e.g., collagen fibers), the sign is related to their long axes. In spherulites, however, the sign is related either to the tangent or to the radius. The orientation of the dye molecules to the reference axis can be inferred from the sign of birefringence. For example, if a collagen fiber exhibits birefringence with negative sign after a staining with toluidine blue, the dye molecules are perpendicularly oriented to the long axis of the fibers. If the orientation of the dye molecules to the chromotrope is known, we can establish the orientation pattern of the macromolecule stained. In the previously mentioned example, the chromotrope is aligned parallel to the long axis of the collagen fibers, because toluidine blue molecules are perpendicularly bound to the polyanions.

For quantification, monochromatic transilluminating light is used. Generally, to have good contrast during the compensation, the retardation value is measured at the wavelength of the complementary color of the polarization color (e.g., in red monochromatic light if the structure is stained with toluidine blue and the birefringence appears in green polarization color). In other words, the monochromatic light used for measurements has the same color as seen in the stained specimen in normal white light (e.g., the metachromasia induced by toluidine blue appears generally in red color).

Retardation values of birefringence are calculated and plotted in histograms, or they are averaged. Histograms are more informative, because they reflect the homogenity or heterogeneity of the orientation pattern. Another possibility of quantification is the recording of anomalous dispersion curves of birefringence. In this case, retardation values are measured and sign of birefringence is determined at different wavelengths, and the retardation values are plotted against the wavelength of the transilluminating light.

The last step of the analysis is the investigation of the dichroism. As has been discussed in Chapter 2, dichroism is a direction-dependent absorption of the polarized light. The analyzer and compensator plate are removed from the microscope, and either the stage or the polarizer is slowly turned. The intensity of the color of the structure is observed in white light. If the vibration direction of polarized light is parallel with the long axes of the dye molecules bound to the chromotrope in an oriented manner, the intensity of the color is higher (the light absorption is greater). In a perpendicular position, the color is pale, because the light absorption is smaller. Dichroism is a very sensitive indicator of an oriented dye-binding, and its qualitiative analysis gives information about the orientation of the dye molecules, and , consequently about the orientation of the chromotrope. Dichroism can be quantified using a cytophotometer. Absorption curves are taken in two positions of the stained structure. The two positions are perpendicular to each other. In one position the vibration plane of the polarized light is parallel to the long axis of the structure, and in the other position they are perpendicular to each other. The two absorption curves are compared. If their difference is great, the degree of the dye orientation (which is supposed to reflect the degree of orientation of the chromotrope) is high. If the two curves are similar, the degree of orientation is low. Examples are shown in Chapter 10.

3. Toluidine Blue

This is the most classical metachromatic dye and the most frequently used cationic topo-optical reagent.[12,13,43,49,50,53-74] It is an inversive topo-optical reaction, and if a poststaining precipitation is used[12] (see Chapter 12), the metachromatic toluidine blue staining induces a strong optical anisotropy (Figures 12, 13). The first observations on the spatial orientation of GAGs were made after toluidine blue staining in polarized light microscope by Romhányi and Jobst,[53] Romhányi,[12] and Vidal.[43] It has been suggested by these authors that sulfated GAGs and collagen fibers are oriented parallel to each other. It should be mentioned here that the same orientation pattern of the sulfated GAGs could be inferred by imbibition analyses of unstained tendon and cartilage tissues using enzymatic extractions to remove GAGs[75-77] (see also Chapter 2).

If the toluidine blue staining is carried out at higher pH (pH 4.5 to 7.2), the orientation of carboxyl group-containing structures of the cell coat can also be investigated.[78-85] In this topo-optical reaction, predominantly the neuraminic acid residues of the different membrane glycoproteins are responsible for the dye-binding. An example is shown in Chapter 2. The toluidine blue staining combined with a poststaining stabilization using potassium ferri-cyanide has been used to show the cell coat by electron microscope.[86] This technique is not reliable, because the toluidine blue binding is vulnerable to the subsequent alcoholic treatment used for dehydration (see Chapter 11), and, furthermore, the electron dense precipitant (potassium ferricyanide) can aspecifically bind to different structures.

Optical anisotropy of nucleic acids can also be induced by this dye, and cell nuclei, chromosomes, and cytoplasmic RNA can be investigated in polarized light.[6,87-91] DNA of nuclei exhibit birefringence only if histones are removed prior to the staining with toluidine blue.

4. Dimethylmethylene Blue

This is also a metachromatic thiazin dye,[92] with two advantages over toluidine blue. The first one is that the metachromatic-orthochromatic colors are always very distinct even using a staining solution of higher pH (pH 5 to 7). The second advantage is that it does not need any poststaining stabilization.[13] It also induces an inversive topo-optical reaction, i.e., it binds perpendicularly to the chromotrope. Intense birefringence can be induced with dimethylmethylene blue (Figures 12,14). It can be used in a topo-optical reactions of GAGs of the extracellular matrix[13] and cell membranes,[93] and as a simple biochemical method to measure sulfated GAGs.[94,95]

5. *N,N′* Diethylpseudoisocyanine

This metachromatic and fluorescent cationic quinoline dye is used in histochemistry of different polyanionic substances.[96-108] Its orthomatic color is red (absorption maximum is at

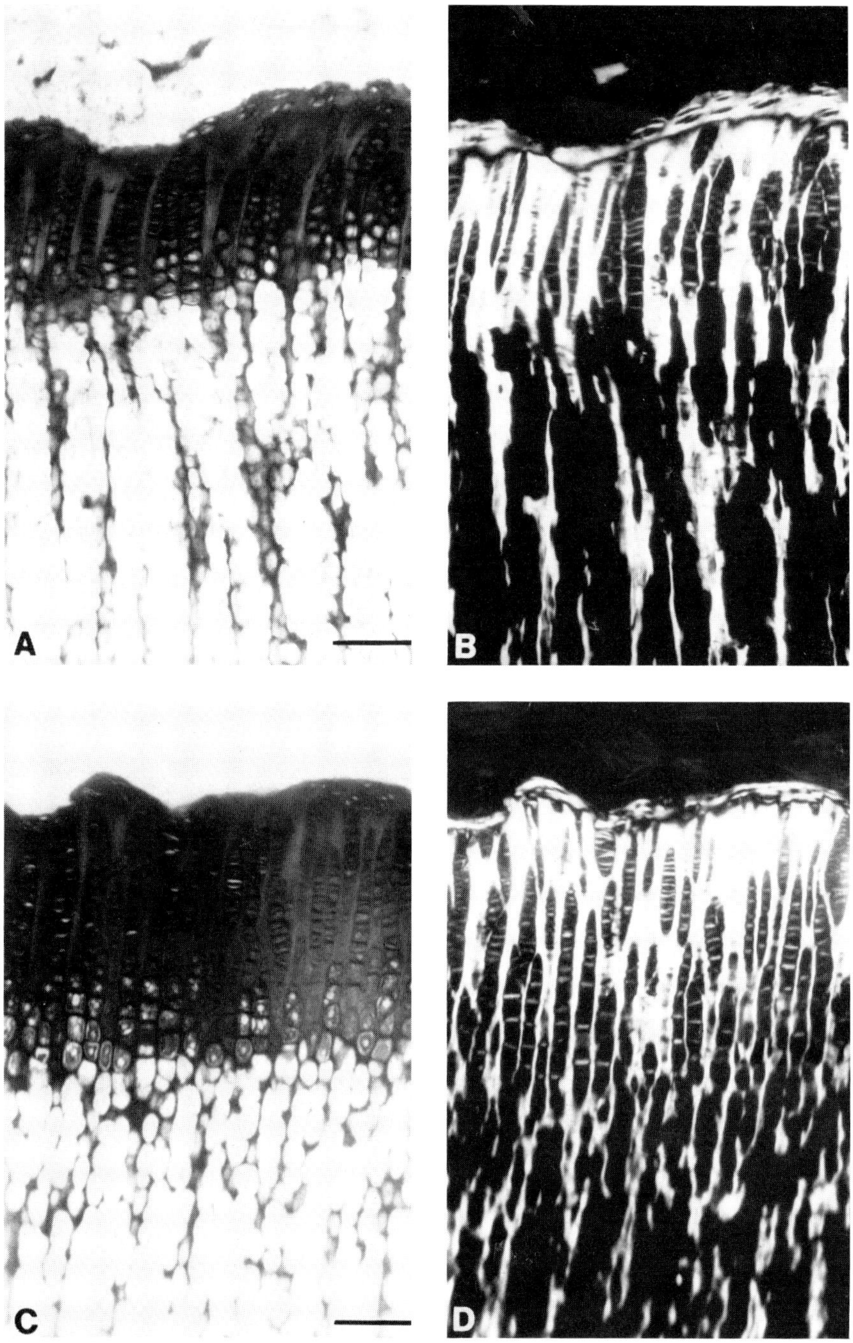

FIGURE 12. Epiphyseal cartilages of CFY-rats after staining with toluidine blue (A, B) and dimethylmethylene blue (C, D), both at pH 2.5, taken in normal (A, C) and polarized light (B, D). Note that the cytoplasm of the chondrocytes is stained but not birefringent. Cartilage matrix where GAGs are oriented exhibits strong optical anisotropy after these stainings. Scales: 100 μm.

λ = 524 nm), and the metachromatic color is violet (the absorption maximum varies between 562 and 581 nm as a function of polyanion added.[13] This dye is bound with its long axis parallel to the chromotrope,[13,37] and can form a multilayer dye aggregate.[35-37] Therefore, it can induce optical anisotropy. It has been shown by Scheuner and Weiss,[109] that neuro-

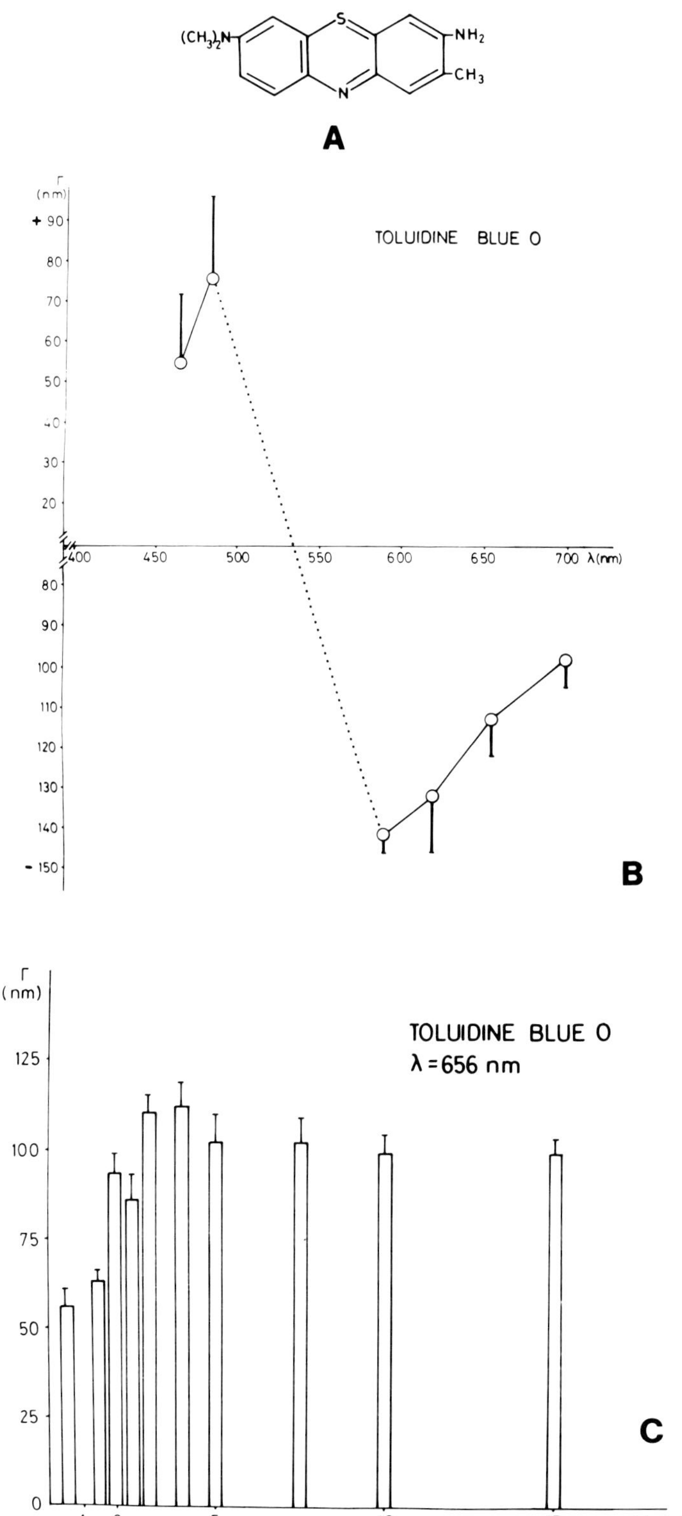

FIGURE 13. (A) Structural formula of toluidine blue. (B) Anomalous dispersion curve of birefringence measured in epiphyseal cartilage matrix stained with toluidine blue at pH 2.5. The sign of birefringence is negative at longer wavelengths of the transilluminating light. The sign of birefringence changes in green light. (C) The change of retardation values measured in epiphyseal cartilage matrix as a function of the staining time. Maximum retardation values can already be measured within 5 min.

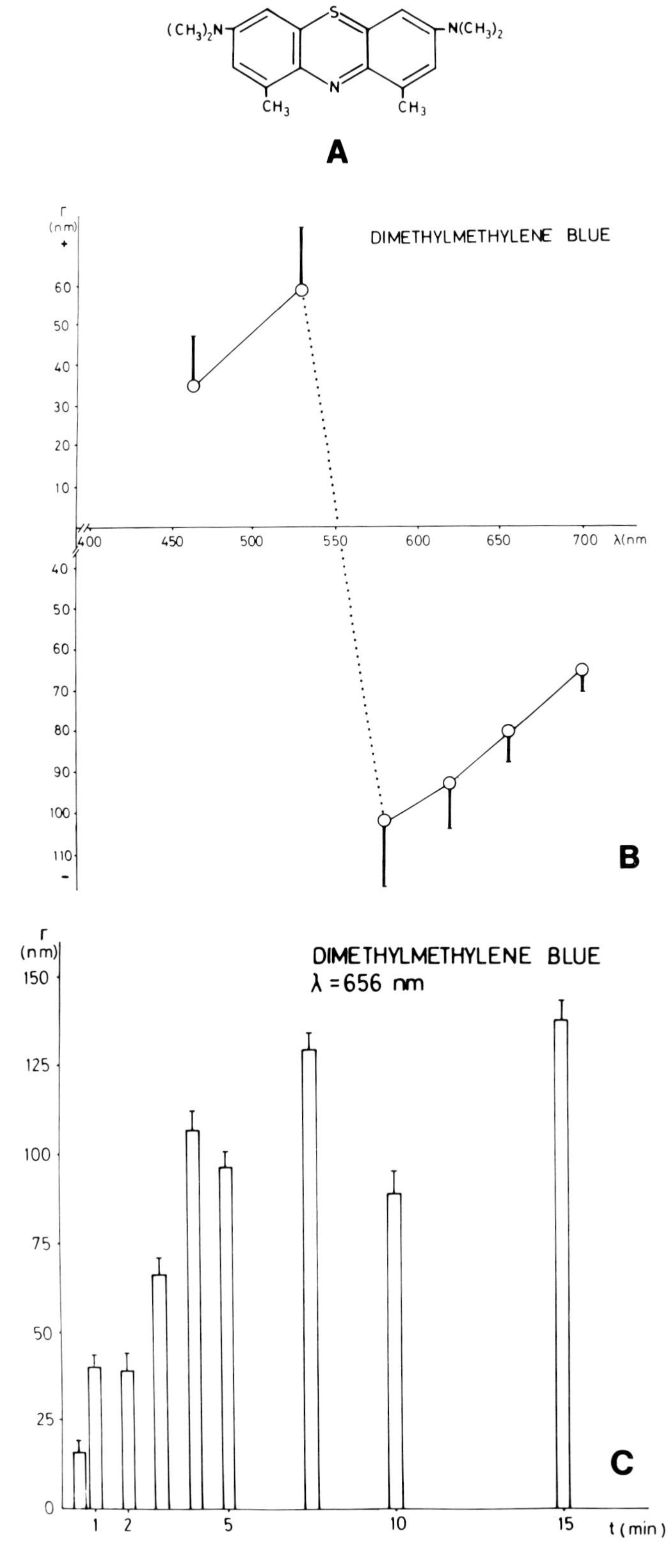

FIGURE 14. Some chemical and optical features of dimethylmethylene blue. (A) Structural formula. (B) Anomalous dispersion of birefringence measured in epiphyseal cartilage matrix. It is similar to that recorded after toluidine blue. (C) Time-dependence of the retardation values. Less than 10 min staining time is enough to obtain maximal birefringence.

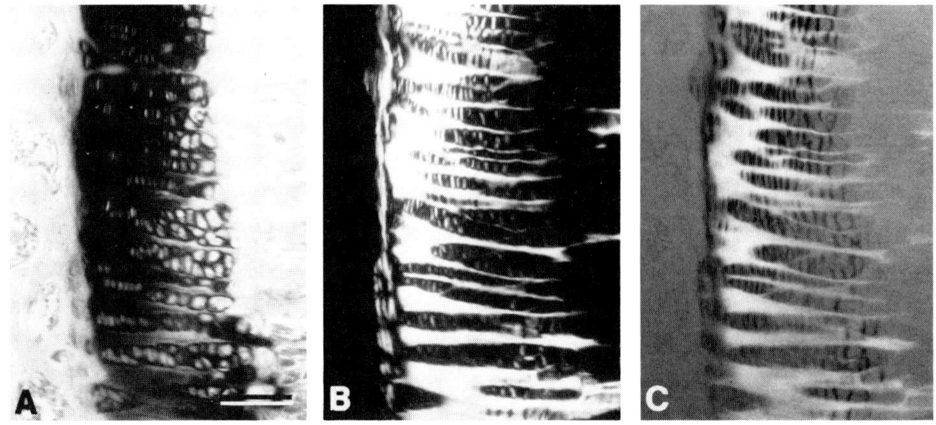

FIGURE 15. Rat epiphyseal cartilage after staining with *N,N'* diethylpseudoisocyanine at pH 2.5 and taken in normal (A) and polarized light without (B) and after (C) compensation. Scale: 100 μm.

secretory droplets are strongly birefringent if the staining with *N,N'*-diethylpseudoisocyanine was preceded by an oxidation of –SH and –SS groups to $-SO_3H$ groups. It has been found in our laboratory, that this dye is an excellent topo-optical reagent of the sulfated GAGs especially in cartilage matrix.[13,110,111] It can be used to enhance the birefringence of polyanions in basement membranes[112] and cell membranes.[93,113] Anomalous dispersion curves of birefringence recorded from dye-smears and stained tissues indicate that the metachromatic aggregate of this dye exhibits a birefringence with positive sign at longer wavelengths of the transilluminating light[13,112,114] (Figures 15, 16). It follows that this is an additive topo-optical reaction. It does not need poststaining stabilization,[13] but a short precipitation with potassium ferricyanide may be used.[112]

6. Rivanol

This cationic acridine dye induces a strong birefringence with negative sign in different connective tissue structures containing sulfated GAGs[12,53,115] (Figures 17, 18), and cell nuclei.[116,117] We have shown a metachromatic shift in UV range if GAGs were added to rivanol solutions.[118] An obvious shortcoming of the topo-optical reaction with rivanol over the other metachromatic stainings is that it stains the tissues in pale yellow color which does not permit normal light-microscopic evaluation of the specimen.

F. REACTIONS WITH PHTHALOCYANIN DYES

It is known that the cationic phthalocyanin dyes like alcian blue, cuprolinic blue, and cupromeronic blue are widely used in electron-microscopic histochemistry of PGs (see Chapter 11). Alcian blue, especially in combination with the CEC technique, is one of the most popular light-microscopic histochemical reagents of the GAGs.[19-21]

It has been shown that GAG-containing tissue constituents, and sulfomucins stained with alcian blue exhibit optical anisotropy[13] and also dichroism[45] (Figures 19, 20). The sign of birefringence is negative, indicating a predominantly axiperpendicular binding of the dye to the chromotrope. Recently, this topo-optical reaction has been used more and more to show the spatial orientation of sulfated GAGs in cartilage and bone.[119-123] Our data, however, show that this dye is not the optimal topo-optical reagent of sulfated GAGs, because the retardation values measured after this staining are much lower as compared to those obtained after toluidine blue dimethylmethylene blue or pseudocyanine stainings (compare Figures 13, 14, 16, 20). In addition, a much longer staining time is needed to saturate the dye-

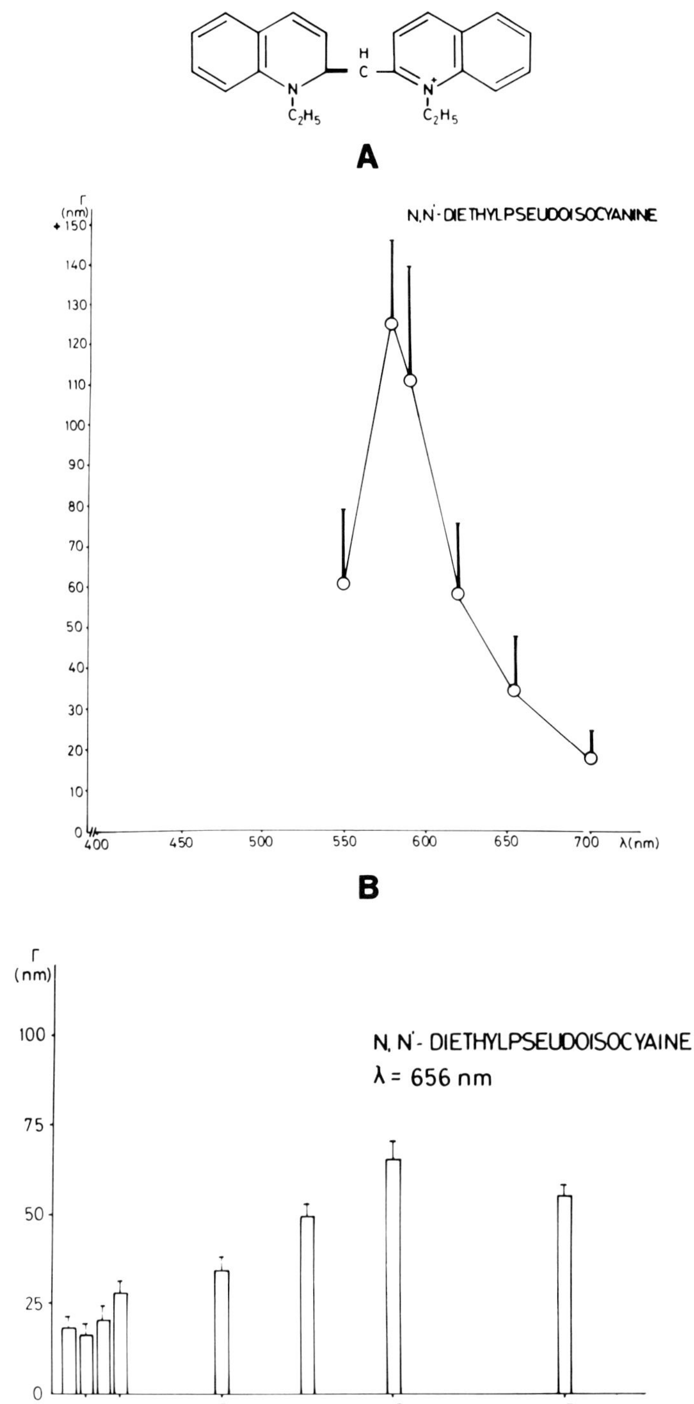

FIGURE 16. Some chemical and optical features of *N,N'*-diethylpseudoisocyanine. (A) Structural formula; (B) anomalous dispersion curve of birefringence; (C) time-dependence of the optical anisotropy. (B) and (C) measurements were carried out in growth plate cartilage matrix stained at pH 2.5. Note the positive sign of birefringence in (B), which can be explained by the localization of cationic group in the dye molecule (A). 10 min staining time is required to obtain maximum retardation values.

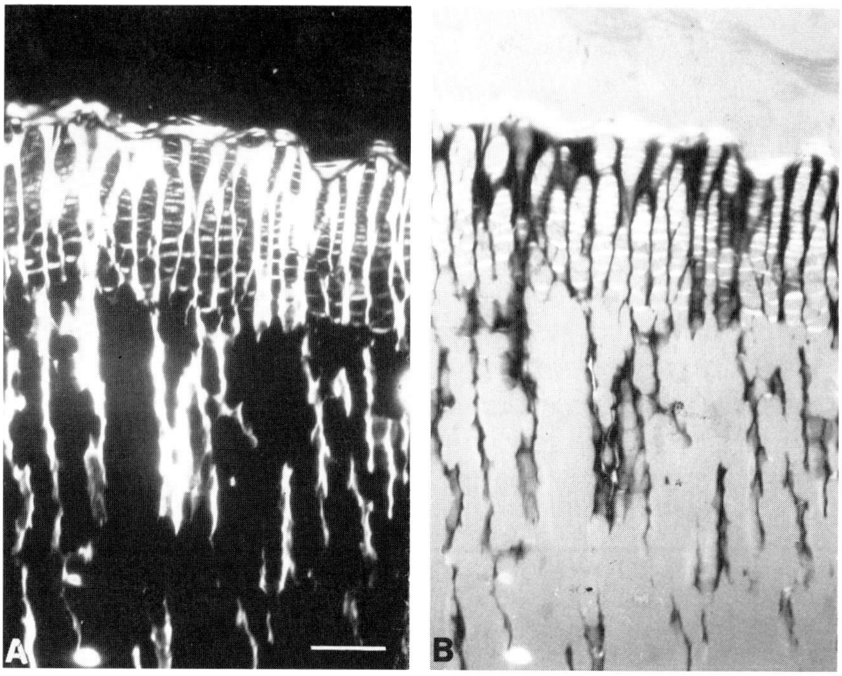

FIGURE 17. Rat epiphyseal cartilage stained with rivanol at pH 2.5 and taken in polarized light without (A) and after compensation (B). Scale: 100 μm.

binding groups than when using the previously mentioned dyes (Figure 20). Both features of the alcian blue topo-optical reaction may be explained by its large size which does not allow a perfect orientation on the chromotrope surface and a rapid penetration of the tissues.

The oriented binding of the alcian blue molecules to polyanionic chromotropes may be due to its ability to form metachromatic aggregates.[45,124] Its metachromatic staining effect is not obvious because the unperturbed orthromatic molecules are observed in disaggregating organic solvents, like dimethylsulphoxide.[124] Single, orthochromatic molecules are not clearly observed in aqueous solution even at low dye concentration (10^{-7} mol/l or less).[124]

Although recognizing that the alcian blue 8GX combined with the CEC principle is a useful and generally accepted histochemical reaction — which also permits the histochemical differentiation of different GAGs — it cannot be recommended as a sensitive topo-optical reagent of sulfated GAG molecules.

Amyloid has been shown to be birefringent after staining with alcian blue.[125] It is not clear, however, which component of the amyloid is responsible for this topo-optical reaction.

We observed that cuprolinic blue can also induce birefringence of negative sign when sulfated GAGs containing tissues are stained (Figures 19, 21). This topo-optical reaction does not show better characteristics than the alcian blue staining.

IV. REACTIONS OF CONNECTIVE TISSUE FIBERS

A. REACTIONS FOR COLLAGEN FIBERS

Polarization microscopy has been used for qualitative and quantitative investigation of the collagen structure in unstained specimens.[126-151] Collagen fibers exhibit a uniaxial birefringence with positive sign. An attractive explanation of the positive intrinsic birefringence of collagen is proposed by Wolman and Kasten[151] (Figure 22). In their model, one of the three helical polypeptide chains is shown with its superhelix superimposed on the primary helix.

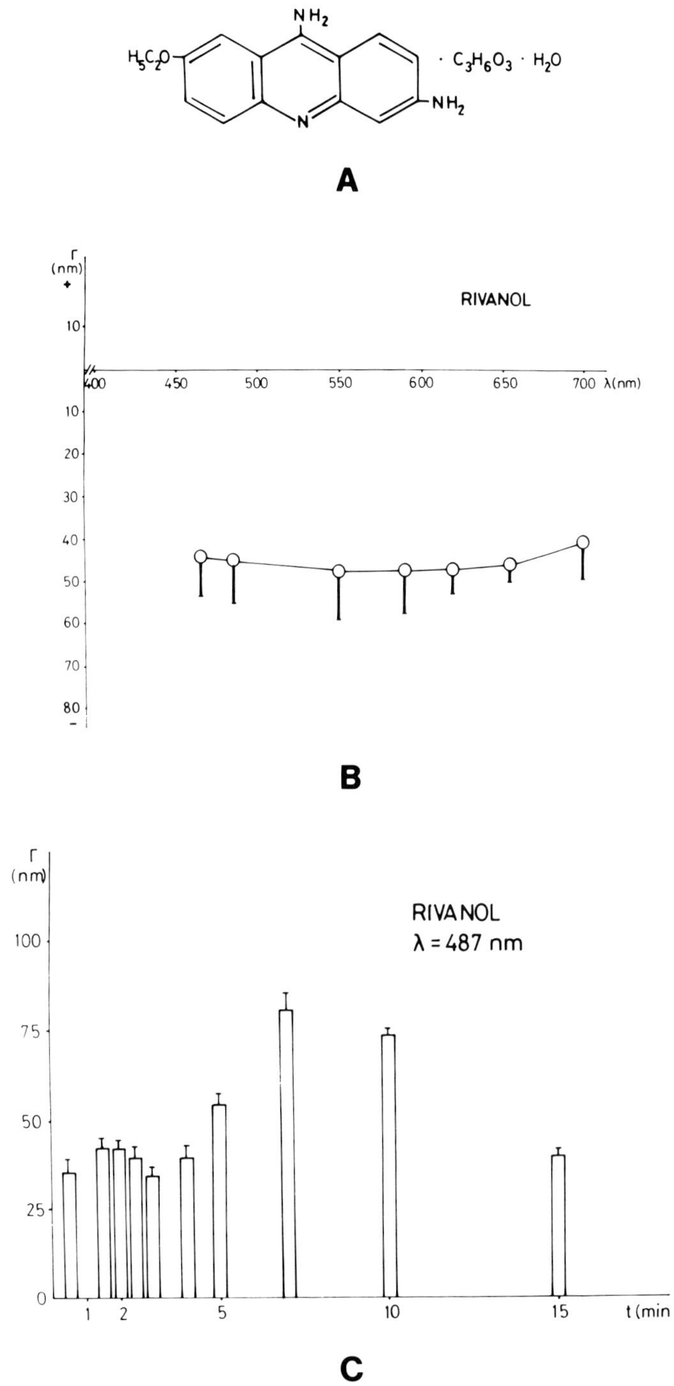

FIGURE 18. Some chemical and optical features of rivanol. (A) Structural formula. (B) Anomalous dispersion curve of birefringence. Compared to other cationic dyes, the birefringence of cartilage matrix stained with this dye does not change as a function of the wavelength of the transilluminating light. The sign of birefringence is negative. (C) Maximum retardation value can be obtained within 10 min staining time.

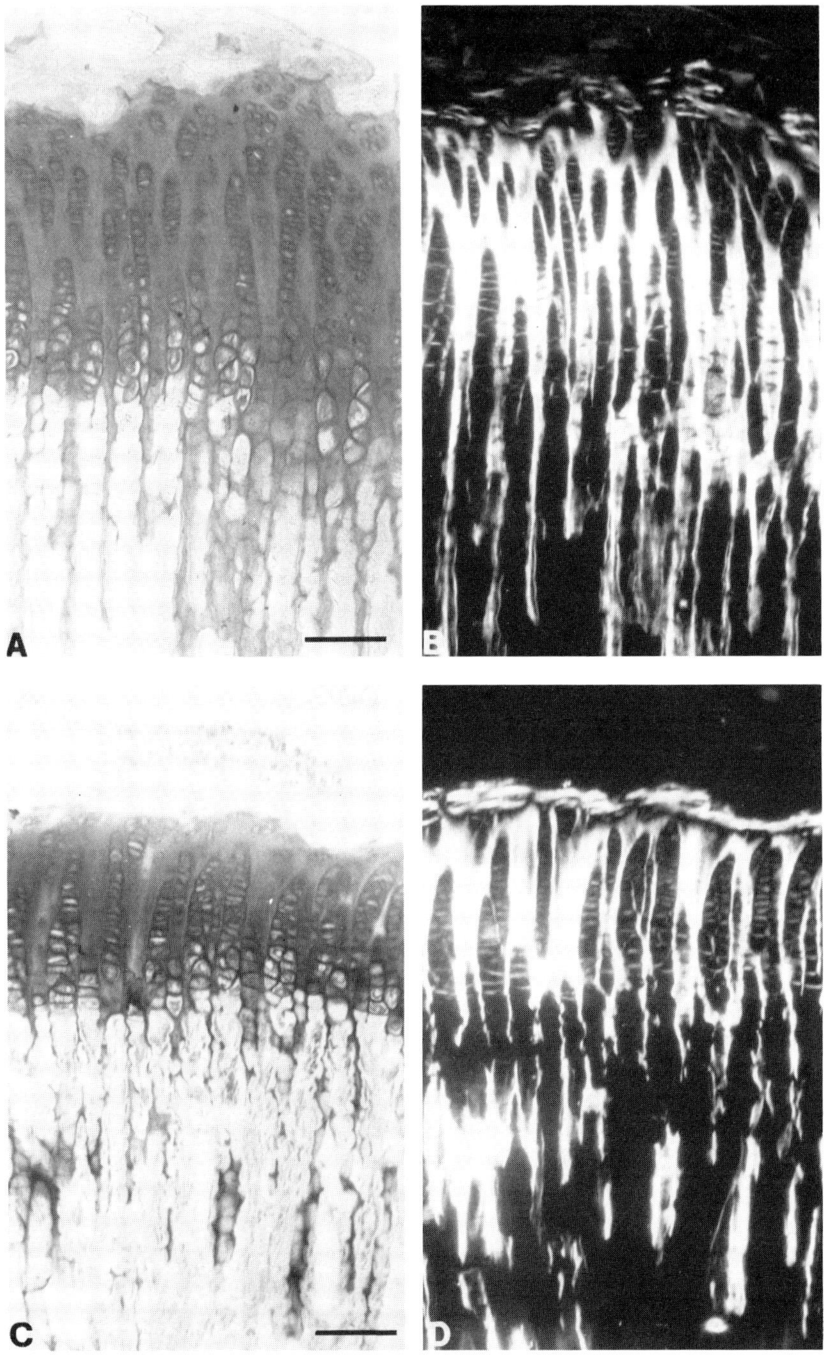

FIGURE 19. Rat epiphyseal cartilage stained with alcian blue 8GX (A, B) and cuprolinic blue (C, D), both at pH 2.5, and taken in normal (A, C) and polarized light (B, D). Scales: 100 μm.

The right-handed supercoiling compensates for the left-handed coiling with a consequent alignment of the polypeptide chains approximately parallel to the molecular axis. This alignment explains why unstained collagen molecules are intensely positively birefringent, and why collagen fibrils and fibers built up from parallel arranged collagen molecules show positive form birefringence.

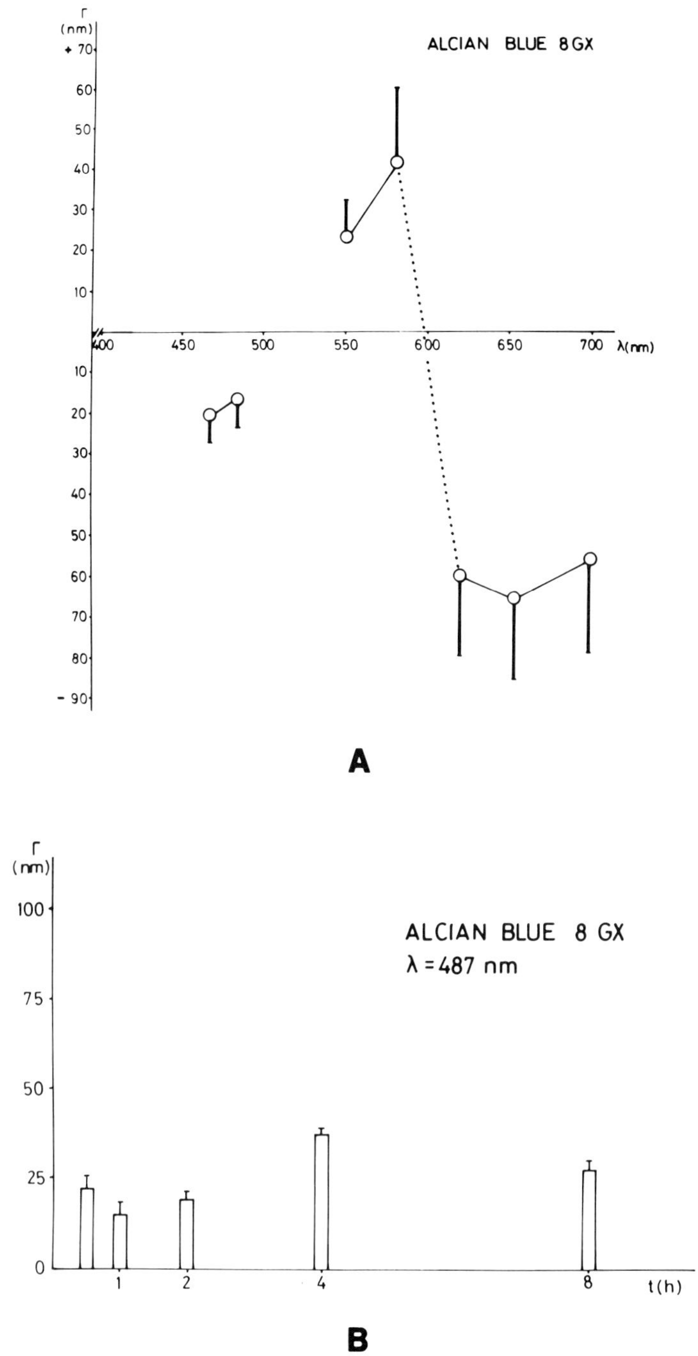

FIGURE 20. Some optical features of the epiphyseal cartilage matrix stained with alcian blue 8GX at pH 2.5. (A) Anomalous dispersion curve of birefringence showing the negative sign of birefringence both at longer and shorter wavelengths of the transilluminating light. (B) The retardation values are considerably lower even after a longer staining time than those obtained on the same specimen stained with toluidine blue, dimethylmethylene blue, *N,N'*-diethylpseudoisocyanine and rivanol. The structural formula of the dye is shown in Chapter 11.

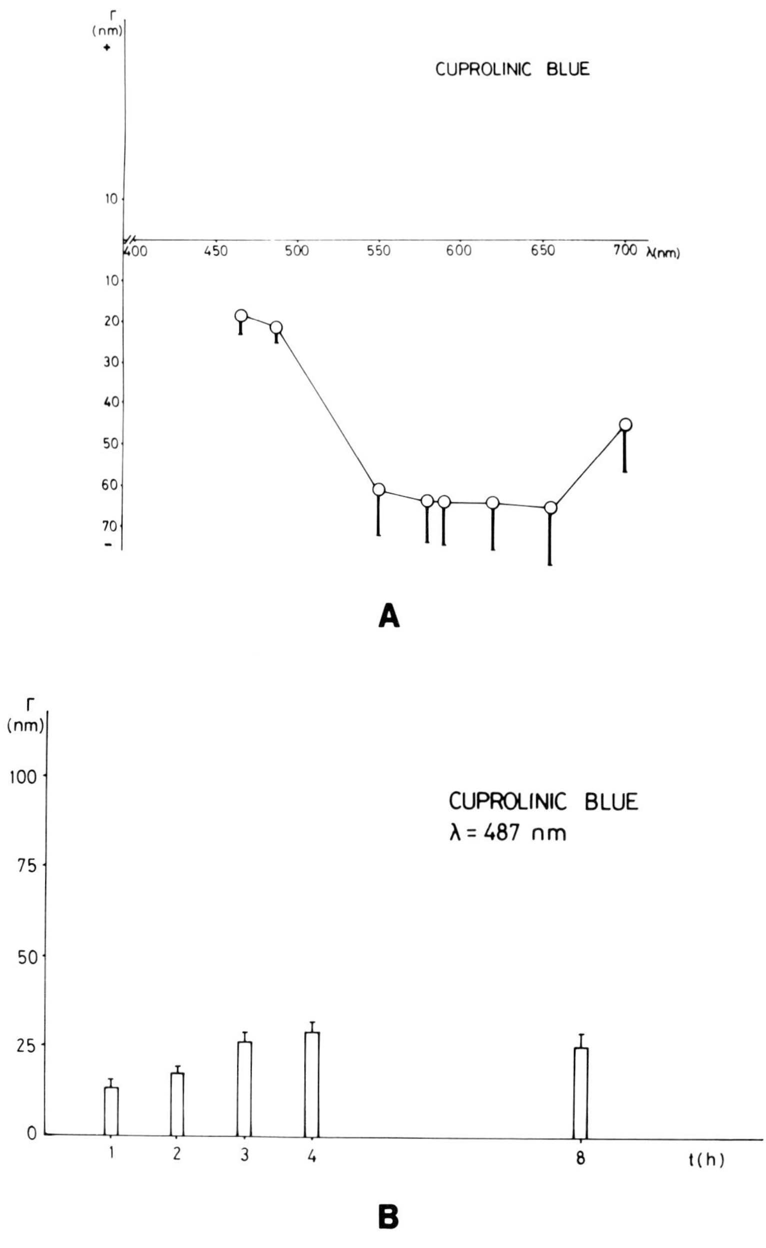

FIGURE 21. Some optical features of a cuprolinic blue topo-optical reaction carried out on rat epiphyseal cartilage. (A) Anomalous dispersion curve of birefringence showing the negative sign of birefringence. (B) Like alcian blue 8GX, cuprolinic blue induces moderate birefringence even after a longer staining time. The structural formula of the dye is shown in Chapter 11.

Polarization microscopic analysis of unstained collagen, however, is not easy and sometimes even impossible in structures where fine collagen fibers are present, and the intensity of their birefringence is low. In such structures, the specific enhancement of birefringence by a topo-optical reaction is a prerequisite of the polarization microscopic investigations. It was realized already in the last century, and since that time that a number of reactions have been suggested to amplify collagen birefringence.

FIGURE 22. A model showing the effect of the left-handed coil and the superimposed right-handed superhelix on the alignment of the polypeptide chain. As the two angles are similar, the collagen chains are aligned parallel to the axis of the superhelix (arrows). The long axis of the superhelix runs in an East-West direction in the figure. This alignment explains why untreated collagen exhibits birefringence with positive sign. (From Wolman, M. and Kasten, F. H., *Histochemistry,* 85, 41, 1986. With permission.)[151]

1. Phenol

This is the most classical topo-optical reagent introduced by Ebner[8] in 1894. Phenol molecules bind to cationic groups of collagen[152,153] and the planar phenol molecules are perpendicularly oriented to the collagen hereby inverting the originally positive sign of collagen fibers to negative (Figure 23). The inversion occurs as a function of the phenol concentration.[154] For quantification of this topo-optical reaction, calculation of an index has been suggested.[155] During this measurement, the retardation value of birefringence of a collagen structure is measured before and after the phenol reaction, and the retardation value determined after the reaction is divided by the value given by unstained collagen.

This is a highly specific topo-optical reaction of collagen fibers. No other birefringent structures are known at the present moment to which the phenol molecules could bind in an oriented manner causing an inversion of the sign of birefringence. There are two drawbacks, however, when working with the phenol reaction. First, it is a colorless reaction, and this feature makes histological investigation difficult. Second — and more important — is the low sensitivity of the phenol reaction as compared to the other collagen-specific topo-optical reactions. In tissues containing loosely oriented and fine collagen fibrils (e.g., in young embryonic hyaline cartilage matrix), the phenol reaction fails to induce birefringence of negative sign but sulfation and picrosirius red F3B topo-optical reactions reveal oriented collagen structure.[156]

2. Sulfation

It has been described by Romhányi et al.[157] that the –OH groups of collagen belonging to OH-proline and OH-lysine can be sulfated using a mild sulfation procedure (see Chapter 12). If vicinal alcoholic–OH groups of different carbohydrates are oxidized by periodic acid and transformed to aldehydes prior to the sulfation, collagen can be selectively sulfated. The sulfate groups are detected by a toluidine blue staining carried out at pH 1 to 1.5. The presence of sulfated GAGs causes an aspecific reaction. In this case, the first step of this topo-optical reaction must be a methylation to remove the sulfate groups of GAGs (see Chapter 12). The collagen-selective sulfation procedure is a sensitive topo-optical reaction of collagen, its sensitivity being about four times higher than that of the phenol reaction.

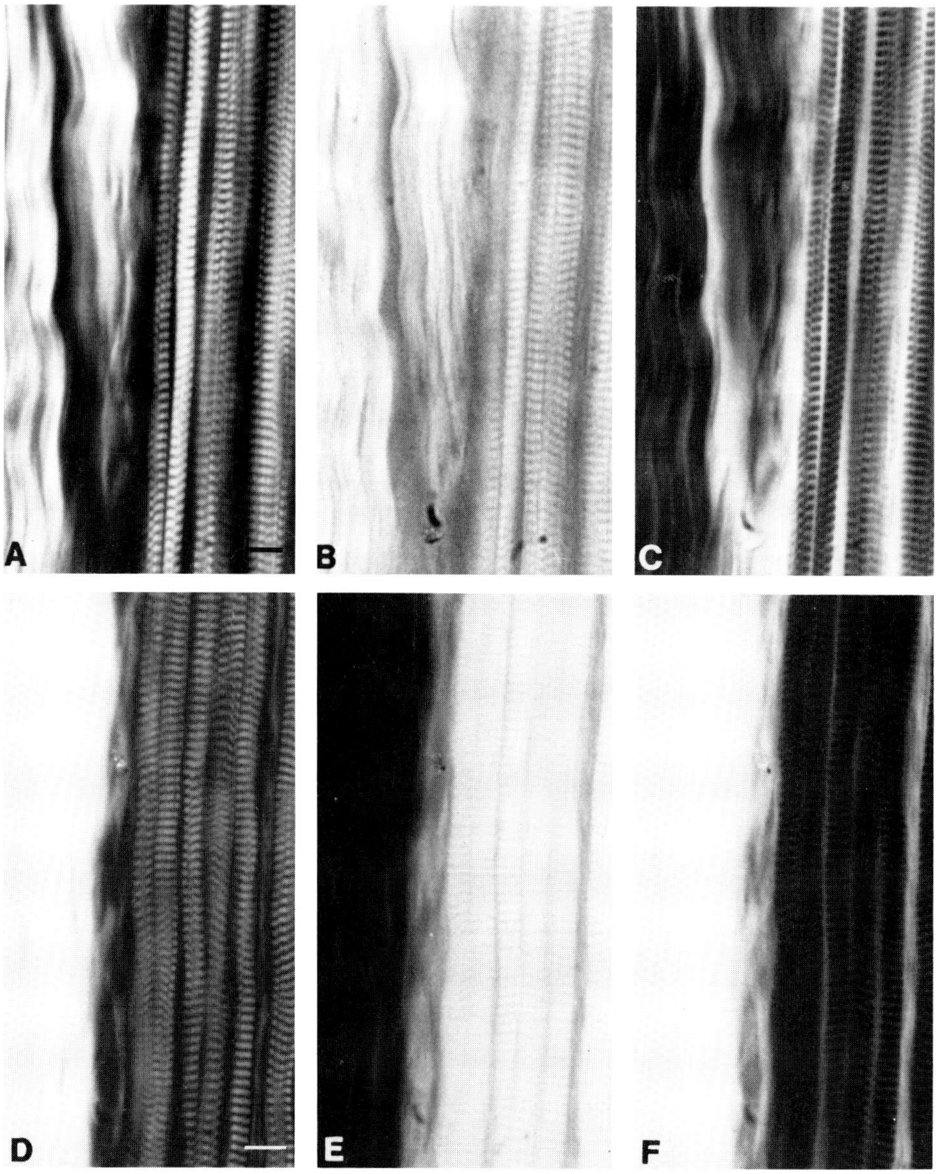

FIGURE 23. Tendon and striated muscle fibers of a young dog (knee joint region). (A, B. C) Unstained section mounted in Canada balsam, and taken in polarized light between crossed polaroids without (A) and after compensation (B, C). Note that the birefringence of collagen (on the left) and muscle fibers are intensified or extinguished together after additive (B) or subtractive compensation (C) indicating the positive sign of birefringence of both of these structures. (D, E, F) Phenol reaction. Polarized light micrographs without (D) and after compensation (E, F). If the birefringence of muscle is intensified by additive compensation, the birefringence of collagen is extinguished (E) and vice versa (F). It follows that the sign of birefringence of the collagen fibers is inverted to negative by the phenol reaction. Scales: 10 μm.

3. Picrosirius Red

Originally the anionic sirius red F3B or F3BA had been proposed to replace the acid fuchsin in the van Gieson procedure to obtain a staining effect where the red color of the collagen was not faded.[158] Constantine and Mowry[159] observed that the optical anisotropy of the collagen is considerably enhanced after a staining with picrosirius red.

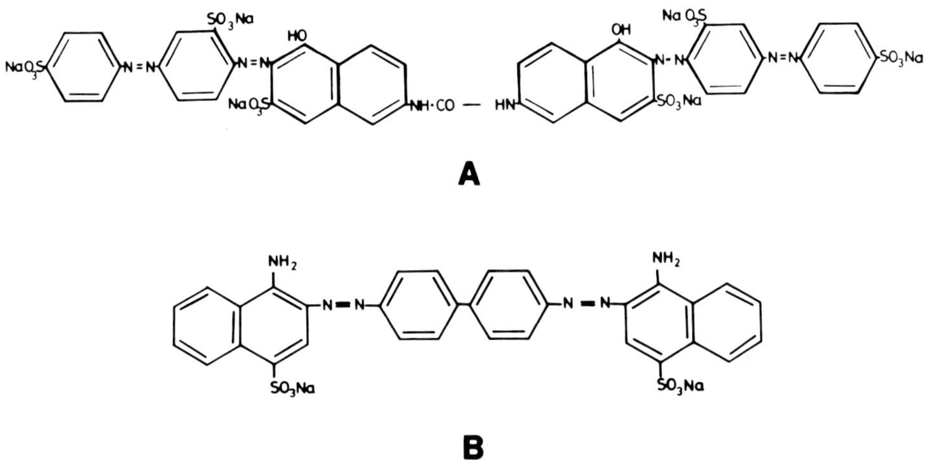

FIGURE 24. Structural formulae of sirius red F3B (A) and congo red (B) molecules according to Lillie.[34] Note the presence of the anionic ($-SO_3^-$) groups. The distribution of the six $-SO_3^-$ groups along the sirius red molecule explains why this dye induces additive topo-optical reaction.

The elongated (4.5 nm long) sirius red F3B molecule (M_r 1373) contains four chromophore azo groups, and six sulfonic acid groups[160,161] (Figure 24). The strongly anionic dye binds to cationic charges present on different tissue constituents by ionic interaction,[162,163] and the dye molecules — due to their charge distribution — are arranged parallel to the long axis of the collagen.[164-166] Consequently, the picrosirius red staining — owing to the non specific ionic interaction of the sirius red molecules to different cationic chromotropes — is an aspecific method when viewed in normal light. The birefringence of positive sign induced by this staining is a specific reaction of the collagen in the extracellular matrix of normal tissues. These limitations — that birefringence induced by picrosirius red is specific for collagen only in ECM and in normal tissues — have to be kept in mind because amyloid, a pathologically occuring and predominantly extracellular material, is also birefringent when stained by this method.[167-170] Furthermore, we found that the intermediate filaments of the cytoskeleton structure exhibit stong birefringence after picrosirius red F3B staining (Figure 25). A similar anisotropic staining of the intermediate filaments has been observed after staining with coomassie blue.[171,172] At present no other structures are known which would give the picrosirius red topo-optical reaction.

The picrosirius red F3B topo-optical reaction attracted much interest and became one of the most popular techniques to show collagen.[173-201] Unfortunately, an erroneous proposal has been made saying that collagen types (types I and III) can be distinguished according to the different polarization colors appearing in different fibers containing genetically distinct collagen types after picrosirius red staining.[162,193,202,203] The same possibility was suggested using another acid dye, solophenyl red 3BL.[204]

It is obvious that the polarization colors are dependent on the thickness of the fibers[205,206] and the sections[207] and not the type of collagen.[208,209] Anyone can be convinced by comparing a 1 μm semifine section to a 8 to 10 μm section of a mature tendon. Tendon collagen contains predominantly type I molecules. According to the original suggestion,[202] it should appear in orange-red color. the collagen fibers of the tendon, however, appear in green in the semifine section, and this color was said to be specific for collagen type III.

The sirius red method is also used for biochemical estimation of collagen.[161,177,186,210-216] It was shown recently, that the dye binding is not stoichiometric, and the same series of studies questioned the electrostatic staining mechanism.[217]

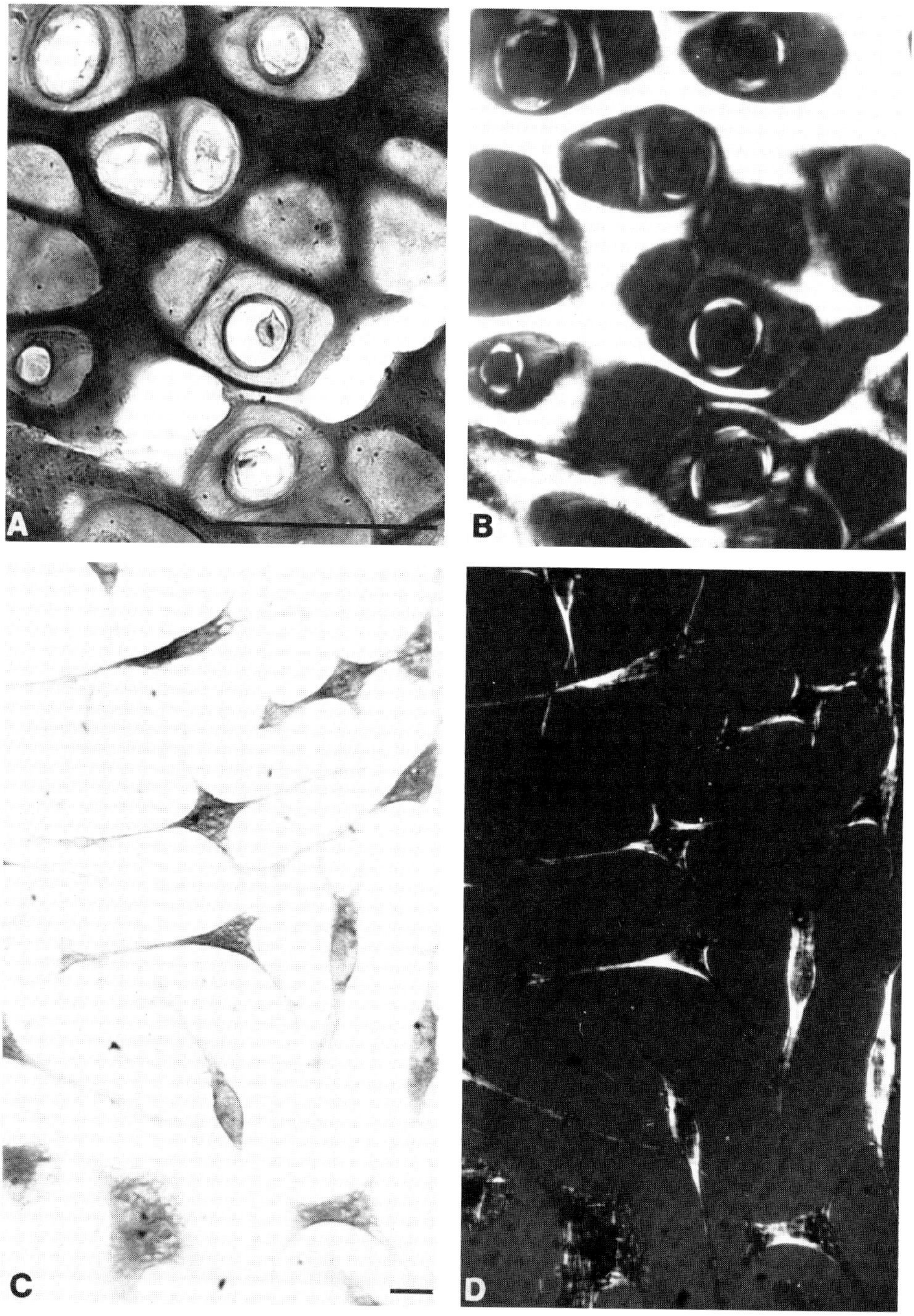

FIGURE 25. Section of human chondrosarcoma (A, B) and mouse L-fibroblasts cultured in monolayer (C, D) after staining with picrosirius red F3B and taken in normal (A, C) and polarized light (B, D). Collagen is stained and birefringent in the cartilage matrix. According to our investigations, the intracytoplasmic birefringence is due to the anisotropic staining of the intermediate filament components of the cytoskeleton. Scales: 100 μm (A), 10 μm (C).

4. Other Reactions

Some other anionic dyes have been recommended to increase the optical anisotropy of the collagen like chlorantine fast red,[218] xylidine ponceau 2R,[219] and congo red.[220] The congo red method — if the sections are mounted in hydrophilic media — is a valuable tool to distinguish amyloid and collagen (see Chapter 12). Metal (silver and gold) impregnations also modify the collagen birefringence.[10,11] Acetylation inverts the sign of birefringence of the collagen fibers[221,222] (see Chapter 12). The toluidine blue staining of collagen at pH 4 to 5[44] has little value because of the simultaneous involvement of anionic groups both of GAGs and collagen in this topo-optical reaction.

B. REACTIONS FOR ELASTIC FIBERS

Aniline induces a birefringence with negative sign of the elastic fibers.[223-227] This topo-optical revealed a three-banded anisotropic pattern of mature elastic fibers suggesting a complex helicoid micellar texture in the fibers which have been considered to be amorphous structures. Unfortunately, the sensitivity of the aniline reaction is low.

The permanganate-bisulfite-toluidine blue (PBT) reaction is a very useful and sensitive topo-optical reaction of elastic fibers[228] (Figure 26). It also detects fine elastic fibers in embryonic tissues.[229] During this reaction, the potassium permanganate is supposed to oxidize the crosslinks of elastin (desmosine, isodesmosine) yielding aldehydes. Anionic bisulfite groups are bound to aldehydes,[32] and bisulfites are stained with toluidine blue. It is an inversive topo-optical reaction. Care must be taken in the blocking of sulfate groups prior to the staining if sulfated GAGs are present in an appreciable amount in the structure subjected to PBT reaction. For this reason, sections must be stained first with toluidine blue at pH 1 to 1.5 (the developer of the PBT reaction), and if metachromasia appears, a methylation must be carried out prior to the PBT reaction, otherwise a non specific reaction will occur.

It is known that young elastic fibers contain free aldehydes. To detect them, only a bisulfite addition + toluidine blue staining is required (BT reaction).[228]

The PBT reaction can also been used to stain RNA if a short (5 to 10 min) oxidation period is chosen.[230] Providing that the rough endoplasmic reticulum of the cell stained by PBT is abundant and its membranes are closely packed (e.g., in plasma cells), in addition to a staining, birefringence also appears.

V. REACTIONS OF CARBOHYDRATES BASED ON OXIDATION BY PERIODIC ACID

A modification of the classical PAS reaction has been developed by Romhányi et al.[33] This is the aldehyde-bisulfite-toluidine blue (ABT) reaction. Its has two related advantages over the PAS technique. First, the ABT reaction results in optical anisotropy if the reacting carbohydrates are spatially oriented. Consequently, this inversive topo-optical reaction can be used in ultrastructural analysis of periodate-reactive carbohydrates. The second advantage of the ABT reaction is that it provides a highly sensitive visualization method of some carbohydrates when evaluated in polarized light. Microorganisms having carbohydrate-rich walls can easily be detected in polarized light,[231-233] and, therefore, the ABT reaction has diagnostic value in histopathology. Carbohydrates of the cell coat can also be visualized after ABT reaction under the polarization microscope (Figure 27).

As has been explained in connection with the collagen-selective sulfation and PBT reactions, the dye-uptake of sulfated GAGs must be controlled and abolished prior to the ABT reaction (see also Chapter 12).

Although the ABT reaction has a well understood mechanism, its interpretation is doubtful. Due to the presence of different periodate-reactive monosaccharides in different oligo- and polysaccharides, it cannot be used as a specific marker of a macromolecule. The

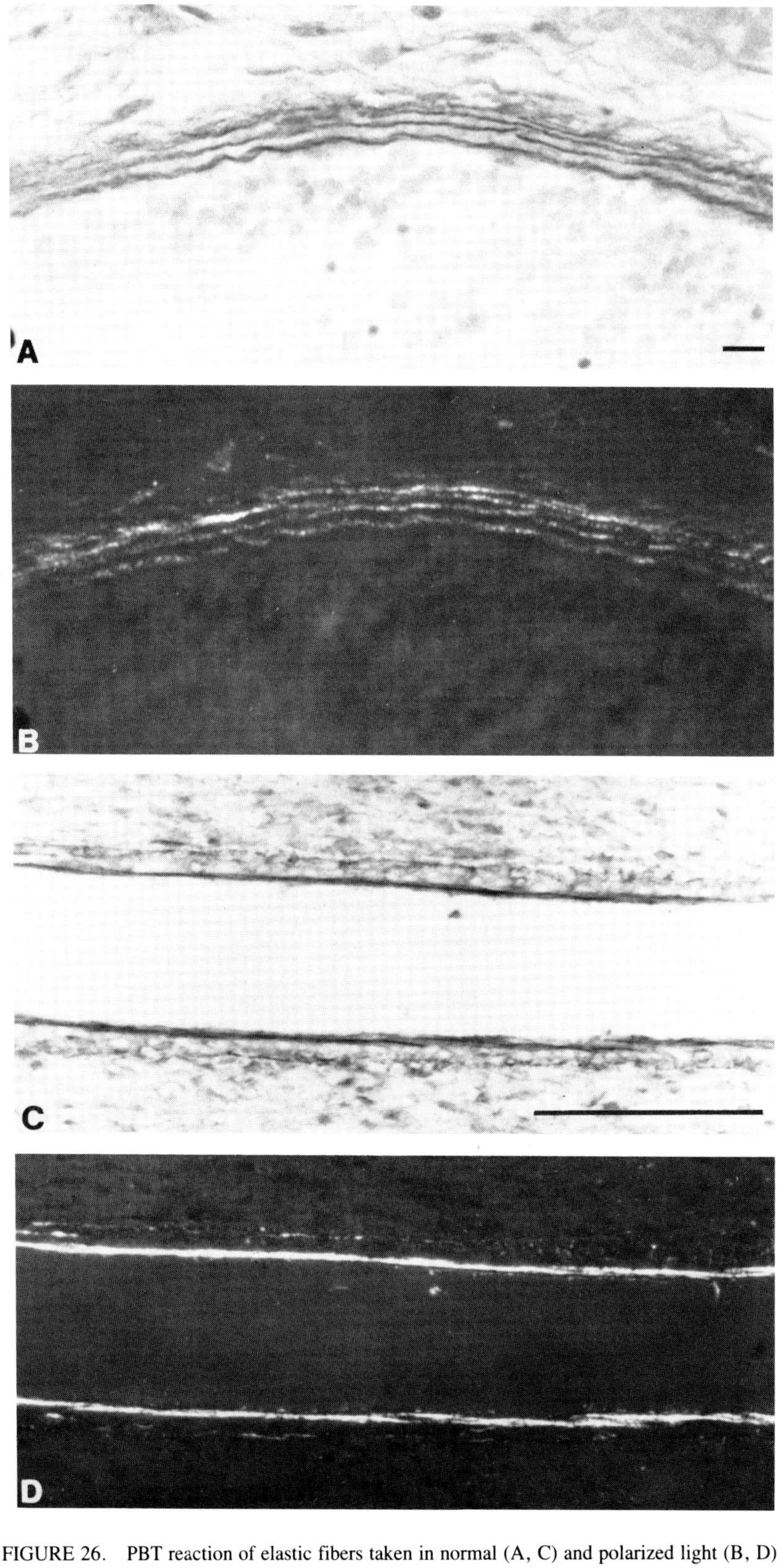

FIGURE 26. PBT reaction of elastic fibers taken in normal (A, C) and polarized light (B, D). (A, B) Portion of cross sectioned aorta wall of a newborn rat. Scale: 10 μm. (C, D) Longitudinal section of a small artery of a 7-month-old human fetus (femoral region). Scale: 100 μm.

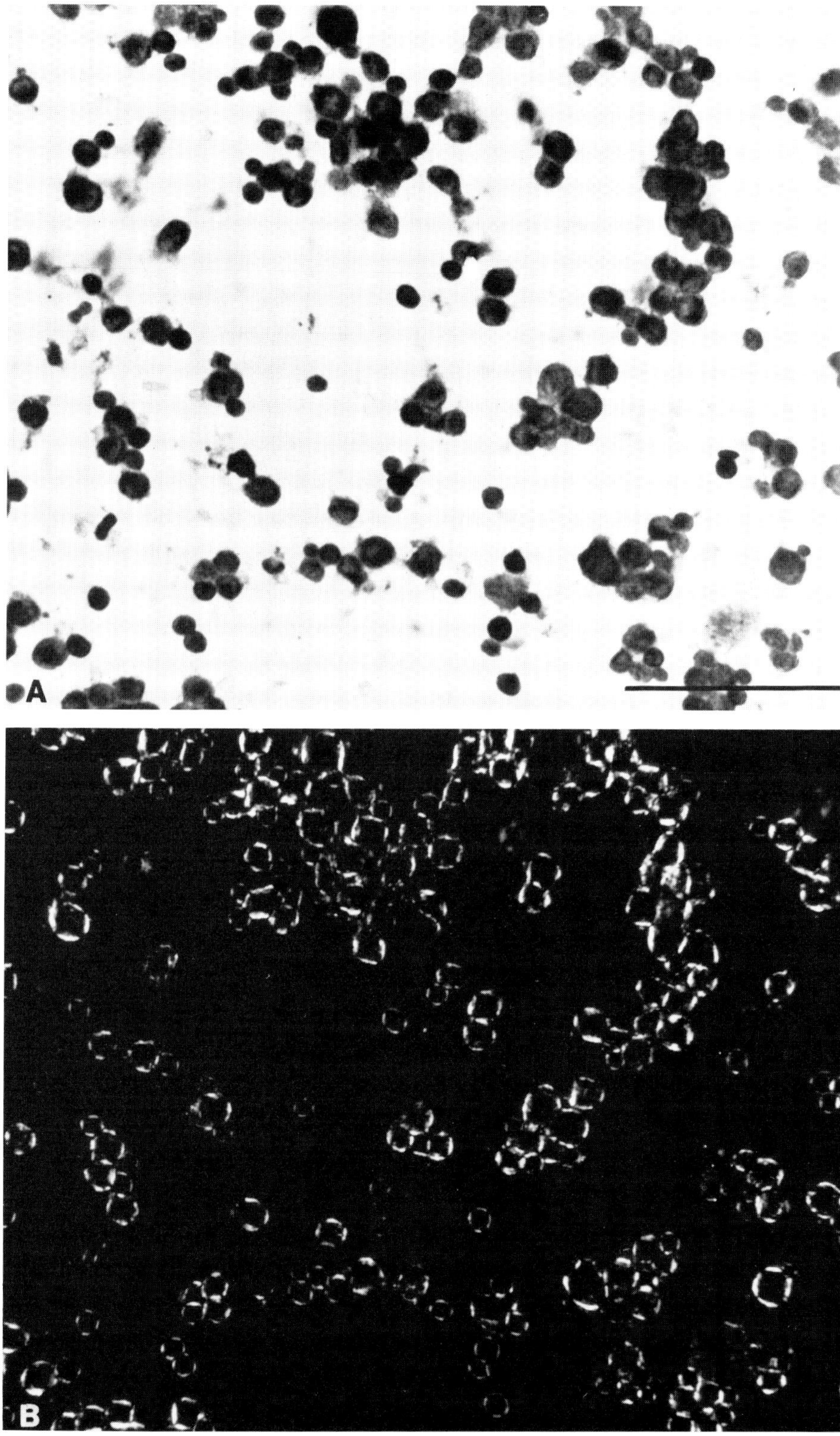

FIGURE 27. Alveolar macrophages isolated from guinea pig after ABT reaction taken in normal (A) and polarized light (B). Note the birefringence of cell membranes indicating the tangential orientation of periodate-reactive carbohydrates. Scale: 50 μm.

only exception is glycogen, because its involvement in the reaction can be controlled by a selective enzymatic digestion.

VI. QUANTIFICATION

There are several possibilities to quantify the topo-optical reactions. Birefringence can be calculated if the retardation value and thickness of the structure are known. The retardation value can easily be measured with a compensator plate. The thickness of the structure can also be determined in some cases, e.g., it can directly be measured if working with isolated fibers. More serious difficulties are encountered, however, if measurements are carried out in sections. If the structure is dense and relatively homogenous — like a differentiated tendon — the measurement of the thickness of the section with interferometry provides a good approximation to the thickness of the birefringent structure (e.g., collagen in tendon). In other cases, the thickness of the section does not correlate with the thickness of the structure, and, therefore, the birefringence cannot be calculated (see also Chapter 12). For example, if the collagen structure of hyaline cartilage matrix is the subject of polarization microscopic analysis, the sum thickness of the collagen fibers present in the area of measurement should be known to calculate the birefringence which is equal to a quotient of the retardation value and the thickness. No techniques are available to measure the thickness of a birefringent structure within a section used for light-microscopic analysis

The other possibility of quantification is the calculation of different anisotropic index values (AI) or quotients (Q).[12,53,155,234,235] The AI and Q formulas have essentially the same content, and they were meant to express in semiquantitative terms the intensity of topo-optical reactions independently of the thickness of the preparation. The calculation of the AI or Q was based on the determination of the retardation value on the same visual field of the same structure before and after staining. The retardation value obtained after the topo-optical reaction is divided by the optical path difference measured prior to the staining. Using this method, however, we often felt that obvious quantitative differences between structures are not reflected by AI or Q values. Therefore, the usefulness of AI was the subject of investigation in our laboratory. In this work, topo-optical reactions specific for sulfated GAGs and collagen were carried out on cartilage tissue, and they were quantified by different methods.[236] Our observations did not warrant the use of AI or Q in the quantitative evaluation of topo-optical reactions at least in cartilage matrix. It was found that such index values did not correspond either to the submicroscopic orientation pattern or to the quantity of the matrix constituent measured.

We use the retardation values of induced birefringence or the anomalous dispersion curves of birefringence to quantify the topo-optical reactions. If the number of measurements is sufficiently high, they can be used to compare semiquantitatively the orientation of different structures (see Chapter 12).

Working with this method, however, the investigator meets the dilemma in some cases: which feature of the structure is responsible for an increase or decrease of the retardation value. There are several alternative possibilities for the explanation. Either the degree of orientation is altered, or the quantity of the substance is changed, or both. In these cases, to find the correct interpretation, the results of the polarization microscopy should be compared with those obtained with cytophotometry, electron microscopy, or biochemistry.

ACKNOWLEDGMENTS

I would like to thank Dr. Róza Ádány for her helpful comments and assistance. I am indebted to my students, Katalin Kocsis and A. Szücs for their help in carrying out the polarization microscopic measurements, and Krisztina Borics for her contribution to development of the CIBR. The birefringence of the cytoskeleton induced by picrosirius red staining in cultured cells was observed during a cooperation with Dr. J. Schlammadinger (Department of Biology, University Medical School, Debrecen). Thanks are due to P. Joó (Department of Biology, University Medical School, Debrecen) for his help in editing this Chapter and for his comments. Chondrosarcoma samples were kindly provided by Dr. J. Frege (Department of Pathology, Cottbus, GDR), and alveolar macrophages were prepared by Dr. Gabriella Fóris (1st Department of Medicine, University Medical School, Debrecen). The scanning densitometry was carried out by Dr. J. Polgár (Department of Clinical Chemistry, University Medical School, Debrecen).

REFERENCES

1. **Ambronn, H.,** *Anleitung zur Benützung des Polarisationsmikroskops bei histologischen Untersuchungen,* Leipzig, 1892.
2. **Ambronn, H. and Frey, A.,** *Das Polarisationsmikroskop. Seine Anwendung in der Kolloidforschung und in der Farberei,* Akademische Verlagsgesellschaft mbH, Leipzig, 1926.
3. **Schmidt, W. J.,** Dichroitische Farbung tierischer und pflanzlicher Gewebe, in *Handbuch der Biologischen Arbeitsmethoden,* Abt. V, Teil 2, Abderhalden, E., Ed., Urban und Schwarzenberg, Berlin, 1932, 1835.
4. **Schmidt, W. J.,** Der Wandel der optischen Anisotropie bei topochemischen Reaktionen histologischer Strukturen, *Ber. d. Oberhes. Ges. f. Naturu. Heilkunde N. F., Naturw. Abtg.,* 23, 56, 1948.
5. **Romhányi, G.,** *On the Ultrastructure of Intercellular Substance of Connective Tissue on the Basis of Polarization Optical Investigations by Means of Topo-Optical Reactions,* D.Sc. Dissertation (in Hungarian), Pécs, 1966.
6. **Romhányi, G.,** Submicroscopic structure of interphase nuclei as revealed by topo-optical staining reaction with cationic dyes, *Acta Morphol. Acad. Sci. Hung.,* 15, 131, 1967.
7. **Romhányi, G.,** Ultrastructure of biomembranes as shown by topo-optical reactions, *Acta Biol. Acad. Sci. Hung.,* 29, 311, 1978.
8. **Ebner, V.,** Über eine optische Reaktion der Bindesubstanzen auf Phenole, *Sitz. Ber. Akad. Wiss. Wien., math. nat. Kl.,* 103, 162, 1894.
9. **Frey, A.,** Die Technik der dichroitischen Metallfarbungen, *Z. Wiss. Mikr.,* 42, 421, 1925.
10. **Mello, M. L. S. and Vidal, B. C.,** Evaluation of dichroism and anomalous dispersion of the birefringence on collagen subjected to metal impregnation, *Ann. Histochim.,* 17, 333, 1972.
11. **Vidal, B. C., Mello, M. L. S., Godo, C., Caseiro, A. C., and Abujadi, J. M.,** Anisotropic properties of silver plus gold-impregnated collagen bundles: ADB and form birefringence curves, *Ann. Histochim.,* 20, 15, 1975.
12. **Romhányi, G.,** Über die submikroskopische strukturelle Grundlage der metachromatischen Reaktion, *Acta Histochem. (Jena),* 15, 201, 1963.
13. **Módis, L.,** *Topo-Optical Investigation of Mucopolysaccharides (Acid Glycosaminoglycans),* Handbuch der Histochemie, Vol.2, Part 4, Graumann, W. and Neumann, K., Eds., G. Fischer Verlag, Stuttgart, 1974.
14. **Goldstein, D. J. and Williams, M. A.,** Quantitative assessment of autoradiographs by photometric reflectance microscopy. An improved method using polarized light, *Histochem. J.,* 6, 223, 1974.
15. **De Waele, M., Renmans, W., Segers, E., Jochmans, K., and Van Camp, B.,** Sensitive detection of immunogold-silver staining with darkfield and epi-polarization microscopy, *J. Histochem. Cytochem.,* 36, 679, 1988.
16. **Gallyas, F.,** A principle for silver staining of tissue elements by physical development, *Acta Morphol. Acad. Sci. Hung.,* 19, 57, 1971.
17. **Gallyas, F. and Wolff, J. R.,** Metal-catalyzed oxidation renders silver intensification selective. Applications for the histochemistry of diaminobenzidine and neurofibrillary changes, *J. Histochem. Cytochem.,* 34, 1667, 1986.
18. **Szirmai, J. A.,** Quantitative approaches in the histochemistry of mucopolysaccharides, *J. Histochem. Cytochem,* 11, 24, 1963.
19. **Scott, J. E. and Dorling, J.,** Differential staining of acid glycosaminoglycans (mucopolysaccharides) by alcian blue staining in salt solutions, *Histochemie,* 5, 221, 1965.
20. **Scott, J. E.,** Affinity, competition and specific interactions in the biochemistry and histochemistry of polyelectrolytes, *Biochem. Soc. Trans.,* 1, 787, 1973.
21. **Scott, J. E.,** Proteoglycan histochemistry — a valuable tool for connective tissue biochemists, *Coll. Relat. Res.,* 5, 541, 1985.

22. **Horobin, R. W. and Goldstein, D. J.,** The influence of salt on the staining of tissue sections with basic dyes: an investigation into the general applicability of the critical electrolyte concentration theory, *Histochem. J.*, 6, 599, 1974.
23. **Goldstein, D. J. and Horobin, R. W.,** Rate factors in staining by alcian blue, *Histochem. J.*, 6, 157, 1974.
24. **Terner, J. Y.,** Histochemical alkylation: a study of methyliodide and its effect on tissues, *J. Histochem. Cytochem.*, 12, 504, 1964.
25. **Ádány, R.,** A selective topo-optical reaction of sulfated glycosaminoglycans, in *VIIth European Symposium on Connective Tissue Research,* Prague, Abstracts, 1980, 262.
26. **Camplejohn, K. L. and Allard, S. A.,** Limitations of safranin O staining in proteoglycan-depleted cartilage demonstrated with monoclonal antibodies, *Histochemistry,* 89, 185, 1988.
27. **Poole, A. R.,** The relationship between toluidine blue staining and hexuronic acid content of cartilage matrix, *Histochem. J.*, 2, 425, 1970.
28. **Ádány, R., Módis, L., and Zsindely, T.,** Sensitization of topo-optical reactions in showing tissue glycosaminoglycans of low local concentration, *Acta Histochem. (Jena),* 67, 247, 1980.
29. **Scott, J. E. and Harbinson, R. J.,** Periodate oxidation of acid polysaccharides. II. Rates of oxidation of uronic acids in polyuronides and acid mucopolysaccharides, *Histochemie.* 19, 155, 1969.
30. **Scott, J. E. and Dorling, J.,** Periodate oxidation of acid polysaccharides. III. A PAS method for chondroitin sulphates and other glycosaminoglycuronans, *Histochemie,* 19, 295, 1969.
31. **Kiviranta, I., Tammi, M., Jurvelin, J., Saamenen, A. -M., and Helminen, H. J.,** Demonstration of chondroitin sulphate and glycoproteins in articular cartilage matrix using periodic acid-Schiff (PAS) method, *Histochemistry,* 83, 303, 1985.
32. **Malinin, G. I.,** Metachromatic staining of sodium bisulfite addition derivatives of glycogen, *J. Histochem. Cytochem,* 18, 834, 1970.
33. **Romhányi, G., Deák, G., and Fischer, J.,** Aldehyde bisulfite-toluidine blue (ABT) staining as a topo-optical reaction for demonstration of linear order of vicinal OH groups in biological structures, *Histochemistry,* 43, 333, 1975.
34. **Lillie, R. D.,** *H.J.Conn's Biological Stains,* 8th ed., Williams and Wilkins, Baltimore, 1969.
35. **Hoppe, W.,** Beziehungen zwischen Struktur und Lichtabsorption der reversiblen Polymerisate aus der Pseudoisocyaninreihe, *Kolloid. Z.*, 109, 21, 1944.
36. **Hoppe, W.,** Röntgenuntersuchungen zur Struktur der reversiblen Polymerisate aus der Pseudoisocyaninreihe, *Kolloid. Z.*, 109, 27, 1944.
37. **Zimmermann, H. and Scheibe, G.,** Zur Konstitution und Lichtabsorption der reversibel polymeren Form des Pseudoisocyanins, *Z. Elektrochem.*, 60, 566, 1956.
38. **Stone, A. L.,** Asymmetric modes of metachromatic binding of dyes to glycosaminoglycans and the influence of polymer conformation on the induced Cotton effects *in vitro*, in *Chemistry and Molecular Biology of the Intercellular Matrix.* Vol.2, Balazs, E. A., Ed., Academic Press, London, 1970, 1067.
39. **Stone, A. L.,** Optical conformation of heparin and heparin complexes with cationic dyes, amines and protein models, *Fed. Proc.*, 36, 101, 1977.
40. **Ballard, R. E., McCaffery, A. J., and Mason, S. F.,** Electronic spectrum, optical activity, and structure of the acridin orange complex with poly-α, L-glutamic acid, *Biopolymers,* 4, 97, 1966.
41. **Húglin, D., Seiffert, W., and Zimmermann, H. W.,** Spektroskopische und thermodynamische Untersuchungen zur Bindung von Azur B an Chondroitinsulfat und zur Bindungsgeometrie des metachromatischen Farbstoffkomplexes, *Histochemistry,* 86, 71, 1986.
42. **Friedrich, K., Hüglin, D., Seiffert, W., and Zimmermann, H. W.,** Modelluntersuchungen zur Struktur des purpuren Farbstoffkomplexes der Giemsa-Farbung, *Histochemistry,* 91, 257, 1989.
43. **Vidal, B. C.,** Pleochroism in tendon and its bearing to acid mucopolysaccharides, *Protoplasma,* 56, 529, 1963.
44. **Mello, M. L. S. and Vidal, B. C.,** Anisotropic properties of toluidine blue-stained collagen, *Ann. Histochim.*, 18, 103, 1973.
45. **Goldstein, D. J. and Horobin, R. W.,** Surface staining of cartilage by alcian blue, with reference to the role of microscopic dye aggregates in histological staining, *Histochem. J.*, 6, 175, 1974.
46. **Sylvén, B.,** Metachromatic dye-substrate interactions, *Quart. J. Microsc. Sci.*, 95, 327, 1954.
47. **Bergeron, J. A. and Singer, M.,** Metachromasy: an experimental and theoretical reevaluation, *J. Biophys. Biochem. Cytol.*, 4, 433, 1958.
48. **Szent-Györgyi, A.,** *Introduction to a Submolecular Biology,* Academic Press, London, 1960, 98.
49. **Scheuner, G. and Hutschenreiter, J.,** Das Auftreten von Metachromasie, Doppelbrechung und Dichroismus durch die Toluidinblaureaktion, *Prog. Histochem. Cytochem.*, 7(2), 1, 1975.
50. **Fischer, J. and Romhányi, G.,** Optical studies on the molecular-sterical mechanism of metachromasia, *Acta Histochem. (Jena),* 59, 29, 1977.
51. **Brewer, D. B.,** Differences in the fine structure of collagen and reticulin as revealed by the polarizing microscope, *J. Pathol. Bacteriol.*, 74, 371, 1957.

52. **Diezel, P. B.,** Die Metachromasie mit verschiedenen Farbstoffen im Mischlicht und im polarisierten Licht, *Acta Histochem. (Jena), Suppl.*, 1, 134, 1958.
53. **Romhányi, G. and Jobst, K.,** Topochemical reactions in polarization microscopy of connective tissue and its intercellular substances, *Ann. Histochim. Suppl.*, 2, 207, 1962.
54. **Missmahl, H. P.,** Metachromasie durch gerichtete Zusammenlagerung von Farbstoffteilchen, erlautert am optischen Verhalten des Toluidinblau, *Histochemie*, 3, 396, 1964.
55. **Missmahl, H. P.,** Birefringence and dichroism of dyes and their significance in the detection of oriented structures, in *Introduction to Quantitative Cytochemistry*, Wied, G. L., Ed., Academic Press, New York, 1966, 539.
56. **Tonna, E. A.,** Changes in the connective tissue framework of long bones during aging. A polarized light study following metachromatic staining, *J. Gerontol.*, 17, 454, 1962.
57. **Tonna, E. A.,** Fracture repair observed under polarized light following metachromatic staining, *Anat. Rec.*, 145, 354, 1963.
58. **Tonna, E. A.,** A histochemical study of the relationship of mucopolysaccharides to connective tissue fibers during osteogenesis and skeletal aging, *Ann. Histochim.*, 8, Suppl., 1, 177, 1963.
59. **Tonna, E. A.,** An evaluation of birefringence of skeletal tissues stained with metachromatic dyes, *J. Roy. Micr. Soc.*, 83, 307, 1964.
60. **Tonna, E. A.,** The connective tissue framework of the femur in mice of different ages, *Anat. Rec.*, 149, 559, 1964.
61. **Tonna, E. A.,** Fracture callus formation in young and old mice observed with polarized light microscopy, *Anat. Rec.*, 150, 349, 1964.
62. **Tonna, E. A. and Hatzel, G.,** Age changes in the distribution of connective tissue elements in human material studied with polarized light microscopy, *J. Gerontol.*, 22, 281, 1967.
63. **Vidal, B. C. and Mello, M. L. S.,** Absorption spectral curves of dichroism on collagen bundles, *Histochemie*, 23, 176, 1970.
64. **Vidal, B. C.,** Anormal dispersion of birefringence, linear dichroism and relationships with ORD (extrinsic Cotton effect), *Histochemie*, 30, 102, 1972.
65. **Vidal, B. C.,** Toluidine blue: Cotton effect-like phenomena produced by crystal aggregates obtained by drying on microscopic slides, *Ann. Histochim.*, 17, 151, 1972.
66. **Vidal, B. C.,** Acid glycosaminoglycans and endochondral ossificaiton: microspectrophotometric evaluation and macromolecular orientation, *Cell. Mol. Biol.*, 22, 45, 1977.
67. **Vidal, B. C. and Mello, M. L. S.,** Proteoglycan arrangement in tendon collagen bundles, *Cell. Mol. Biol.*, 30, 195, 1984.
68. **Franz, H., Scheuner, G., and Hutschenreiter, J.,** Modellversuche zur Ursache der Toluidinblaumetachromasie, *Acta Histochem. (Jena)*, 41, 120, 1971.
69. **Heuft, G. and Böhm, N.,** Altersabhängigkeit des Toluidinblau-Dichroismus und der Anisotropie der Fasern in der Wachstumszone menschlicher Rippen, *Histochemistry*, 59, 45, 1978.
70. **Mello, M. L. S., Godo, C., Vidal, B. C., and Abujadi, J. M.,** Changes in the macromolecular orientation of collagen fibers during the process of tendon repair in the rat, *Ann. Histochim.*, 20, 145, 1975.
71. **Mello, M. L. S., Vidal, B. C., de Carvalho, A. C., and Caseior-Filho, A. C.,** Change with age of anisotropic properties of collagen bundles, *Gerontology*, 25, 2, 1979.
72. **Romhányi, G. and Deák, G.,** The submicroscopic structure of collagen fibers as revealed by topo-optical staining reactions with cationic dyes, *Acta Morphol. Acad. Sci. Hung.*, 15, 147, 1967.
73. **Scheuner, G., Franz, H., and Glöckner, R.,** Zur Beziehung zwischen Metachromasie und optischer Anisotropie, *Acta Histochem. (Jena)*, 36, 411, 1970.
74. **Scheuner, G., Hutschenreiter, J., and Franz, H.,** Modelle zur Erklarung des Zusammenhang zwischen Metachromasie, Doppelbrechung und künstlichem Dichroismus an biologischen Objekten, *Acta Histochem. (Jena)*, 40, 23, 1971.
75. **Vidal, B. C.,** The part played by the mucopolysaccharides in the form birefringence of collagen, *Protoplasma*, 59, 472, 1964.
76. **Vidal, B. C.,** The part played by proteoglycans and structural glycoproteins in the molecular orientation of collagen bundles, *Cell. Mol. Biol.*, 26, 415, 1980.
77. **Vidal, B. C. and Vilarta, R.,** Articular cartilage: collagen type II-proteoglycan interactions. Availability of reactive groups. Variation in birefringence and differences as compared to collagen I. *Acta Histochem. (Jena)*, 83, 189, 1988.
78. **Romhányi, G. and Deák, G.,** Ultrastructure of the ergastoplasm of pancreas acinar cells as revealed by topo-optical staining reactions, *Acta Biochim. Biophys. Acad. Sci. Hung.*, 2, 115, 1967.
79. **Romhányi, G. and Deák, G.,** Dependence of topo-optical staining reaction of the ergastoplasm of the liver cells on the redox state, *Acta Histochem. (Jena)*, 33, 308, 1969.
80. **Romhányi, G. and Molnár, L.,** Optical polarization indicates linear arrangement of rhodopsin oligosaccharide chain in rod disk membranes of frog retina, *Nature*, 249, 486, 1974.

81. **Romhányi, G., Molnár, L., and Németh, Á.**, Ultrastructural differences in cell membranes of erythrocytes myeloid, and lymphoid cells shown by topo-optical reactions, *Histochemistry*, 39, 261, 1974.
82. **Romhányi, G.**, Ultrastructure of biomembranes as shown by topo-optical reactions, *Acta Biol. Acad. Sci. Hung.*, 29, 311, 1978.
83. **Romhányi, G., Fischer, J., and Corless, J. M.**, Orientation of acidic polysaccharides and rhodopsin oligosaccharides in frog retinal rod outer segment, *Histochemistry*, 56, 65, 1978.
84. **Makovitzky, J. and Geyer, G.** Untersuchungen über die anisotrope Toluidinblaufarbung der Glykokalyx, *Histochemistry*, 50, 261, 1977.
85. **Geyer, G. and Makovitzky, J.**, Erythrocyte membrane topo-optical staining reflects glycoprotein conformational changes, *J. Microsc.*, 119, 407, 1980.
86. **Geyer, G., Linss, W., and Makovitzky, J.**, Electron microscopic toluidine blue staining of erythrocyte membrane, *Anat. Anz.*, 143, 291, 1978.
87. **Romhányi, G. and Jobst, K.**, Polarisationsmikroskopische Untersuchungen über saurebedingte intramolekulare Strukturanderungen der Desoxyribonukleinsaure, *Acta Histochem. (Jena)*, 3, 308, 1956/57.
88. **Jobst, K. and Kellermayer, M.**, Submicroscopic structure and dry weight of isolated thymus nuclei following trypsin and salt treatment. Polarization optical, interference microscopic and cytophotometric studies. *Acta Morphol. Acad. Sci. Hung.*, 15, 221, 1967.
89. **Jobst, K. and Kellermayer, M.**, On the submicroscopic structure of chromosomes, *Acta Biol. Acad. Sci. Hung.*, 22, 25, 1971.
90. **Kellermayer, M. and Jobst, K.**, Ion-dependent anisotropy of deoxyribonucleoprotein structures in tissue cultures, *Exp. Cell Res.*, 63, 204, 1970.
91. **Vidal, B. C.**, Birefringence measurements on toluidine blue-stained chromosomes during mitosis, *Ann. Histochim.*, 17, 145, 1972.
92. **Toepfer, K.**, Spektralphotometrische und histochemische Eigenschaftern von 1:9 Dimethyl-Methylenblau als metachromatischer Farbstoff, *Histochemie*. 21, 64, 1970.
93. **Makovitzky, J.**, Untersuchungen über die anisotrope 1:9-Dimethylmethylenblau-und *N,N'*-Diaethylpseudoisocyaninchloridfarbung der Glykokalyx, *Histochemistry*, 59, 55, 1978.
94. **Farndale, R. W., Sayers, C. A., and Barrett, A. J.**, A direct spectrophotometric microassay for sulfated glycosaminoglycans in cartilage cultures, *Connect. Tissue Res.*, 9, 247, 1982.
95. **Templeton, D. M.**, The basis and applicability of the dimethylmethylene blue binding assay for sulfated glycosaminoglycans, *Connect. Tissue Res.*, 17, 23, 1988.
96. **Schiebler, T. H.**, Darstellung vom B-Zellen in Pankreasinseln und von Neurosekret mit Pseudoisocyanin, *Naturwissenschaften*, 45, 214, 1958.
97. **Schiebler, T. H. and Schiessler, S.**, Über den Nachweis vom Insulin mit den metachromatisch reagierenden Pseudoisocyaninen, *Histochemie*, 1, 445, 1959.
98. **Sterba, G.**, Histochemischer und biochemischer Nachweis der Identitat zwischen der als Neurosekret nachweisbaren Substanz und dem Tragerprotein der Oxytocine, *Gen. Comp. Endocrinol.*, 3, 733, 1963.
99. **Sterba, G. and Hoheisel, G.**, Darstellung des neurosekretorischen Systems der Insekten mit Pseudoisocyanin, *Z. Mikrosk. Anat. Forsch.*, 72, 31, 1964.
100. **Sterba, G.**, Grundlagen des histochemischen und biochemischen Nachweises von Neurosekret (= Tragerprotein der Oxytocine) mit Pseudoisocyanin, *Acta Histochem. (Jena)*, 17, 268, 1964.
101. **Kuhlman, D.**, Fluoreszenzmikroskopische Untersuchungen mit Pseudoisocyaninchlorid am Schlundring einiger stylommatophorer Pulmonaten, *Z. Zellforsch.*, 113, 344, 1971.
102. **Harzer, K. and Benz, H. U.**, Quantitative Metachromasie mit Pseudoisocyanin: eine neue Methode zur Bestimmung von Sulfatiden sowie ihre Anwendung bei der Diagnose der metachromatischen Leukodystrophie (Sulfatid-Lipidose), *Z. Klin. Chem. Klin. Biochem.*, 11, 471, 1973.
103. **Vilter, V., Tremblay, P., and Charton, G.**, Détection histophysique des polyanions tissulaires par la métachromasie bathochrome de la pseudo-isocyanine, *C. R. Soc. Biol. (Paris)*, 167, 802, 1973.
104. **Vilter, V., Da Lage, C., and Gaudin-Harding, F.**, Identification spectrophotometrique *in vitro* des glycosaminoglycannes sulfatés par la métachromasie bathochrome de la pseudo-isocyanine, *C. R. Acad. Sci. (Paris)*, 279, 815, 1974.
105. **Vilter, V. and Da Lage, C.**, Interet de l'histospectrophotometrie pour l'etude des glycosaminoglycannes et des polyglycannes anioniques apres coloration a la pseudo-isocyanine, *Bull. Ass. Anat. (Nancy)*, 59, 779, 1975.
106. **Zimmermann, P.**, Methodische Modifikationen und eine neue Technik zur Darstellung des neurosekretorischen Apparates und der Neuroglia bei Wirbellosen, *Z. Wiss. Mikr.*, 68, 154, 1967.
107. **Reutter, K., Takahashi, H., and Graumann, W.**, Histologische und spektralphotometrische Untersuchungen zur Bestimmung der "Starke der Metachromotropie" mit Dichlorpseudoisocyanin, *Histochemie*, 24, 187, 1970.
108. **Scheibe, G. and Zanker, V.**, Physikochemische Grundlagen der Metachromasie, *Acta Histochem. (Jena)*, *Suppl.*, 1, 6, 1958.

109. **Scheuner, G. and Weiss, J.**, Polarisationsoptischer Nachweis von Neurosekret mit *N,N′*-diethylpseudoisocyaninchlorid, *Histochemie,* 11, 325, 1967.

110. **Módis, L., Módis-Süveges, I., and Conti, G.**, Quantitative polarisationsoptische Untersuchungen an Mucopolysacchariden, *Acta Histochem. (Jena) Suppl.*, 12, 169, 1972.

111. **Módis, L., Hadházy, Cs., László, M. B., Kostenszky, K. S., and Földes, I.**, Proteoglycan biosynthesis is stimulated by D-penicillamine in chondrifying high density cell cultures, *Exp. Pathol.*, 35, 159, 1988.

112. **Scheuner, G. and Flemming, G.**, Zum Einsatz von N,N′-Diaethylpseudoisocyaninchlorid für den polarisationsoptischen Nachweis von Polyanionen im histologischen Schnitt, *Acta Histochem. (Jena),* 64, 249, 1979.

113. **Makovitzky, J.**, Polarization optical analysis of blood cell membranes, *Prog. Histochem. Cytochem.*, 15/3, 1, 1984.

114. **Vidal, B. C.**, Anormal dispersion of birefringence, linear dichroism and relationships with ORD (extrinsic Cotton effect), *Histochemie,* 30, 102, 1972.

115. **Scheuner, G.**, Zum polarisationsoptisch-histochemischen Nachweis gerichtet angeordneter saurer Gruppen mit Akridinderivaten, *Acta Histochem. (Jena),* 31, 36, 1968.

116. **Surján, L. and Sebök, J.**, Increase in intranuclear birefringence during chromatin activation reaction, *Exp. Cell Res.*, 78, 241, 1973.

117. **Surján, L., Juhász, J., and Sebök, J.**, Lymphocyte activation and the increase of nuclear birefringence, *Acta Histochem. (Jena).* 64, 139, 1979.

118. **Módis, L., Módis-Süveges, I., and Conti, G.**, Sulla reazione metachromatica del rivanolo, *Riv. Istoch. Norm. Pat.*, 17, 3, 1971.

119. **Kent, G. N., Dodds, R. A., Klenerman, L., Watts, R. W. E., Bitensky, L,. and Chayen, J.**, Changes in crystal size and orientation of acidic glycosaminoglycans at the fracture site in fractured necks of femur, *J. Bone Joint Surg. (Br.),* 65, 189, 1983.

120. **Dunham, J., Shackleton, D. R., Nahir, A. M., Billingham, M. E. J., Bitensky, L., Chayen, J., and Muir, H.**, Altered orientation of glycosaminoglycans and cellular changes in the tibial cartilage in the first two weeks of experimental canine osteoarthritis, *J. Orthop. Res.*, 3, 258, 1985.

121. **Dunham, J., Shackleton, D. R., Billingham, M. E. J., Bitensky, L., Chayen, J., and Muir, H. I.**, A reappraisal of the structure of normal canine articular cartilage, *J. Anat.*, 157, 89, 1988.

122. **Ferris, B. D., Klenerman, L., Dodds, R. A., Bitensky, L., and Chayen, J.**, Altered organization of non-collagenous bone matrix in osteoporosis, *Bone,* 8, 285, 1987.

123. **Skerry, T. M., Bitensky, L., Chayen, J., and Lanyon, L. E.**, Loading-related reorientation of bone proteoglycan *in vivo.* Strain memory in bone tissue? *J. Orthop. Res.*, 6, 547, 1988.

124. **Scott, J. E.**, Histochemistry of alcian blue. I. Metachromasia of alcian blue, astrablue and other cationic phthalocyanin dyes, *Histochemie,* 21, 277, 1970.

125. **Stiller, D. and Katenkamp, D.**, Demonstration of orderly arranged acidic groups in amyloid by alcian blue, *Histochemistry,* 39, 163, 1974.

126. **Valentin, G.**, *Die Untersuchung der Pflanzen und der Tiergewebe in polarisiertem Lichte,* Wilhelm Engelmann, Leipzig, 1861.

127. **Benninghoff, A.**, Form und Bau der Gelenkknorpel in ihren Beziehungen zur Funktion. II. Der Aufbau des Gelenkknorpels in seinen Beziehungen zur Funktion, *Z. Zellforsch.*, 2, 783, 1925.

128. **Benninghoff, A.**, Der funktionelle Bau des Hyalinknorpels, *Ergebn. Anat. Entwickl.-Gesch.*, 26, 1, 1925.

129. **Bucher, O.**, Beitrag zum funktionellen Bau des hyalinen Knorpels (auf Grund von Untersuchungen im polarisierten Lichte), *Z. Zellforsch.*, 32, 281, 1942.

130. **Banga, I., Baló, J., and Szabó, D.**, Submicroscopic structure of collagen fibers: their contraction and relaxation, *Acta Morphol. Acad. Sci. Hung.*, 6, 391, 1956.

131. **Szabó, D.**, Special aspects in the study of connective-tissue fibres in polarized light, *Acta Morphol. Acad. Sci. Hung.*, 7, 15, 1956.

132. **Lindner, J.**, Special polarisation-optical examination on connective tissue fibers altered experimentally, *Ann. Histochim. Suppl.*, 2, 193, 1962.

133. **Hutschenreiter, J. and Scheuner, G.**, Untersuchungen zur Eigen-dopplebrechung des Kollagens, *Acta Histochem. (Jena),* 35, 337, 1970.

134. **Hutschenreiter, J. and Scheuner, G.**, Histophysikalische Bestimmung des submikroskopischen Lückenvolumens am Kollagen, *Exp. Pathol.*, 6, 167, 1972.

135. **Hutschenreiter, J. and Scheuner, G.**, Zum Quellungverhalten des Kollagens in Abhängigkeit von der Zeit, *Exp. Pathol.*, 6, 170, 1972.

136. **Kaplan, D. and Bettelheim, F. A.**, On the birefringence of bovine cornea, *Exp. Eye Res.*, 13, 219, 1972.

137. **Bairati, A., Petruccioli, M. G., and Tarelli, L. T.**, Birefringence of collagen fibrils after glutaraldehyde treatment, *J. Submicrosc. Cytol.*, 4, 89, 1972.

138. **Wolman, M. and Gillman, T.**, A polarized light study of collagen in dermal wound healing. *Br. J. Exp. Pathol.*, 53, 85, 1972.

139. **Gathercole, L. J., Keller, A., and Shah, J.S.**, The periodic wave pattern in native tendon collagen: a correlation of polarizing with scanning electron microscopy, *J. Microsc.*, 102, 95, 1974.
140. **Laufer, M., Ashkenazi, C., Katz, D., and Wolman, M.**, Orientation of collagen in wound healing, *Br. J. Exp. Pathol.*, 55, 233, 1974.
141. **Nicholls, S. P., Gathercole, L. J., Keller, A., and Shah, J. S.**, Crimping in rat tail tendon collagen: morphology and transverse mechanical anisotropy, *Int. J. Biol. Macromol.*, 5, 283, 1983.
142. **Nicholls, S. P., Gathercole, L. J., and Shah, J. S.**, Morphology of human palmaris longus tendon, *Ann. Rheum. Dis.*, 43, 477, 1984.
143. **Niven, H., Baer, E., and Hiltner, A.**, Organization of collagen fibers in rat tail tendon at the optical microscope level. *Coll. Relat. Res.*, 2, 131, 1982.
144. **Meyer, W., Neurand, K., Radke, B.**, Collagen fiber arrangement in the skin of the pig, *J. Anat.*, 134, 139, 1982.
145. **McBride Jr., D. J., Hahn, R. A., and Silver, F. H.**, Morphological characterization of tendon development during chick embryogenesis: measurement of birefringence retardation, *Int. J. Biol. Macromol.*, 7, 71, 1985.
146. **Bigi, A., Ripamonti, A., Roveri, N., Compostella, L., Roncon, L., and Schivazappa, L.**, Structure and orientation of collagen fibers in human mitral valve, *Int. J. Biol. Macromol.*, 4, 387, 1982.
147. **Boyde, A., Bianco, P., Portigliatti Barbos, M., and Ascenzi, A.**, Collagen orientation in compact bone: I. A new method for the determination of the proportion of collagen parallel to the plane of compact bone sections, *Metab. Bone Dis. Relat. Res.* , 5, 299, 1984.
148. **Yarker, Y. E., Aspden, R. M., and Hukins, D. W. L.**, Birefringence of articular cartilage and the distribution of collagen fibril orientations, *Connect. Tissue Res.*, 11, 207, 1983.
149. **Whittaker, P., Boughner, D. R., Perkins, D. G., and Canham, P. B.**, Quantitative structural analysis of collagen in chordae tendineae and its relation to floppy mitral valves and proteoglycan infiltration, *Br. Heart J.*, 57, 264, 1987.
150. **Whittaker, P., Schwab, M. E., and Canham, P. B.**, The molecular organization of collagen in saccular aneurysms assessed by polarized light microscopy, *Connect. Tissue Res.*, 17, 43, 1988.
151. **Wolman, M. and Kasten, F. H.**, Polarized light microscopy in the study of the molecular structure of collagen and reticulin, *Histochemistry*, 85, 41, 1986.
152. **Hutschenreiter, J. and Scheuner, G.**, Angaben zur submikroskopischen Struktur des Kollagens — eine Grundlage für die Interpretation entsprechender polarisationsoptischer und histochemischer Befunde, *Z. Mikrosk. Anat. Forsch.*, 90, 969, 1976.
153. **Hutschenreiter, J. and Scheuner, G.**, Zur polarisationsoptischen Erfassung funktioneller Gruppen des Kollagens durch die Phenolreaktion, *Verh. Anat. Ges.*, 71, 1443, 1977.
154. **Neumark, T.**, Phenol reaction of collagen fibres and acid mucopolysaccharide content of connective tissue, *Acta Morphol. Acad. Sci. Hung.*, 14, 59, 1966.
155. **Missmahl, H. P.**, Quantitative Auswertung der Phenolreaktion an der kollagenen Faser, *Acta Histochem. (Jena), Suppl.*, 4, 188, 1964.
156. **Földes, I., Módis, L., Antalffy, J., and Ádám, I.**, Ultrastructure of extracellular matrix of embryonic chick limb bud cartilage, *Acta. Biol. Acad. Sci. Hung.*, 31, 81, 1980.
157. **Romhányi, G., Bukovinszky, A., and Deák, G.**, Sulfation as a collagen-specific reaction. The ultrastructure of sulfate collagen, basement membranes, and reticulin fibers as shown by topo-optical staining reactions, *Histochemie*, 36, 125, 1973.
158. **Sweat, F., Puchtler, H., and Rosenthal, S. I.**, Sirius red F3BA as a stain for connective tissue, *Arch. Pathol.*, 78, 69, 1964.
159. **Constantine, V. S. and Mowry, R. W.**, The selective staining of human dermal collagen. II. The use of picrosirius red F3BA with polarization microscopy, *J. Invest. Dermatol.*, 50, 419, 1968.
160. **Lillie, R. D. and Fullmer, H. M.**, *Histopathologic Technic and Practical Histochemistry*, 4th ed., McGraw-Hill, A Blakiston Publication, New York, 1976, 287.
161. **López-De León, A. and Rojkind, M.**, A simple micromethod for collagen and total protein determination in formalin-fixed paraffin-embedded sections, *J. Histochem. Cytochem.*, 33, 737, 1985.
162. **Junqueira, L. C. U., Bignolas, G., and Brentani, R. R.**, Picrosirius staining plus polarization microscopy, a specific method for collagen detection in tissue sections, *Histochem. J.*, 11, 447, 1979.
163. **Junqueira, L. C. U., Montes, G. S., and Krisztán, R. M.**, The collagen of the vertebrate peripheral nervous system. *Cell Tissue Res.*, 202, 453, 1979.
164. **Vickerstaff., T.**, *The Physical Chemistry of Dyeing*, Oliver and Boyd, London, 1954.
165. **Puchtler, H., Waldrop, F. S., and Valentine, L. S.**, Polarization microscopic studies of connective tissue stained with picrosirius red F3BA, *Beitr. Pathol. Anat.*, 150, 174, 1973.
166. **Vidal, B. C., Mello, M. L. S., and Pimentel, E. R.**, Polarization microscopy and microspectrophotometry of sirius red, picrosirius and chlorantine fast red aggregates and their complexes with collagen, *Histochem. J.*, 14, 857, 1982.

167. **Cooper, J. H.,** An evaluation of current methods for the diagnostic histochemistry of amyloid, *J. Clin. Pathol.,* 22, 410, 1969.
168. **Brigger, D. and Muckle, T. J.,** Comparison of sirius red and congo red as stains for amyloid in animal tissues, *J. Histochem. Cytochem.,* 23, 84, 1975.
169. **DeLellis, R. S. and Bowoling, M. C.,** The use of sirius red and congo red staining in routine histopathology, *Hum. Pathol.,* 1, 655, 1970.
170. **Llewellyn, B. D,** An improved sirius red method for amyloid, *J. Med. Lab. Technol.,* 27, 308, 1970.
171. **Pena, S. D. J.,** A new technique for visualization of the cytoskeleton in cultured fibroblasts with coomassie blue R250, *Cell Biol. Int. Rep.,* 4, 149, 1980.
172. **Pena, S. D. J., Opas, M., Turksen, K., Kalnins, V. I., and Carpenter, S.,** Immunocytochemical studies of intermediate filament aggregates and their relationship to microtubules in cultured skin fibroblasts from patients with giant axonal neuropathy, *Eur. J. Cell Biol.,* 31, 227, 1983.
173. **Jackson, J. G., Puchtler, H., and Sweat, F.,** Investigation of staining, polarization and fluorescence-microscopic properties of pseudo-elastic fibers in the renal arterial system, *J. Roy. Micr. Soc.,* 88, 473, 1968.
174. **Puchtler, H., Waldrop, F. S., and Valentine, L. S.,** Fluorescence microscopic distinction between elastin and collagen, *Histochemie,* 35, 17, 1973.
175. **Carrasco, H. F., Montes, G. S., Krisztán, R. M., Shigihara, K. M., Carneiro, J., and Junqueira, L. C. U.,** Comparative morphologic and histochemical studies on the collagen of vertebrate arteries, *Blood Vessels,* 19, 28, 1982.
176. **Junqueira, L. C. U., Montes, G. S., Almeida Neto, E., Baross, C., and Tedesco-Marchese, A. J.,** The collagen of permanently damaged nerves in human leprosy, *Int. J. Lepr.,* 48, 291, 1980.
177. **Junqueira, L. C. U., Bignolas, G., Mourao, P. A. S., and Bonetti, S. S.,** Quantitation of collagen-proteoglycan interaction in tissue sections, *Connect. Tissue Res.,* 7, 91, 1980.
178. **Junqueira, L. C. U., Zugaib, M., Montes, G. S., Toledo, O. M. S., Krisztán, R. M., and Shigihara, K. M.,** Morphologic and histochemical evidence for the occurence of collagenolysis and for the role of neutrophilic polymorphonuclear leukocytes during cervical dilation, *Am. J. Obstet. Gynec.,* 138, 273, 1980.
179. **Junqueira, L. C. U., Montes, G. S., Kaupert, D., Shigihara, K. M., Bolonhani, T. M., and Krisztán, R. M.,** Morphologic and histochemical studies on the collagen of neurinomas, neurofibromas, and fibromas, *J. Neuropathol. Exp. Neurol.,* 40, 123, 1981.
180. **Junqueira, L. C. U., Montes, G. S., Martins, J. E. C., and Joazeiro, P. P.,** Dermal collagen distribution. A histochemical and ultrastructural study, *Histochemistry,* 79, 397, 1983.
181. **Junqueira, L. C. U., Toledo, O. M. S., and Montes, G. S.,** Histochemical and morphological studies on a new type of acellular cartilage, *Basic Appl. Histochem.,* 27, 1, 1983.
182. **Junqueira, L. C. U., Montes, G. S., Toledo, O. M. S., Bexiga, S. R. R., Gordilho, M. A., and Brentani, R. R.,** Evidence for collagen molecular orientation in basement membranes, *Histochem. J.,* 15, 785, 1983.
183. **Junqueira, L. C. U., Toledo, O. M. S., and Montes, G. S.,** Correlation of specific sulfated glycosaminoglycans with collagen types I, II, and III, *Cell Tissue Res.,* 217, 171, 1981.
184. **Junqueira, L. C. U. and Roscoe, J. T.,** Reduced collagen content and fiber bundle disorganization in skin biopsies of patients with Ehlers-Danlos syndrome, *Histochem. J.,* 17, 1197, 1985.
185. **Junqueira, L. C. U.,** A study on human osteosarcoma collagen by the histochemical picrosirius -polarization method, *J. Pathol.,* 148, 189, 1986.
186. **Junqueira, L. C. U., Montes, G. S., Toledo, O. M. S., and Joazeiro, P. P.,** Morphological, histochemical, and biochemical observations on the connective tissue matrix of *in situ* and isolated hepatic granulomas in experimental murine schistosomiasis, *Ann. Trop. Med. Parasitol.,* 80, 27, 1986.
187. **Montes, G. S., Kristztán, R. M., Shigihara, K. M., Tokoro, R., Mourao, P. A. S., and Junqueira, L. C. U.,** Histochemical and morphological characterization of reticular fibers, *Histochemistry,* 65, 131, 1980.
188. **Montes, G. S. and Junqueira, L. C. U.,** Biology of collagen, *Rev. Can. Biol. Exp.,* 41, 143, 1982.
189. **Montes, G. S., Bezerra, M. S. F., and Junqueira, L. C. U.,** Collagen distribution in tissues, in *Ultrastructure of the Connective Tissue Matrix,* Ruggeri, A. and Motta, P. M., Eds., Martinus Nijhoff, Boston, 1984, chap. 3.
190. **Montes, G. S., Krisztán, R. M., and Junqueira, L. C. U.,** Preservation of elastic system fibers and of collagen molecular arrangement and stainability in an egyptian mummy, *Histochemistry,* 83, 117, 1985.
191. **Zambrano, N. Z., Montes, G. S., Shigihira, K. M., Sanchez, E. M., and Junqueira, L. C. U.,** Collagen arrangement in cartilage, *Acta Anat.,* 113, 26, 1982.
192. **Bonafe, J. L., Pieraggi, M. T., and Rochiccioli, P.,** Etude du collagéne dermique par la coloration du rouge sirius au cours de maladies congénitales ou acquises du tissu conjonctif, *Ann. Dermatol. Vénéréol.,* 108, 633, 1981.

193. **Pieraggi, M. T. and Bonafe, J. L.**, Appréciation de l'effet de la D-pénicillamine sur la synthese et le maintien de la structure collagene, grace a la coloration histologique par le rouge sirius examinée en lumiere polarisée. Discussion se la valeur de cette coloration en pathologie du collagene, *Ann. Pathol.*, 2, 155, 1982.
194. **Greenberg, S. R.**, A new method for the histological study of collagen, *Lab. Med.*, 17, 474, 1986.
195. **Dziedzic-Goclawska, A., Rozycka, M., Czyba, J. C., Sawicki, W., Moutier, R., Lenczowski, S., and Ostrowski, K.**, Applicaiton of the optical fourier transform for analysis of the spatial distribution of collagen fibers in normal and osteopetrotic bone tissue, *Histochemistry*, 74, 123, 1982.
196. **Flint, F. O. and Firth, B. M.**, Histochemical demonstration of collagen in comminuted meat products, *Analyst*, 108, 757, 1983.
197. **Flint, F. O. and Pickering, K.**, Demonstration of collagen in meat products by an improved picro-sirius red polarisation method., *Analyst*, 109, 1505, 1984.
198. **Canham, P. B., Smith, J. F. H., and Starkey, J.**, Orientation of collagen in the brain artery using polarized light microscopy, *Mikroskopie*, Suppl. 37, 246, 1980.
199. **Weber, K. T.**, Cardiac interstitium in health and disease: the fibrillar collagen network, *J. Am. Coll. Cardiol.*, 13, 1637, 1989.
200. **Whittaker, P., Boughner, D. R., and Kloner, R. A.**, Analysis of healing after myocardial infarction using polarized light microscopy, *Am. J. Pathol.*, 134, 879, 1989.
201. **Rosenquist, T. H. and Módis, L.**, Spatial disorder of collagen III in the great vessels, associated with congenital heart defects, *Anat. Rec.*
202. **Junqueira, L. C. U., Cossermelli, W., and Brentani, R. R.**, Differential staining of collagen type I, II and III by sirius red and polarization microscopy, *Arch. Histol. Jpn.*, 41, 267, 1978.
203. **Józsa, L,. Réffy, A., and Bálint, J. B.**, Polarization and electron microscopic studies on the collagen of intact and ruptured human tendons, *Acta Histochem. (Jena)*, 74, 209, 1984.
204. **Ogbuihi, S., Müller, Z., and Zink, P.**, Zur quantitativen polarisationsmikroskopischen Darstellung von Kollagen Typ I und Typ III an histologischen Paraffinschnitten, *Z. Rechtsmed.*, 100, 101, 1988.
205. **Pérez-Tamayo, R. and Montfort, I.**, The susceptibility of hepatic collagen to homologous collagenase in human and experimental cirrhosis of the liver. *Am. J. Pathol.*, 100, 427, 1980.
206. **Szendröi, M., Vajta, G., Kovács, L., Schaff, Zs., and Lapis, K.**, Polarization colours of collagen fibers: a sign of collagen production activity in fibrotic processes, *Acta Morphol. Hung.*, 32, 47, 1984.
207. **Junqueira, L. C. U., Montes, G. S., and Sanchez, E. M.**, The influence of tissue section thickness on the study of collagen by the picrosirius-polarization method, *Histochemistry*, 74, 153, 1982.
208. **Piérard, G. E. and Lapiere, C. M.**, The colors of collagen appearing in polarized light after staining with picrosirius red are unrelated to the molecular type of collagen, *Am. J. Dermatopathol.*, 9, 172, 1987.
209. **Piérard, G. E.**, Sirius red polarization method is useful to visualize the organization of connective tissues but not the molecular composition of their fibrous polymers, *Matrix*, 9, 68, 1989.
210. **Junqueira, L. C. U. and Montes, G. S.**, Biology of collagen-proteoglycan interaction, *Arch. Histol. Jpn.*, 46, 589, 1983.
211. **Armendáriz-Borunda, J. and Rojkind, M.**, A simple quantitative method for collagen typing in tissue samples: its application to human liver with schistosomiasis, *Coll. Relat. Res.*, 4, 35, 1984.
212. **Kuttan, R. and Di Ferrante, N.**, Sirius-red-collagen interaction: a method for the measurement of collagen and bacterial collagenase activity. *Biochem. Int.*, 1, 455, 1980.
213. **Munoz, M. L., Calderón, J., and Rojkind, M.**, The collagenase of entamoeba histolytica, *J. Exp. Med.*, 155, 42, 1982.
214. **Linares, H. A.**, Measurement of collagen-proteoglycan interaction in hypertrophic scars, *Plast. Reconstr. Surg.*, 71, 817, 1983.
215. **Yue, B. Y. J. T., Sugar, J., and Schrode, K.**, Collagen staining in corneal tissues, *Curr. Eye Res.*, 5, 559, 1986.
216. **Junqueira, L. C. U., Bignolas, G., and Brentani, R. R.**, A simple and sensitive method for the quantitative estimation of collagen, *Anal. Biochem.*, 94, 96, 1979.
217. **Puchtler, H. Meloan, S. N., and Waldrop, F. S.**, Are picro-dye reactions for collagen quantitative? Chemical and histochemical considerations, *Histochemistry*, 88, 243, 1988.
218. **Vidal, B. C., Mello, M. L. S., and Pimentel, E. R.**, Polarization microscopy and microspectrophotometry of sirius red, picrosirius and chlorantine fast red aggregates and their complexes with collagen, *Histochem. J.*, 14, 857, 1982.
219. **Vidal, B. C.**, Dichroism in collagen bundles stained with xylidine ponceau 2R, *Ann. Histochim.*, 15, 289, 1970.
220. **Romhányi, G., Deák, G., and Bukovinszky, A.**, Collagen-specific topo-optical staining reaction with congo red and its ultrastructural interpretation, *Acta Morphol. Acad. Sci. Hung.*, 18, 261, 1970.
221. **Romhányi, G.**, Über die Ultrastruktur und die enzymatische Abbaubarkeit des Acetylkollagen, *Nova Acta Leopoldina*, 36, 27, 1971.

222. **Romhányi, G., Bukovinszky, A., and Deák, G.,** On the ultrastructure and proteolytic susceptibility of acetylcollagen as studied by topo-optical reactions, *Acta Morphol. Acad. Sci. Hung.*, 20, 49, 1972.
223. **Romhányi, G.,** Über dei submikroskopischen Strukturen der elastischen Fasern, *Acta Morphol. Acad. Sci. Hung.*, 5, 311, 1955.
224. **Romhányi, G.,** Submicroscopic structure of elastic fibers as observed in the polarization microscope, *Nature*, 182, 930, 1958.
225. **Romhányi, G.,** On the submicroscopic structure of the elastic fibers of the bovine ligamentum nuchae as revealed by the polarization microscope, *Acta Morphol. Acad. Sci. Hung.*, 13, 397, 1964.
226. **Romhányi, G.,** The characteristic three-banded birefringence of *ligamentum nuchae* elastic fibres indicates ordered micellar texture, *Histochemistry*, 77, 133, 1983.
227. **Romhányi, G.,** Specific topo-optical reactions of connective tissue elements and their ultrastructural interpretation, *Connect. Tissue Res.*, 15, 13, 1986.
228. **Fischer, J.,** Ultrastructure of elastic fibers as shown by polarization optics after the topo-optical permanganate-bisulfute-toluidine blue (PBT) reaction, *Acta Histochem. (Jena)*. 65, 87, 1979.
229. **Rosenquist, T. H., Beall, A., Módis, L., and Fishman, R.,** Impaired elastic matrix development in the great arteries after ablation of the cardiac neural crest, *Anat. Rec.*, 226, 347, 1990.
230. **Fischer, J.,** Optical studies on the molecular arrangement of RNA in tissues with a selective topo-optical reaction of RNA, *Histochemistry*, 59, 325, 1979.
231. **Fischer, J.,** Demonstration of microorganisms in tissues by the ABT and KOH-ABT topo-optical reactions, *Acta Morphol. Acad. Sci. Hung.*, 24, 203, 1976.
232. **Fischer, J.,** Optical polarization reveals different ultrastructural molecular arrangement of polysaccharides in the yeast cell walls, *Acta Biol. Acad. Sci. Hung.*, 28, 49, 1977.
233. **Fischer, J. and Emödy, L.,** Molecular order of carbohydrate components in cell walls of bacteria, fungi, and algae according to the topo-optical reaction of the vicinal OH groups, *Acta Microbiol. Acad. Sci. Hung.*, 23, 97, 1976.
234. **Scheuner, G. and Hutschenreiter, J.,** *Polarisationsmikroskopie in der Histophysik*, VEB G. Thieme, Leipzig, 1972.
235. **Hutschenreiter, J. and Scheuner, G.,** Polarisationsoptisches Verfahren zur quantitativen Bestimmung gerichtet angeordneter saurer Gruppen, *Acta Histochem. (Jena)*, 45, 144, 1973.
236. **Ádány, R., Módis, L., and Balogh, G.,** A critical re-evaluation of the anisotropic index used in polarization microscopy: a study on epiphyseal cartilage, *Acta Histochem. (Jena)*, 65, 243, 1979.

Chapter 4

STRUCTURE OF THE HYALINE CARTILAGE MATRIX

TABLE OF CONTENTS

I. INTRODUCTION

The mature hyaline cartilage tissue is characterized by predominance of the extracellular matrix over the cells. For example, the total cell volume occupies only about 2% of the tissue volume in adult human articular cartilage. The properties of this tissue, therefore, are established by the properties of the matrix.

Due to the high frequency of degenerative and inflammatory joint diseases in human medicine, much effort has been directed towards a detailed understanding of the structure, composition, molecular anatomy, and function of the cartilage matrix. The essence of recent progress in this field is summarized in outstanding monographs.[1-6]

Although our knowledge of cartilage biochemistry has been rapidly increasing, the structure of the cartilage matrix is not yet well understood. This — among others — may be due to the limitations of the morphological techniques. For example, the old terminology: ''amorphous'' cartilage matrix created in last century survived and was widely used even in the contemporary literature. The major message of this chapter is to show that this definition cannot be held anymore, because at least some macromolecular constituents form a spatially oriented system in most of the matrix compartments.

II. MACROMOLECULAR CONSTITUENTS OF THE HYALINE CARTILAGE MATRIX

Chemically, most of the cartilage matrix macromolecules are complexes of carbohydrates and proteins. There are three major classes of the cartilage matrix constituents: (1) collagens; (2) proteoglycans and glycosaminoglycans; (3) noncollagenous glycoproteins. Topographically, we distinguish cell-surface associated, fibrillar, perifibrillar, and interfibrillar components. In this short overview the chemical classification will be followed.

In adult cartilage matrix the water content is about 70 to 75%, and about 70% of the dry substance represents collagen, 20% proteoglycans and 10% other proteins frequently referred as matrix proteins.

At present, the following 5 of the 14 identified collagen types are known to be present in the cartilage: types II, VI, IX, X and XI.[7-22] Four of them (types II, IX, X and XI) are said to be cartilage-specific, but they are also found in other tissues such as vitreous and notochord. The major collagen of the hyaline cartilage is the type II, and it forms fibrils. From the minor cartilage collagens (types VI, IX, X and XI), types IX and XI collagens are also involved in the organization of collagen fibrils: type IX collagen molecules are covalently cross-linked to type II collagen and they lie periodically along the surface of type II fibrils, i.e., a collagenous and noncollagenous domain of this molecule periodically project out from the fibrillar surface endowed with a bottlebrush structure, while the exact location of type XI in the fibril is not known.[11,14,15] Type X collagen is synthesized only by the hypertrophic chondrocytes and supposedly it is involved in the calcification of the cartilage matrix.[17-20] Type XI collagen is associated with the type II fibrils and it may interact with proteoglycans.[21] According to immunohistochemical studies, collagen type XI is preferentially located pericellularly and it is supposed to be involved in the organization of the pericellular matrix.[22] Using mechanically isolated chondrons as experimental models,[23-25] collagen types VI and IX have also been shown in the pericellular matrix.[26,27] An increase of minor collagens (types IX and XI) has been demonstrated in osteoarthrotic cartilage.[28] In sera of patients with rheumatic diseases antibodies against collagen types I, II, IX, and XI could be detected.[29] In mice and rats, a so-called ''collagen-induced arthritis'' could be brought about that is considered as an experimental model of human rheumatoid arthritis.[30]

While collagen fibrils contribute predominantly to tensile strength of the cartilage, the other major macromolecular species of the matrix, the proteoglycan (PG) class, provides resistance to compression of the loaded cartilage. There are at least five different types of PGs found in cartilage matrix. Their common structural feature is that they contain a protein core and at least one glycosaminoglycan (GAG) chain (for review, see references.[31-36] The most characteristic PG of the cartilage matrix is the large aggregating PG, it represents about 80% of the PG of hyaline cartilage.[36] The other PGs isolated from cartilage matrix are the following: an aggregating PG of smaller size with higher keratan sulfate content than the large aggregating PG, a small dermatan sulfate PG, and a heparan sulfate PG.

The most abundant type of PG in normal and differentiated hyaline cartilage matrix is the large aggregating PG. Its protein core comprising about 2000 amino acids can be divided into structurally and functionally different domains: the aminoterminal end forms two globular domains which contain no or few GAG chains but a number of *N*-linked oligosaccharides.[36] This domain specifically interacts with hyaluronic acid and link protein forming the characteristic PG aggregate (Figure 1). To the next major domain predominantly keratan sulfate chains and *O*-linked oligosaccharides are attached. The longest part of the molecule is the chondroitin sulfate binding region, but it also contains keratan sulfate chains and *O*-linked oligosaccharides. There is a globular protein domain at the carboxyterminal end, which binds *N*-linked oligosaccharides but devoid GAGs.

The GAG side chains form covalent bonds with preferential amino acid residues of the core protein through an oligosaccharide structure. The amino acid residue and the oligosaccharide together constitute the linkage region. The sequence of the linkage region in the case of chondroitin sulfate is: serine-D-xylose-D-galactose-D-galactose-D-glucoronic acid.

The GAG side chains attached to the linear core protein have a helical conformation (for review, see Módis).[32] The most important common feature of GAGs is their strong anionic character, which is due to the carboxyl of their hexuronic acid and to the sulfate groups of their hexosamine residues. Due to this feature, GAGs can interact with microcations (Na^+, K^+, Ca^{2+}, Mg^{2+}, etc.) and polycations such as collagen. Numerous morphological methods make use of the cation binding capacity of GAGs: GAGs (and PGs) can easily be detected using cationic reagents including dyes (see Chapters 3 and 11). GAGs and PGs are able to bind high amounts of water. The high fixed charge density of the PGs leads to a large osmotic swelling pressure by the Donnan effect. PGs being in free solution occupy about five times the volume they have in the tissue. This swelling is counteracted by the collagen structure and enables the tissue to resist compressive load.

The content of chondroitin and keratan sulfates in large PGs varies with age and also with the localization within different matrix compartments.[37-40] Keratan sulfate is found in tissues and parts of a tissue where the O_2 supply is low (e.g., in the deep zones of cartilage).[41]

PGs are found in different locations within the hyaline cartilage matrix. One population is present at the surface of chondrocytes[42-45] resulting in a high local concentration of PGs in the pericellular matrix.[46,47] Another population is located on the surface of collagen fibrils,[48-54] these PGs are found to be attached at the d and e bands of the fibrils electron microscopically.[52,53] It is supposed that this PG population is involved in the fibrillogenesis and regulation of collagen fibril diameter.[48,49] PGs and collagen are known to be in interaction with each other. The predominant type of their interactions is electrostatic in nature.[55-59] We have experimental evidence proving that GAGs and collagen fibrils are arranged in an axiparallel manner (see later in this chapter), and their electrostatic interaction is the major factor which is responsible for this orientation pattern (see Chapter 8).

It is known that cartilage PGs may be potential autoantigens in humans and in experimental animals, and they may induce a joint disease resembling to the human rheumatoid arthritis.[60-62]

A number of noncollagenous proteins (cartilage matrix proteins) have been identified biochemically.[63-68] They are glycoproteins in nature, and according to their functions, they can be divided into different subclasses. The link proteins stabilize the proteoglycan aggre-

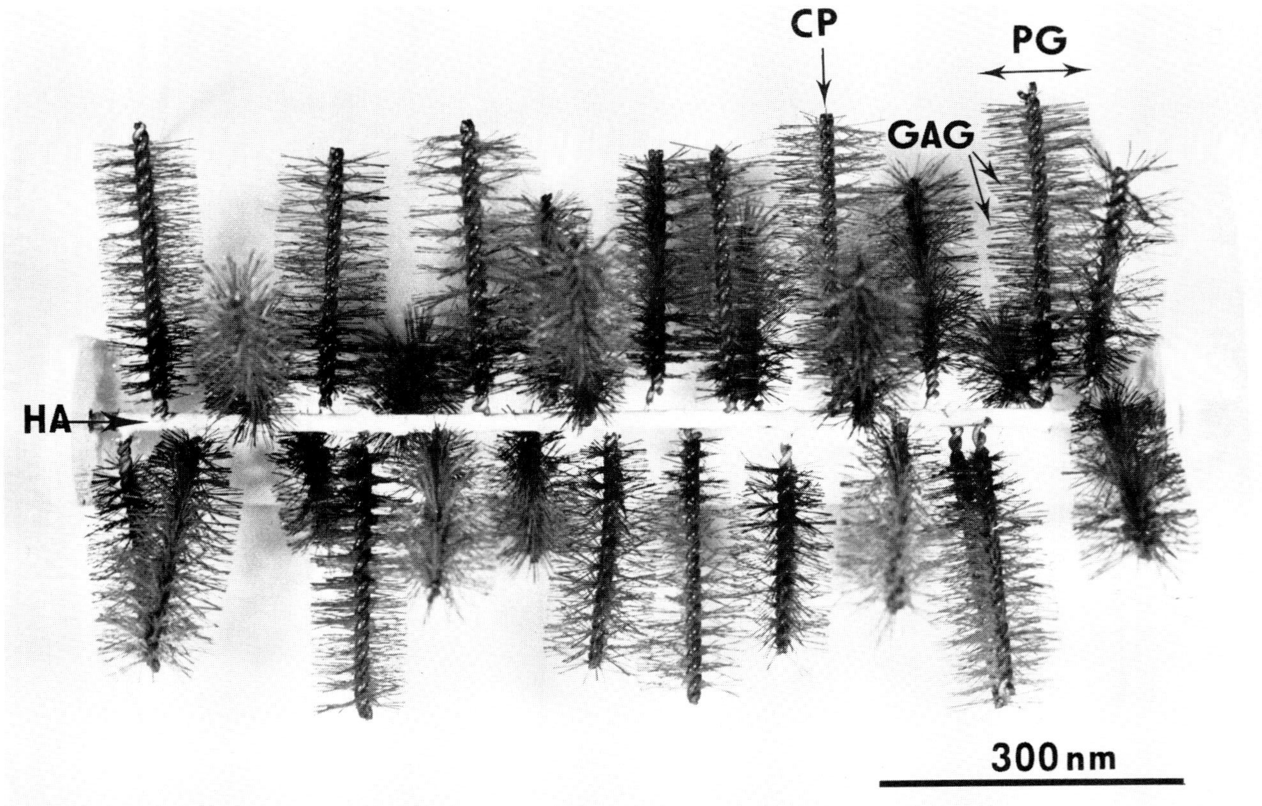

FIGURE 1. Schematic representation of the shape and approximate size of a cartilage proteoglycan (PG) aggregate. CP: core protein; GAG: glycosaminoglycan side chains (chondroitin sulfates and — near to the hyaluronic acid binding region — keratan sulfates); HA: hyaluronic acid. The link proteins which stabilize the binding of PG to HA are not shown. This model suggests that both GAGs and PGs possess an extended configuration. This is due to the high anionic charge density of the GAGs. The HA backbone — depending on the number of PGs of bottlebrush shape bound to it — may be flexible.

gates,[11,66] more of the cartilage matrix proteins are supposed to modulate collagen fibril formation (e.g., fibromodulin[67] and 54 kDa collagen binding protein),[68] while a third group is involved in the attachment of chondrocytes to the matrix (e.g., 38 kDa and 58 kDa proteins,[67] fibronectin.[69-72] Chondrocalcin, which has been found identical with the C-propeptide of type II procollagen, is a calcium-binding protein and it is found in developing cartilage matrix where mineralization occurs.[73-74]

The different cartilage matrix components may be bound to the chondrocyte membrane in different ways. One possible way of interaction of PGs with the cell membrane is their attachment to the hyaluronic acid which is bound to specific receptor of the chondrocyte.[75-77] It is worth mentioning here that the hyaluronic acid receptor is associated with the microfilament system of the cytoskeleton.[76] Collagen receptors also shown on the chondrocyte surface, they are 34 kDa anchorin CII molecules.[78,79] Chondronectin is supposed to bind collagen to the cell membrane in cartilage tissue,[80,81] although the role of fibronectin in this process cannot be excluded.[69-72]

III. ULTRASTRUCTURE OF THE HYALINE CARTILAGE MATRIX AND ITS DIFFERENTIATION

The fine structure of the cartilage matrix has been intensively studied both with electron microscope[2,5,50-54,82-85] and polarization optics.[86-102] Owing to its significance in osteoarthritis and other human joint diseases, the articular cartilage was the major target of the investigations.

Both electron and polarization microscopy revealed the arrangement of collagen fibrils in detail, while little is known about the exact location of different glycoconjugates.

Benninghoff[86,87] was the first, who — based on polarization microscopic studies — proposed that the collagen fibrils were arranged in an oriented manner in the differentiated articular cartilage matrix. According to his description, the collagen fibrils run parallel to the articular surface in the superficial (tangential) zone, while they are perpendicularly arranged to the surface in the deep (radial) zone, and the collagen fibrils — interconnecting the tangential and radial zones — form Gothic arcades. The arcades are crossing each other, and, therefore, they extinguish the birefringence of each other. Due to this, the matrix of the intermediate zone of articular (and other mature hyaline) cartilage appears to be isotropic under a polarization microscope. According to polarization microscopic analysis of different hyaline cartilages stained with picrosirius red for collagen, the Benninghoff's model can be used for the description of collagen orientation in other cartilages, too.[102]

Scanning and transmission electron microscopic studies have generally confirmed Benninghoff's description[86,87] of the radial and tangential zones, but some studies failed to support the arcade-like arrangement of collagen in the intermediate zone.[82,83]

Using polarization microscopy combined with topo-optical reactions for sulfated GAGs, induced anisotropy could be observed in the cartilage matrix indicating that sulfated GAGs are spatially oriented.[94-101] According to our observations, the preferential orientation axis of the sulfated GAGs is the long axis of the collagen fibrils.[94-96,99] This conclusion has been supported by using other experimental methods.[98,101] Vidal and Vilarta[93] analyzed rat articular cartilage matrix without any staining under the polarization microscope. They found that the digestion with testicular hyaluronidase diminished the optical anisotropy (quantified in retardation values) of different matrix zones with 15 to 31% compared to untreated specimens. They concluded that this phenomenon is due to the axiparallel orientation of "acid proteoglycans" and collagen fibers. Obviously, these authors recorded the contribution of chondroitin sulfate molecules to the birefringence of the cartilage matrix. Their conclusion — modifying the "acid proteoglycans" to chondroitin sulfates — is a significant evidence proving that the results obtained after topo-optical reactions are not due to the dye-orientation alone but they reflect the spatial orientation pattern of the sulfated GAGs.

Electron microscopic investigations of cartilage matrix stained with cationic probes for PGs revealed that a number of partially collapsed PG particles are periodically and perpendicularly attached to the collagen fibril surfaces[50-54] (see also Chapter 11). The GAG side chains of these PG molecules are extended *in vivo* (and collapsed during the tissue processing for electron microscopy), and the GAGs of PGs oriented more or less perpendicularly to the collagen must run parallel to the long axis of the collagen fibrils. In this sense, the electron microscopic findings confirm the polarization microscopic observations. Electron microscopic investigations, however, have demonstrated an interfibrillary located and randomly distributed PG population, too (see Chapter 11). These molecules can never contribute to the birefringence of the cartilage matrix. In this respect, electron microscopy complemented the polarization microscopic results.

According to our studies, the spatial orientation of GAGs and PGs can be detected even at the early stage of cartilage differentiation. It is known, that the first stage of cartilage matrix differentiation is the deposition of a primary matrix in the presumptive cartilaginous areas. This is rich in hyaluronic acid,[103] and contains fibronectin, collagen type I, and some PGs.[104,105] The secondary cartilage matrix — which appears after expression of the cartilage phenotype — is characterized by the appearance of collagen type II fibrils and high amounts of cartilage PGs.[104,105]

Using topo-optical reactions for sulfated GAGs, we could demonstrate an induced birefringence both in the precartilaginous and cartilaginous areas at early stages of limb differentiation in chick[104] and human embryos (Figure 2). It is interesting to note that no oriented collagen structure can be detected in the matrix areas exhibiting induced birefringence of GAGs except the regions of the more mature cartilage[104] (Figure 2). The relative sign of birefringence induced by stainings for GAGs indicates that these molecules are tangentially arranged around the chondrocytes and chondroblasts. These reactions proved to be susceptible to a digestion with testicular hyaluronidase indicating that chondroitin sulfates are the major GAGs of the PG which are stained in our experiments.[104] Using staining techniques for PGs, electron-microscopic investigations revealed that PGs are present in two different locations in early embryonic cartilage: they cover the outer surface of the chondrocyte membrane, and their long axes are predominantly in perpendicular orientation related to the tangent of the cell membrane (see Chapter 11). The other group is comprised of PGs randomly distributed in the matrix. It is known that different PG types appear in the embryonic cartilage,[106,107] but our technique is not suitable for their distinction. It is reasonable to suppose, however, that the cell surface-bound oriented PGs are responsible for the optical anisotropy induced by staining specifically for sulfated GAGs. Thus, we can conclude that the chondrocytes are surrounded by a pericellular matrix containing tangentially arranged chondroitin sulfates and radially oriented PG core proteins at the beginning of cartilage matrix differentiation. The formation of this oriented PG structure precedes the deposition of the oriented collagen structure. It is possible that the spatially oriented PGs promote collagen fibrillogenesis and their oriented deposition.

The same results could be obtained by investigation of chondrifying high density cultures of limb bud mesenchymal cells isolated from chicken embryos of stages 23 to 24.[108,109] In this model, the formation of an oriented sulfated GAG structure precedes the oriented collagen deposition, and GAGs — like the *in vivo* models — are circularly arranged around the chondrocytes (Figures 3, 4). This optical reaction could be abolished by an antecedent digestion with testicular hyaluronidase or chondroitinase AC, it is likely that the pericellularly located PGs responsible for our topo-optical reactions are either large aggregating PGs or intermediate-size PGs or both, and not dermatan sulfate or heparan sulfate PGs.[110]

At the later stages of embryonic cartilage differentiation, cartilage matrix trabecules develop between the cartilage cells which are characterized by deposition of oriented collagen fibrils and sulfated GAGs (predominantly chondroitin sulfate) molecules along the long axes

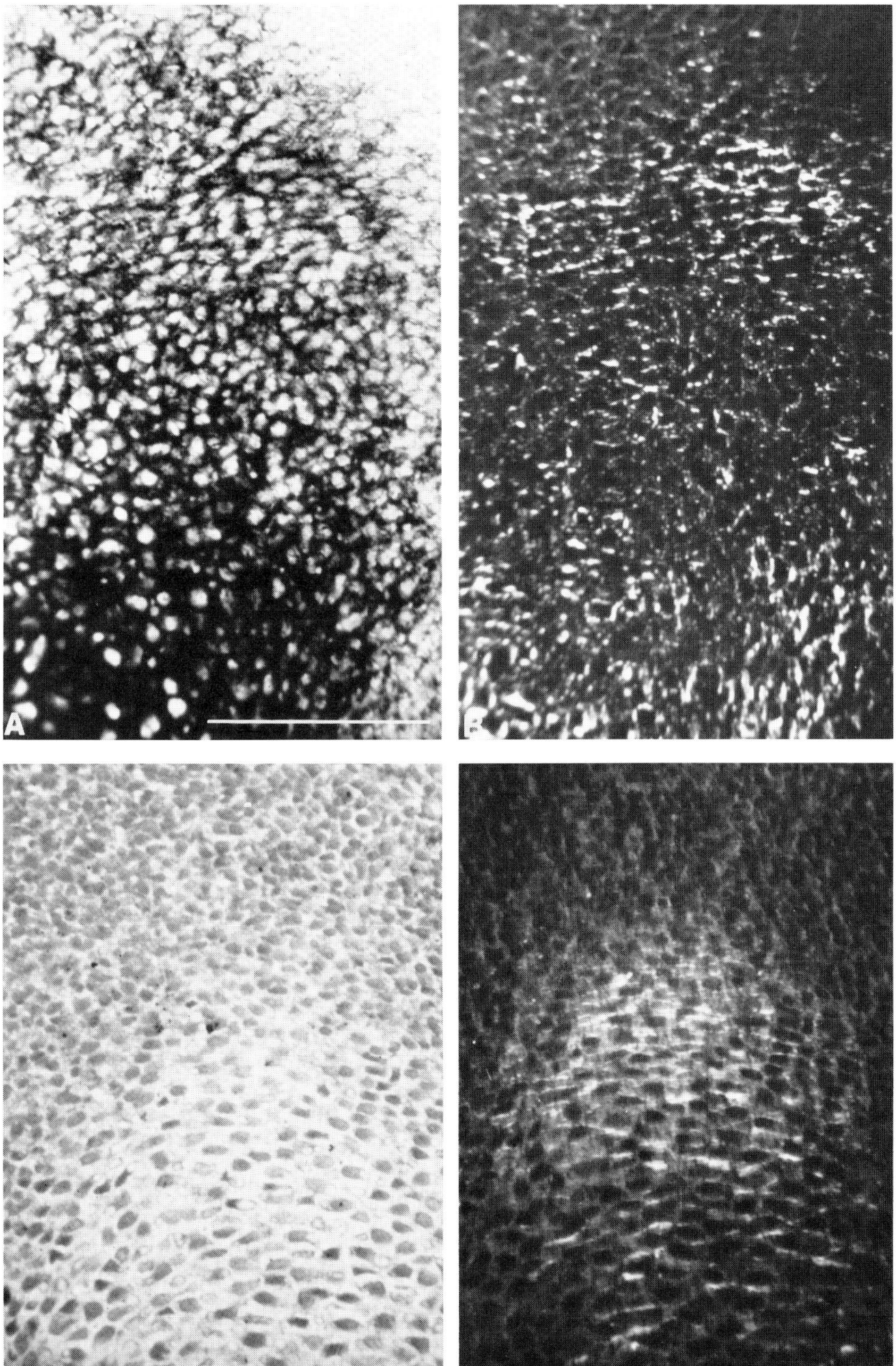

FIGURE 2. Condylar portion of the femur anlage of a 5-week-old human embryo stained with *N,N'*-diethylpseudoisocyanine for sulfated GAGs (A, B) and with picrosirius red F3B for collagen (C, D). The picrosirius red staining was carried out after a digestion with testicular hyaluronidase. (A) and (C) are taken in normal light (A in monochromatic light at λ = 590 nm), B and D are photographed in polarized light. Spatially oriented sulfated GAGs are present in the matrix of cartilaginous and precartilaginous areas (A and B), while oriented collagen structures can only be observed where the cartilage phenotype is expressed (D). Scales: 100 μm.

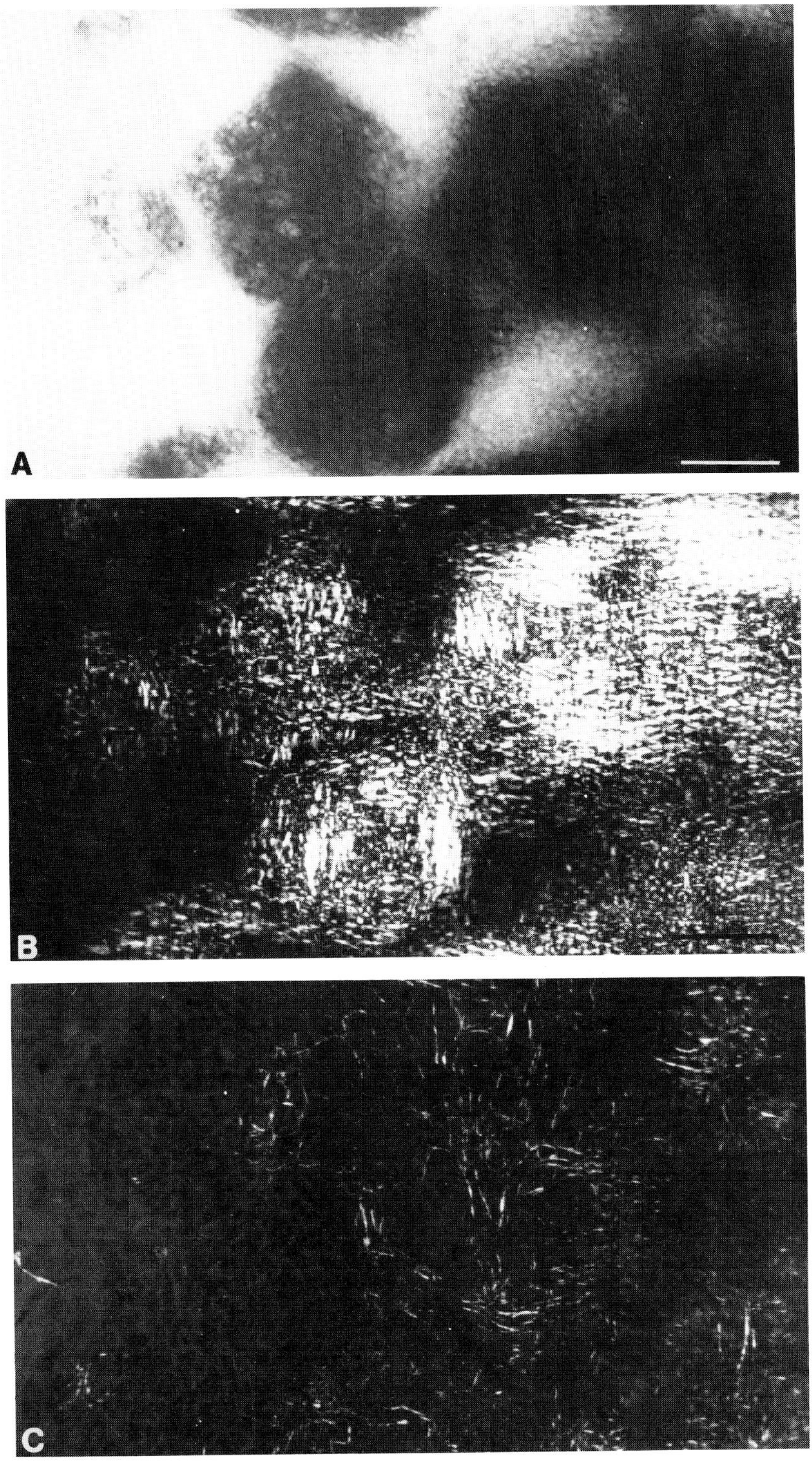

FIGURE 3. Four-day-old chondrifying high density cultures of mesenchymal cells isolated from stages 23 to 24 chick embryonic limb buds. (A, B:) *N,N'*-diethylpseudoisocyanine staining for sulfated GAGs; (C) collagen-selective sulfation (see Chapter 12). (A) normal, (B, C:) polarized light micrographs. Spatially oriented sulfated GAGs are detected in the matrix of cartilage nodules (A, B), while oriented collagen structure is revealed by polarization microscopy only in the central parts of the cartilage nodules. Scale: 100 μm.

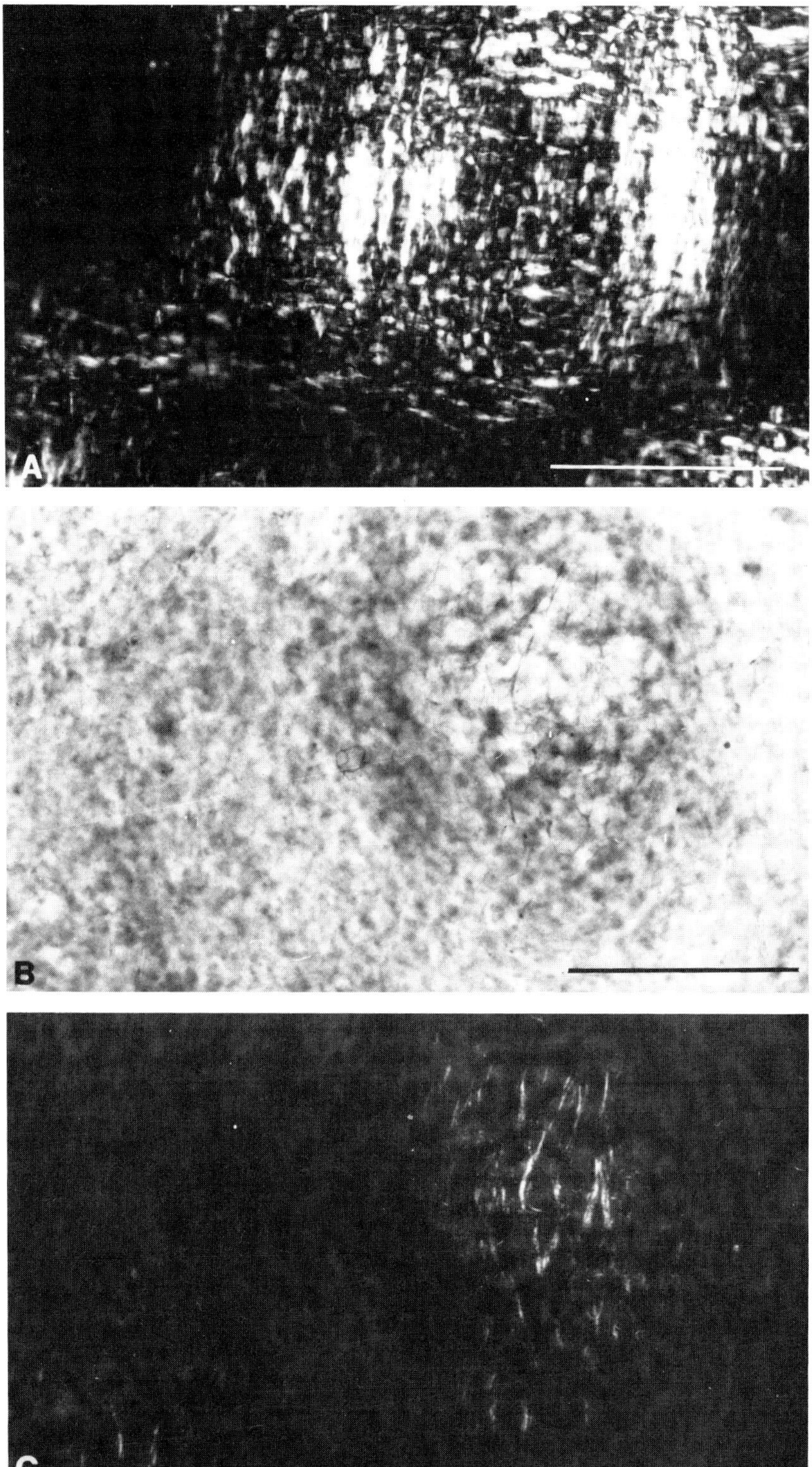

FIGURE 4. Cartilage nodules of 4-day-old chondrifying high density cultures after N,N′-diethyl-pseudoisocyanine staining (A) and collagen-selective sulphation (B, C). (A, C) polarized, (B) normal light micrographs. Spatially oriented sulfated GAG structure can be observed around the chondrocytes throughout the nodule, induced birefringence of collagen is restricted to the central part of the nodule. Scales: 100 μm.

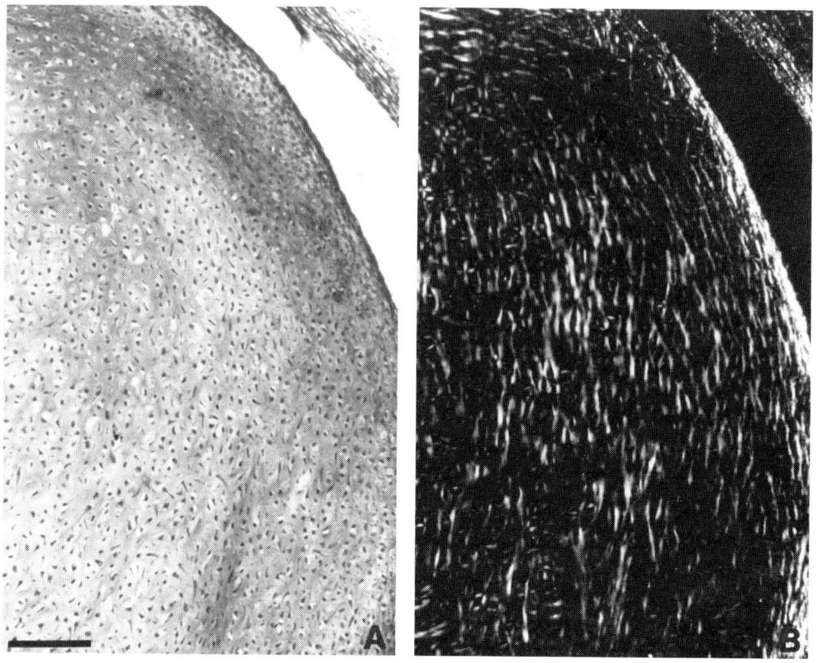

FIGURE 5. Portion of femoral condyle of an 11-week-old human embryo after testicular hyaluronidase digestion + picrosirius red F3B staining taken in normal (A) and polarized light (B). Note the intense birefringence of the cartilage matrix trabecules especially near to the articular surface. Scale: 100 μm.

of the trabecules. These two constituents are ordered parallel to each other. This orientation pattern is probably due to their electrostatic interactions. At this stage of development, the birefringence induced by topo-optical reactions specific for sulfated GAGs and collagen is considerably increased comparing to the early stages.[95,104] The induced birefringence is more intensive near the articular surface when it is formed. Both collagen and GAGs are arranged parallel to the surface, indicating that the tangential zone is formed at the early stage of development (Figure 5). At the same time, the pericellular orientation of PGs and GAGs also persist. Electron microscopical investigation of embryonic and fetal cartilages revealed that PGs can be found in three different locations: in pericellular, perifibrillar, and interfibrillar spaces (see Chapter 11). Those occuring in pericellular and perifibrillar spaces are oriented perpendicularly to the cell surface and fibrillar surface, respectively. Presumably, the sulfated GAGs of these PGs are responsible for the induced topo-optical reactions. It is possible that while the pericellularly located oriented PGs are involved in the collagen fibrogenesis and ordered deposition of the collagen fibrils, the perifibrillar PGs play a role in regulation of the collagen fibril diameter. The interfibrillary localized PGs are randomly distributed.

In postnatal life, territorial and interterritorial matrix areas are separated in the middle and deep zones of human articular cartilage. Collagen and GAGs are circularly oriented in this matrix compartment (Figures 6 to 8). On the other hand, GAGs and collagen fibrils are arranged longitudinally in the interterritorial matrix trabecules, Using energy dispersive X-ray microanalytical techniques, we found a higher local concentration of sulfated GAGs in the territorial matrix as compared to the interterritorial matrix (see also Chapter 11). Significantly higher retardation values could be measured after topo-optical reactions specific for sulfated GAGs and collagen in the territorial matrix than in the interterritorial areas (Figures 9-11). It follows that a highly oriented, densely packed, basket-like matrix structure develops in chondrons which supposedly protects the cartilage cells against mechanical stresses and contributes to the resiliency of the articular cartilage structure. A similar — although much narrower — structure ("perilacunar rim") separating the territorial and interterritorial matrices was anticipated by Szirmai.[112] We think that the intense induced birefringence seen throughout the territorial matrix can be well explained by the densely packed fine collagen fibrils observed electron microscopically in isolated chondrons[24] and in situ.[85] We do not know at this moment that oriented deposition of minor collagens contribute to the intense induced birefringence of the territorial collagen.

It is reasonable to suppose that the strong optical anisotropy recorded after GAG-specific topo-optical reactions is either due to the high local concentration of PGs, or to a high degree of spatial orientation of their GAG side chains in the territorial matrix,[111] or both.

These data suggest that the original Benninghoff proposal[86,87] about the orientation pattern of articular cartilage matrix can be supplemented and developed in the following points: (1) The chondrocytes are surrounded by circularly oriented and densely packed collagen fibrils in the territorial matrix within the intermediate and deep zones. (2) Sulfated GAGs of PGs associated with the cell membrane and collagen fibrils are also spatially oriented (Figure 12). (3) The oriented microstructure of the cartilage matrix gradually develops (Figure 13).

In connection with our findings that oriented GAGs and PGs could be demonstrated in hyaline cartilage matrix (and also in other connective tissues), we want to mention here that Weiss[113,114] has raised this possibility several decades ago expressing the following opinion: "...hyaline ground substance in reality possesses definite structural order analogous to crystallinity." This prediction did not have any experimental evidences at that time. Our other observation was that the oriented deposition of PGs and their GAGs precedes the oriented deposition of collagen in cartilage matrix can be regarded as experimental evidence of another hypothesis forwarded by Karl Meyer,[115] the great pioneer of mucopolysaccharide/glycosaminoglycan biochemistry:...(GAG molecules)..."are a code of extracellular information for the arrangement and organization of the fibrillar proteins, chiefly collagen fibrils."

IV. CONCLUDING REMARKS

Based on our present knowledge including our own observations (see also Chapter 11) we propose the following hypothesis to describe the ultrastructure of the hyaline cartilage matrix (Figures 12 to 14).

The cartilage matrix is a heterogeneous system, in which microscopic and submicroscopic compartments can be distinguished. The compartments are different from each other chemically, structurally, and certainly, functionally. The microscopic compartments are the zones, and within the zones are the territorial and interterritorial areas. Both sulfated GAGs and collagen exhibit a circular orientation pattern in the territorial matrix, while these constituents are longitudinally arranged in the interterritorial matrix. The territorial matrix possesses a much more compact structure than the interterritorial matrix, and it is also richer in PGs. The submicroscopic compartments are the pericellular, fibrillar, perifibrillar, and interfibrillar spaces. Oriented PGs and GAGs can be detected in the pericellular and perifibrillar areas where PGs are perpendicularly, and GAGs are tangentially arranged. Supposedly, various types of cell-matrix interactions contribute to a much more intricate structure of the pericellular matrix as it can be observed morphologically. Probably, the pericellular cartilage matrix (like other tissues) is in contact with the cytoskeleton. During the differentiation, the formation of the spatially oriented pericellular PG structure precedes the oriented deposition of the collagen. Territorial matrix with circularly oriented GAG and collagen structure appears first in early postnatal life in human articular cartilage.

Further efforts are needed to refine, correct, or reject different aspects of this hypothetical model.

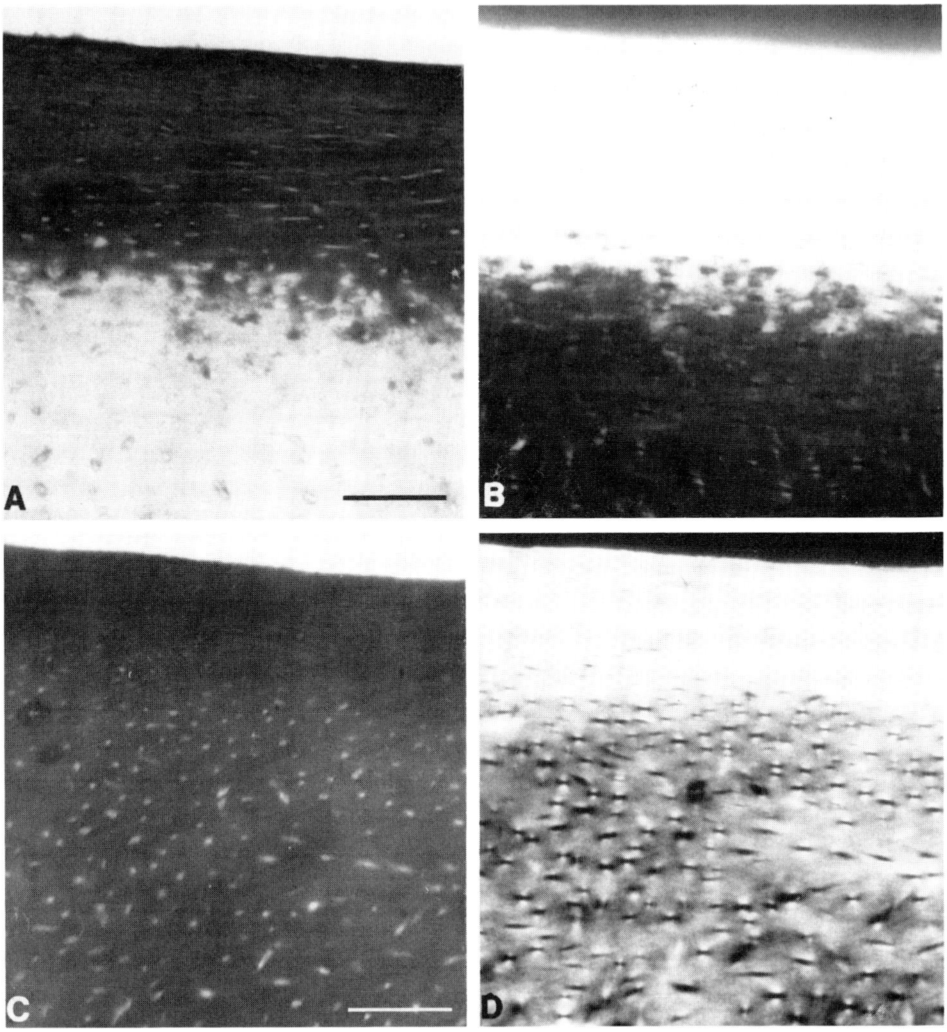

FIGURE 6. Articular cartilage (femur condyle) of a 1-day-old child after picrosirius red F3B staining (A, B), and after digestion with hyaluronidase + picrosirius red F3B staining (C, D, E) for collagen, as well as after staining with *N,N'*-diethylpseudoisocyanine for sulfated GAGs (F). (A, C, F) normal, (B, D, E) polarized light micrographs. Note that collagen can only be stained in the tangentional zone with picrosirius red F3B if the staining is not combined with antecedent enzymatic removal of GAGs (A, B). After digestion both the staining and birefringence is increased in deeper regions, too, (C, D) indicating that the testicular hyaluronidase removed GAGs masking the dye-binding groups of the collagen (the presence of sulfated GAGs is shown in (F). If the specimen seen in (D) is turned under the microscope until the articular surface is approximately parallel with the axis of the polarizer, the birefringence of the interterritorial matrix is more or less extinguished while the spherulite structure of the territorial matrix becomes obvious (E). Scales: 100 μm.

ACKNOWLEDGMENTS

The author is indebted to Dr. Cs. Hadházy (Department of Anatomy, Debrecen) for providing the high density cultures of chondrifying mesenchymal cells for polarization microscopic investigations. The skillful help of Cs. Király and E. Szabó (from the same department) in quantification of topo-optical reactions is highly appreciated.

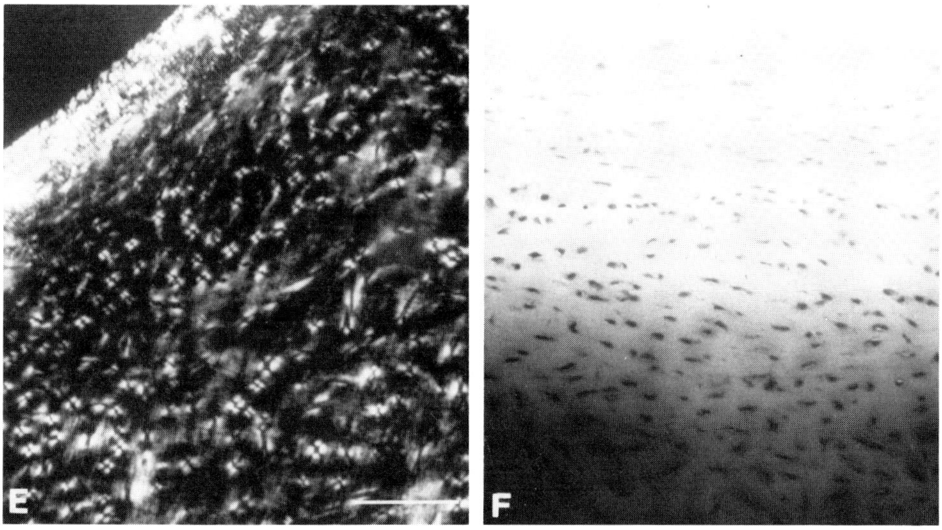

FIGURE 6, E and F.

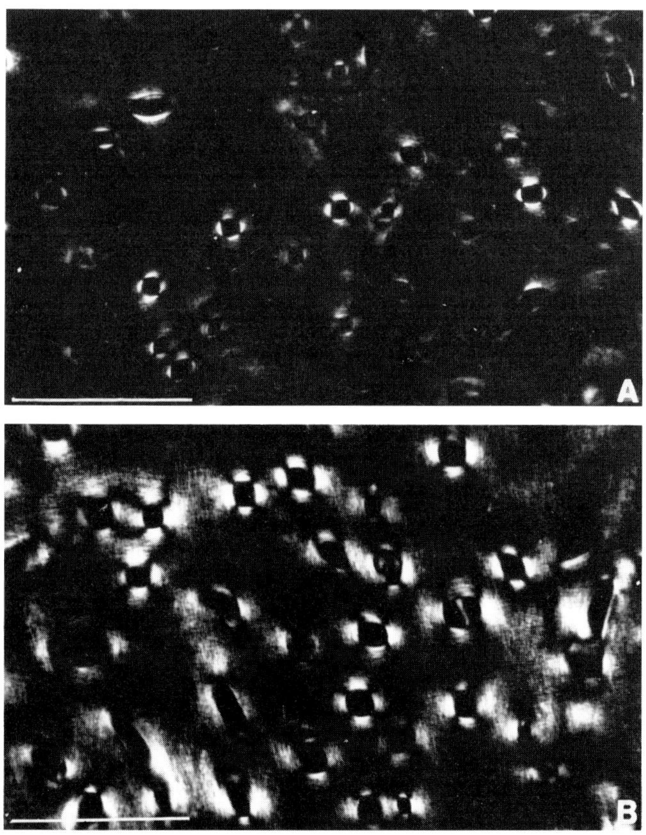

FIGURE 7. Polarized light micrographs taken from middle zone of femur condyle cartilage of a 1-year-old child after picrosirius red F3B staining without (A) and after digestion with testicular hyaluronidase (B). The birefringent zones are thicker in territorial spherulites after the digestion because more dye binding group is accessible to the dye molecules on the collagen demasked. The appearance of the spherulites after this staining in polarized light indicates a circular orientation pattern of collagen in the territorial matrix. Scales: 100 μm.

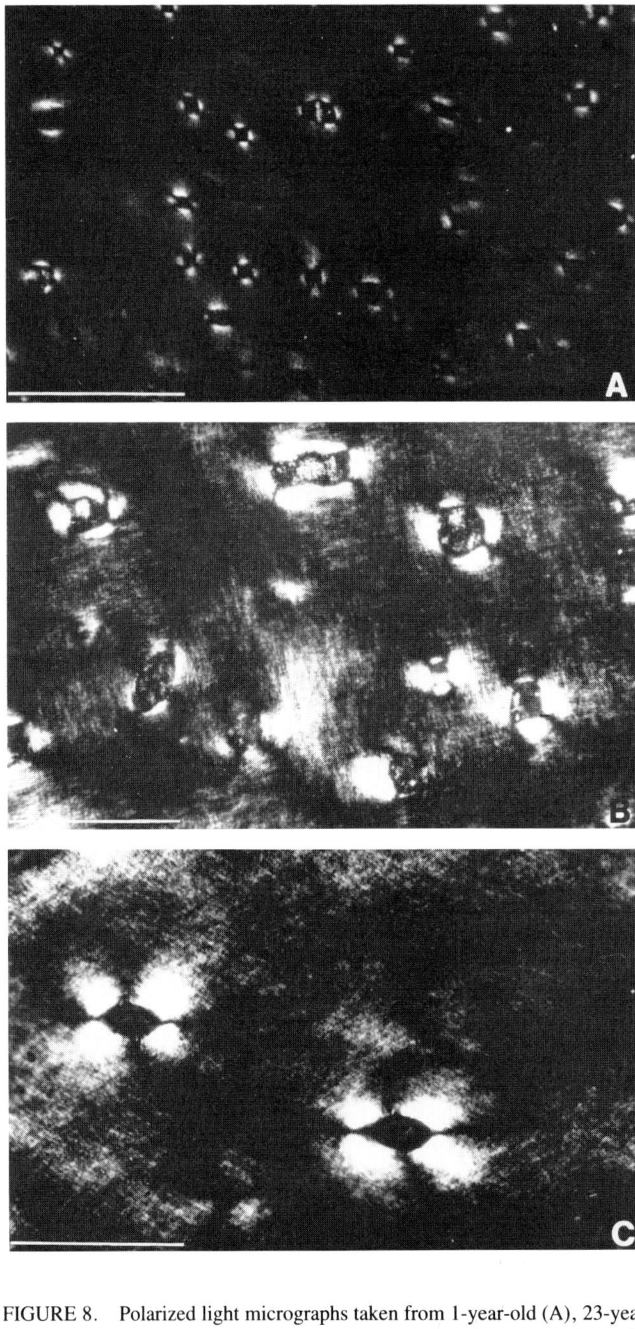

FIGURE 8. Polarized light micrographs taken from 1-year-old (A), 23-year-old (B), and 69-year-old (C) human articular cartilages (femur condyles, middle zones) after staining with toluidine blue at pH 2 for sulfated GAGs. The sections are rotated under the microscope until the birefringence of the interterritorial matrix is extinguished. The spherulite structure of the territorial matrix areas now can easily be recognized. This optical phenomenon indicates that sulfated GAGs are circularly arranged in this matrix compartment. Scales: 100 μm.

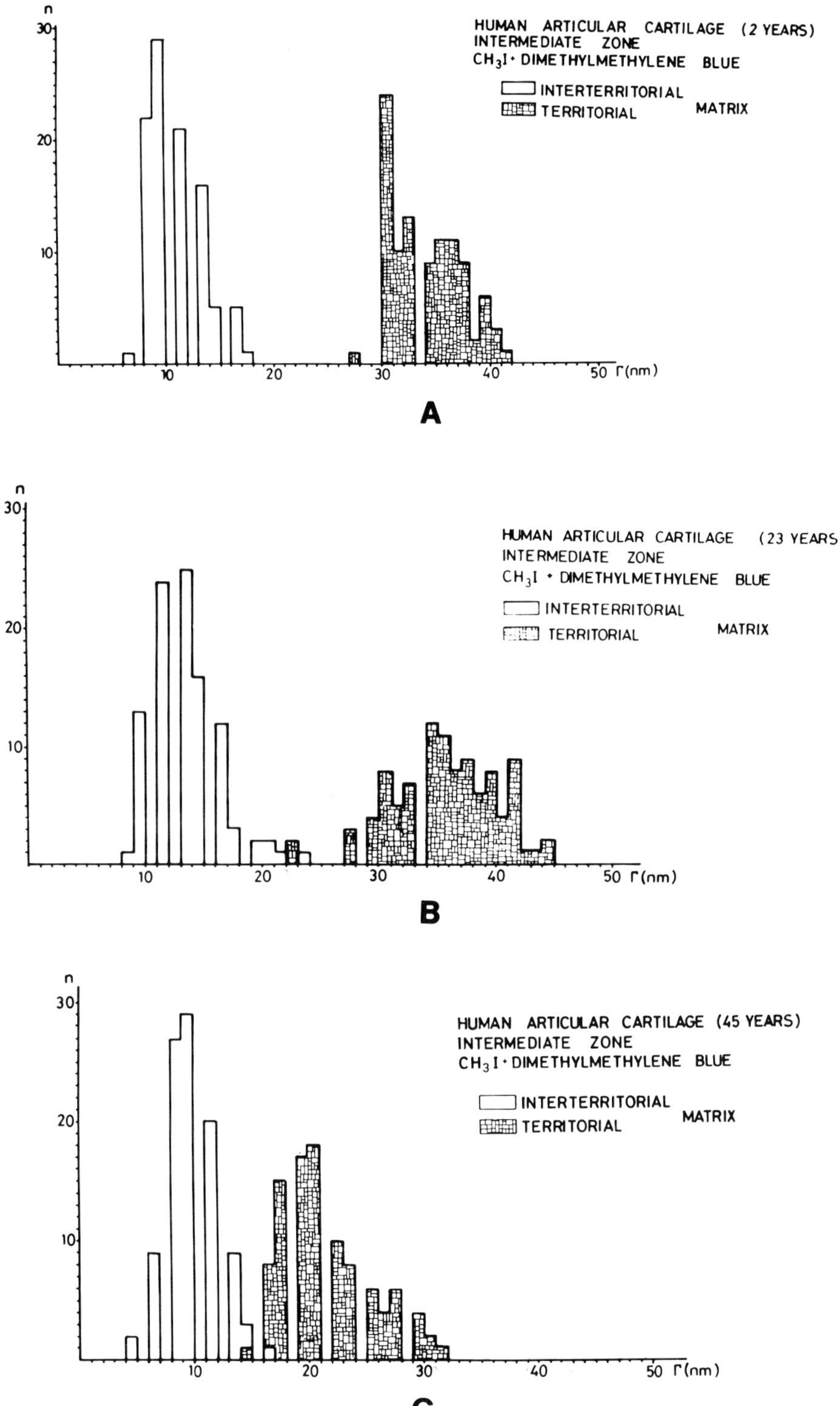

FIGURE 9. Histograms showing the distribution of retardation values of birefringence measured in territorial and interterritorial matrix of the intermediate zones of 2- (A), 23- (B) and 45-year-old (C) human articular cartilages (femur condyles) after a topo-optical reaction specific for sulfated GAGs (see Chapter 12). The territorial matrix areas exhibit significantly higher retardation values than the interterritorial matrix. It follows that oriented GAGs are more densely packed in the territorial matrix.

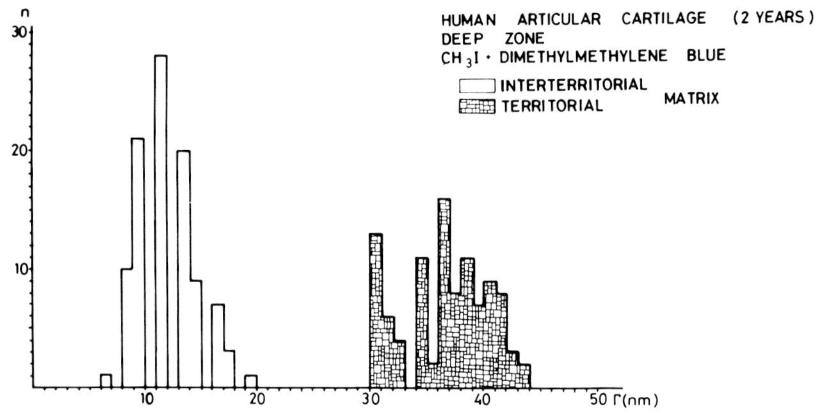

A

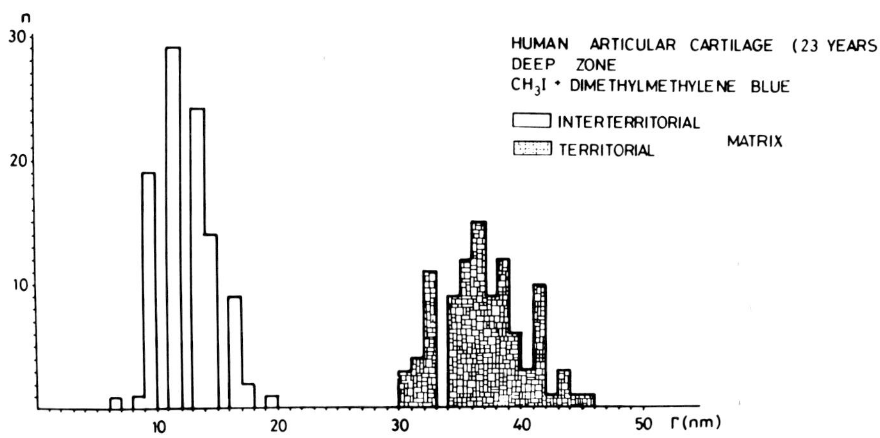

B

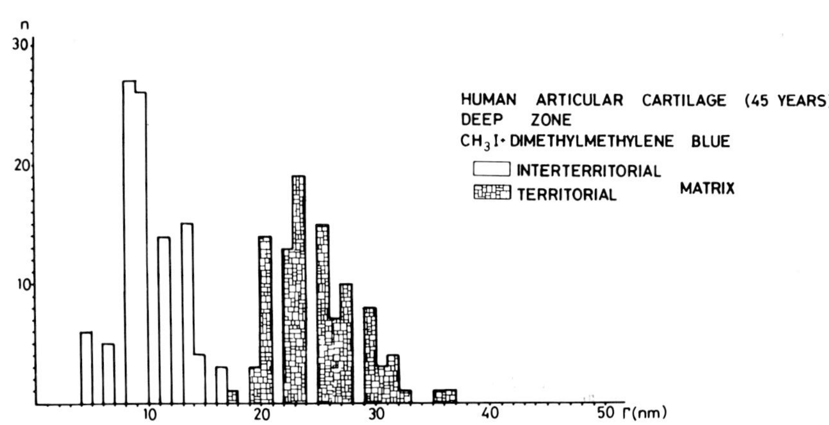

C

FIGURE 10. Distribution of retardation values of birefringence recorded from the deep zones of 2- (A), 23- (B) and 45-year-old (C) human articular cartilages after a sulfated GAG-specific topo-optical reaction. The retardation values are significantly higher in the territorial matrix compared with the interterritorial matrix within the deep zone, too.

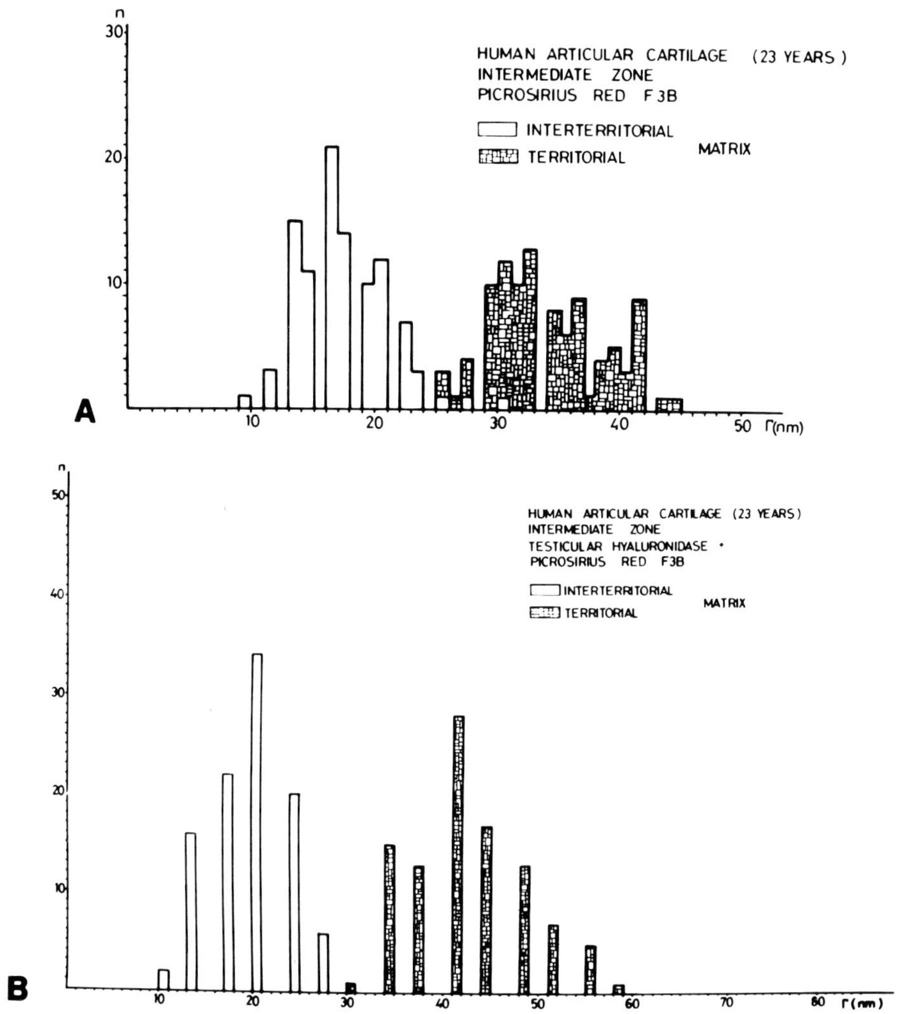

FIGURE 11. Distribution of retardation values of birefringence measured in different matrix compartments of 23-year-old human articular cartilage stained for collagen with picrosirius red F3B (A) and testicular hyaluronidase digestion + picrosirius red F3B (B, C). Note that higher retardation values are recorded if the staining is carried out after the digestion (A, B). The optical anisotropy is more intense in the matrix compartments of the deep zone than in those of the intermediate zone (B, C) indicating a higher level of collagen orientation in the matrix of the deep zone.

C

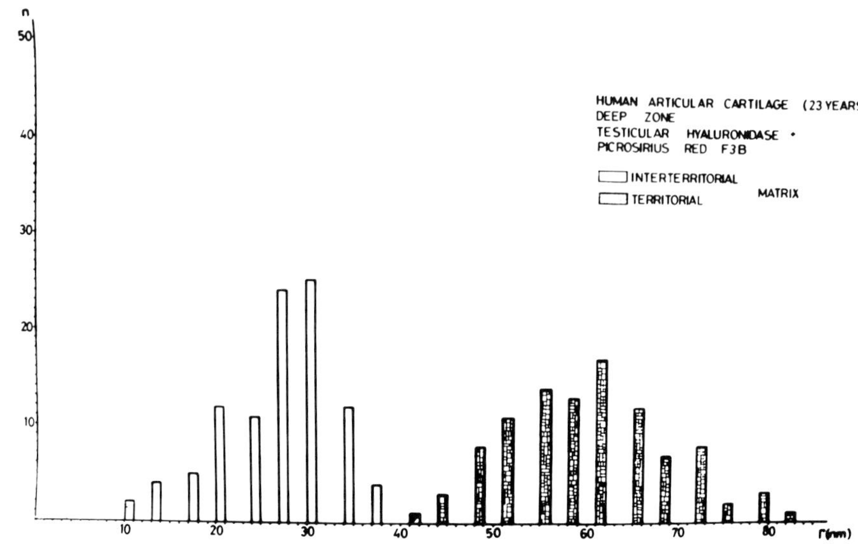

FIGURE 11C.

A B

FIGURE 12. Schematic diagram illustrating the arrangement of collagen in differentiated articular cartilage matrix (A). In (B), an enlarged chondron is depicted between the interterritorial matrix trabecules containing longitudinally oriented fibrils. The collagen orientation pattern in the interterritorial matrix is identical to that of proposed by Benninghoff.[86,87] According to our observations, the territorial matrix contains circularly oriented collagen fibrils in the middle and deep zones. Sulfated GAG molecules follow the orientation of collagen except the tangential zone where their local concentration is low.

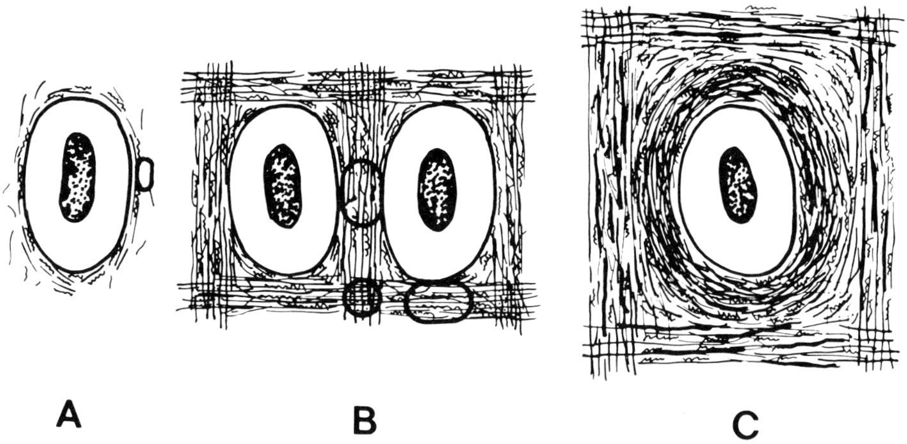

FIGURE 13. Diagrams showing the major orientation patterns of collagen fibrils and the axiparallelly arranged sulfated GAGs around the cells at different stages of human articular cartilage matrix differentiation. (A) Early embryonic stage. GAGs (and somewhat later also collagen fibrils) are tangentially oriented around the cell membrane. The preferential orientation of these components is also shown by an index ellipsoid. (B) Later embryonic stage. Matrix trabecules with longitudinally oriented collagen fibrils and GAG molecules develop, but the pericellular matrix persists. Where collagen fibrils — running in different matrix trabecules — cross each other, the optical anisotropy is extinguished (illustrated by a circle). (C) Postnatal stage. Territorial matrix with circularly oriented and densely packed collagen fibrils and GAGs develops and is interposed between pericellular and interterritorial matrix compartments (From Módis, L., Ádány, R., and Lakatos, I., *Acta Histochem. (Jena), Suppl.*, 26, 305, 1982. With permission.)

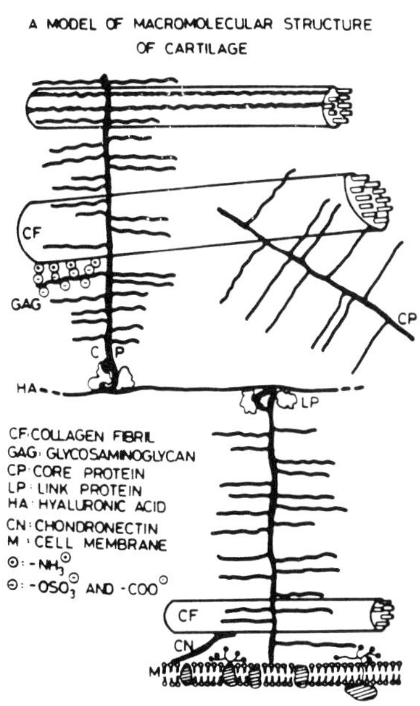

FIGURE 14. A simplified scheme showing the macromolecular organization of the hyaline cartilage matrix. Large proteoglycans with bottlebrush structure are present in three different localizations: a group of them is perpendicularly oriented to the cell membrane in the pericellular space, their GAGs are tangentially arranged around the chondrocyte membrane; another population is perpendicularly attached to the collagen fibrils, their GAGs are in electrostatic interaction with the collagen, and, due to this, they run more or less parallel with the surface of the collagen fibrils; a third group comprises PGs randomly distributed in the interfibrillar space (on the right side of the Figure).

REFERENCES

1. **Sokoloff, L., Ed.,** *The Joints and Synovial Fluid,* Vols. 1, 2, Academic Press, New York, 1978.
2. **Stockwell, R. R.,** *Biology of Cartilage Cells,* Cambridge University Press, Cambridge, 1979.
3. **Freeman, M. A. R., Ed.,** *Adult Articular Cartilage,* 2nd ed., Pitman Medical, London, 1979.
4. **Hall, B. K., Ed.,** *Cartilage,* Vols. 1, 2, 3, Academic Press, New York, 1983.
5. **Kuettner, K. E., Schleyerbach, R., and Hascall, V. C., Eds.,** *Articular Cartilage Biochemistry,* Raven Press, New York, 1986.
6. **Helminen, H. J., Kiviranta, I., Saamanen, A-M., Tammi, M., Paukkonen, K., and Jurvelin, J., Eds.,** *Joint Loading. Biology and Health of Articular Structures,* Wright, Butterworth, Bristol, 1987.
7. **Herbage, D.,** Les collagenes de l'articulation, *Ann. Biol. Clin.,* 44, 152, 1986.
8. **Piez, K. A.,** Collagen types: a review, *J. Cell Biochem.,* Suppl., 10, 101, 1986.
9. **Eyre, D. R., Wu, J.-J., and Apone, S.,** Growing family of collagens in articular cartilage: identification of 5 genetically distinct types, *J. Rheumatol.,* Suppl., 14, 25, 1987.
10. **Ayad, S., Evans, H., Weiss, J. B., and Holt, L.,** Type VI collagen but not type V collagen is present in cartilage, *Coll. Relat. Res.,* 4, 165, 1984.
11. **Mayne, R.,** Cartilage collagens. What is their function, and are they involved in articular disease? *Arthritis Rheum.,* 32, 241, 1989.
12. **Bruckner, P., Vaughan, L., and Winterhalter, K. H.,** Type IX collagen from sternal cartilage of chicken embryo contains covalently bound glycosaminoglycans, *Proc. Natl. Acad. Sci. U S.A.* 82, 2608, 1985.
13. **Van der Rest, M., Mayne, R., Ninomiya, Y., Seidah, N. G., Chretien, M., and Olsen, B. R.,** The structure of type IX collagen, *J. Biol. Chem.,* 260, 220, 1985.
14. **Van der Rest, M. and Mayne, R.,** Type IX collagen proteoglycan from cartilage is covalently cross-linked to type II collagen, *J. Biol. Chem.,* 263, 1615, 1988.
15. **Eyre, D. R., Apone, S., Wu, J.-J., Ericsson, L. H., and Walsh, K. A.,** Collagen type IX: evidence for covalent linkages to type II collagen in cartilage, *FEBS Lett.,* 220, 337, 1987.
16. **Müller-Glauser, W., Humbel, B., Glatt, M., Strauli, P., Winterhalter, K. H., and Bruckner, P.,** On the role of type IX collagen in the extracellular matrix of cartilage: type IX collagen is localized to intersections of collagen fibrils, *J. Cell Biol.,* 102, 1931, 1986.
17. **Gibson, G. J. and Flint, M. H.,** Type X collagen synthesis by chick sternal cartilage and its relationship to endochondral development, *J. Cell Biol.,* 101, 277, 1985.
18. **Gibson, G. J., Bearman, C. H., and Flint, M. H.,** The immunoperoxidase localization of type X collagen in chick cartilage and lung, *Coll. Relat. Res.,* 6, 163, 1986.
19. **Cancedda, R., Capasso, O., Castagnola, P., Descalzi-Cancedda, F., and Quarto, N.,** Deposition of type X collagen in the cartilage extracellular matrix, *J. Cell. Biochem.,* 28, 7, 1985.
20. **Schmid, T. M. and Linsenmayer, T. F.,** Developmental acquisition of type X collagen in the embryonic chick tibiotarsus, *Dev. Biol.,* 107, 373, 1985.
21. **Smith, G. N., Williams, J. M., Jr., and Brandt, K. D.,** Interaction of proteoglycans with the pericellular (1α, 2α, 3α) collagens of cartilage, *J. Biol. Chem.,* 260, 10761, 1985.
22. **Smith, G. N., Hasty, K. A., and Brandt, K. D.,** Type XI collagen is associated with the chondrocyte surface is suspension culture, *Matrix,* 9, 186, 1989.
23. **Poole, C. A., Flint, M. H., and Beaumont, B. W.,** Morphology of the pericellular capsule in articular cartilage revealed by hyaluronidase digestion, *J. Ultrastruct. Res.,* 91, 13, 1985.
24. **Poole, C. A., Flint, M. H., and Beaumont, B. W.,** Chondrons in cartilage: ultrastructural analysis of the pericellular microenvironment in adult human articular cartilages, *J. Orthop. Res.,* 5, 509, 1987.
25. **Poole, C. A., Flint, M. H., and Beaumont, B. W.,** Chondrons extracted from canine tibial cartilage: preliminary report on their isolation and structure, *J. Orthop. Res.,* 6, 408, 1988.
26. **Poole, C. A., Ayad, S., and Schofield, J. R.,** Chondrons from articular cartilage: I. Immunolocalization of type VI collagen in the pericellular capsule of isolated canine tibial chondrons, *J. Cell Sci.,* 90, 635, 1988.
27. **Poole, C. A., Wotton, S. F., and Duance, V. C.,** Localization of type IX collagen in chondrons isolated from porcine articular cartilage and rat chondrosarcoma, *Histochem. J.,* 20, 567, 1988.
28. **Németh-Csóka, M. and Mészáros, T.,** The importance of minor collagenous chains in the articular cartilage, *Acta Biol. Hung.,* 35, 163, 1984.
29. **Charriere, G., Hartman, D. J., Vignon, E., Ronziere, M.-C., Herbage, D., and Ville, G.,** Antibodies to types I, II, IX, and XI collagen in the serum of patients with rheumatic diseases, *Arthritis Rheum.,* 31, 325, 1988.
30. **Wooley, P. H., Dillon, A. M., Luthra, H. S., Stuart, J. M., and David, C. S.,** Genetic control of type II collagen-induced arthritis in mice: factors influencing disease susceptibility and evidence for multiple MHC-associated gene control, *Transplant. Proc.,* 15, 180, 1983.
31. **Hardingham, T. and Muir, H.,** The specific interaction of hyaluronic acid with cartilage proteoglycans, *Biochim. Biophys. Acta.* 279, 401, 1972.

32. **Módis, L.**, The molecular structure of the interfibrillar matrix in connective tissue, *Acta Biol. Acad. Sci. Hung.*, 29, 197, 1978.
33. **Muir, H.**, Proteoglycans as organizers of the intercellular matrix, *Biochem. Soc. Trans.*, 11, 613, 1983.
34. **Handley, C. J., Lowther, D. A., and McQuillan, D. J.**, Mini-review: the structure and synthesis of proteoglycans of articular cartilage, *Cell Biol. Int. Rep.*, 9, 753, 1985.
35. **Poole, A. R.**, Proteoglycans in health and disease: structures and functions, *Biochem. J.*, 236, 1, 1986.
36. **Lohmander, S.**, Proteoglycans of joint cartilage. Structure, function, turnover and role as markers of joint disease, *Bailliere's Clin. Rheumatol.*, 2, 37, 1988.
37. **Eggli, P., Herrmann, W., Hunziker, E. B., and Schenk, R. K.**, Matrix compartments in the growth plate of the proximal tibia of the rats, *Anat. Rec.*, 211, 246, 1985.
38. **Schenk, R. K., Eggli, P. S., and Hunziker, E. B.**, Articular cartilage morphology, in *Articular Cartilage Biochemistry*, Kuettner, K. E., Schleyerbach, R., and Hascall, V. C., Eds., Raven Press, New York, 1986, 3.
39. **Stockwell, R. A. and Scott, J. E.**, Observations on the acid glycosaminoglycan (mucopolysaccharide) content of the matrix of aging cartilage, *Ann. Rheum. Dis.*, 24, 341, 1965.
40. **Scott, J. E.** Distribution of acid mucopolysaccharides in cartilage, in *Current Developments in Rheumatology*, Lövtren, O., Boström, H., Olhagen, B., and Sannerstedt, R., Eds., Lindgren, Mölndal, Stockholm, 1973, 10.
41. **Scott, J. E. and Haigh, M.**, Keratan sulphate and the ultrastructure of cornea and cartilage: a 'stand-in' for chondroitin sulphate in conditions of oxygen lack?, *J. Anat.*, 158, 95, 1988.
42. **Hunziker, E. B., Herrmann, W., and Schenk, R. K.**, Ruthenium hexammine trichloride (RHT)-mediated interaction between plasmalemmal components and pericellular matrix proteoglycans is responsible for the preservation of chondrocytic plasma membranes *in situ* during cartilage fixation, *J. Histochem. Cytochem.*, 31, 717, 1983.
43. **Heinegard, D. and Sommarin, Y.**, Macromolecules and interactions at the chondrocyte cell surface, *J. Clin. Chem. Clin. Biochem.*, 24, 937, 1986.
44. **Ratcliffe, A., Fryer, P. R., and Hardingham, T. E.**, The distribution of aggregating proteoglycans in articular cartilage, *J. Histochem. Cytochem.*, 32, 193, 1984.
45. **Glant, T. T. and Mikecz, K.**, Antigenic profile of human, bovine, and canine articular chondrocytes, *Cell Tissue Res.*, 224, 359, 1986.
46. **Mitchell, N. and Shepard, N.**, Pericellular proteoglycan concentrations in early degenerative arthritis, *Arthritis Rheum.*, 24, 958, 1981.
47. **Módis, L., Lustyik, Gy., Ádány, R., and Zs.-Nagy, I.**, Energy dispersive X-ray microanalysis of sulfated glycosaminoglycans in cartilage matrix stained with alcian blue 8GX, *Bas. Appl. Histochem.*, 32, 415, 1988.
48. **Katz, E. P., Wachtel, E. J., and Maroudas, A.**, Extrafibrillar proteoglycans osmotically regulate the molecular packing of collagen in cartilage, *Biochim. Biophys. Acta*, 882, 136, 1986.
49. **Kuijer, R., van de Stadt, R. J., De Koning, M. H. M. T., and van der Korst, J. K.**, Influence of constituents of proteoglycans on type II collagen fibrillogenesis, *Coll. Relat. Res.*, 5, 379, 1985.
50. **Dell'Orbo, C., Quacci, D., and Ruggeri, A.**, Nouve tecniche istochimiche per la dimostrazione dei proteoglicani estracellulari al microscopio elettronico, *Riv. Istoch. Norm. Pat.*, 21, 193, 1977.
51. **Ruggeri, A., Dell'Orbo, C., and Quacci, D.**, Electron microscopic visualizaiton of proteoglycans with alcian blue, *Histochem. J.*, 7, 187, 1975.
52. **Orford, C. R. and Gardner, D.**, Proteoglycan association with collagen d band in hyaline articular cartilage, *Connect. Tissue Res.*, 12, 345, 1984.
53. **Scott, J. E. and Haigh, M.**, Proteoglycan-collagen interactions in intervertebral disc. A chondroitin sulfate proteoglycan associates with collagen fibrils in rabbit annulus fibrosus at the d-e bands, *Biosci. Rep.*, 6, 879, 1986.
54. **Morocutti, M., Marchini, M., Colombo, B., and Ruggeri, A.**, Heterogeneity of proteoglycan particles in thin sections and replicas of human articular cartilage, *Acta Biol. Hung.*, 35, 233, 1984.
55. **Mathews, M. B.**, The interaction of collagen and acid mucopolysaccharides. A model for connective tissue, *Biochem. J.*, 96, 710, 1965.
56. **Laurent, T. C.**, Interaction between proteins and glycosaminoglycans, *Fed. Proc.*, 36, 24, 1977.
57. **Ruggeri, A. and Benazzo, F.**, Collagen-proteoglycan interaction, in *Ultrastructure of the Connective Tissue Matrix*, Ruggeri, A. and Motta, P. M., Eds., Martinus Nijhoff, Boston, 1984, chap. 6.
58. **Scott, J. E.**, Proteoglycan-collagen interactions, *CIBA Symposia*, 124, 104, 1986.
59. **Scott, J. E.**, Proteoglycan—fibrillar collagen interactions, *Biochem. J.*, 252, 313, 1988.
60. **Glant, T.**, Induction of cartilage degradation in experimental arthritis produced by allogeneic and xenogeneic proteoglycan antigens, *Connect. Tissue Res.*, 9, 137, 1982.
61. **Glant, T. T., Mikecz, K., Arzomanian, A., and Poole, A. R.**, Proteoglycan-induced arthritis in BALB/c mice: clinical features and histopathology, *Arthritis Rheum.*, 30, 201, 1987.
62. **Mikecz, K., Glant, T. T., Buzás, E., and Poole, A. R.**, Cartilage proteoglycans as potential autoantigens in humans and in experimental animals, *Agents Actions*, 22, 63, 1988.

63. **Fife, R. S. and Brandt, K. D.,** Identification of a high-molecular-weight (>400,000) protein in hyaline cartilage, *Biochim. Biophys. Acta,* 802, 506, 1984.
64. **Paulsson, M. and Heinegard, D.,** Noncollagenous cartilage proteins. Current status of an emerging research field, *Coll. Relat. Res.,* 4, 219, 1984.
65. **Heinegard, D., Larsson, T., Sommarin, Y., Franzén, A., Paulsson, M., and Hedbom, E.,** Two novel matrix proteins isolated from articular cartilage show wide distributions among connective tissues, *J. Biol. Chem.,* 261, 13866, 1986.
66. **Roughley, P. J., Poole, A. R., and Mort, J. S.,** The heterogeneity of link proteins isolated from human articular cartilage proteoglycan aggregates, *J. Biol. Chem.,* 257, 11908, 1982.
67. **Heinegard, D. and Oldberg, A.,** Structure and biology of cartilage and bone noncollagenous macromolecules, *FASEB J.,* 3, 2042, 1989.
68. **Chandrasekhar, S., Laurie, G. W., Cannon, F. B., Martin, G. R., and Kleinman, H. K.,** In vitro regulation of cartilage matrix assembly by a M_r 54.000 collagen-binding protein, *Proc. Natl. Acad. Sci. U.S.A.,* 83, 5126, 1986.
69. **Dessau, W., Sasse, J., Timpl, R., Jilek, F., and van der Mark, K.,** Synthesis and extracellular deposition of fibronectin in chondrocyte cultures. response to the removal of extracellular cartilage matrix, *J. Cell Biol.,* 79, 342, 1978.
70. **Clemmensen, I., Holund, B., Johansen, N., and Andersen, R. B.,** Demonstration of fibronectin in human articular cartilage by an indirect immunoperoxidase technique, *Histochemistry,* 76, 51, 1982.
71. **Glant, T. T., Hadházy, Cs., Mikecz, K., and Sipos, A.,** Appearance and persistence of fibronectin in cartilage. Specific interaction of fibronectin with collagen type II, *Histochemistry,* 82, 149, 1985.
72. **Kulyk, W., M., Upholt, W. B., and Kosher, R., A.,** Fibronectin gene expression during limb cartilage differentiation, *Development,* 106, 449, 1989.
73. **Poole, A. R., Pidoux, I., Reiner, A., Choi, H., and Rosenberg, L. C.,** Association of an extracellular protein (chondrocalcin) with the calcification of cartilage in endochondral bone formation, *J. Cell Biol.,* 98, 54, 1984.
74. **Hinek, A., Reiner, A., and Poole, A. R.,** The calcification of cartilage matrix in chondrocyte culture: studies of the C-propeptide of type II collagen (chondrocalcin), *J. Cell Biol.,* 104, 1435, 1987.
75. **Knudson, C. B. and Toole, B. P.,** Changes in the pericellular matrix during differentiation of limb bud mesoderm, *Dev. Biol.,* 112, 308, 1985.
76. **Lacy, B. E. and Underhill, C. B.,** The hyaluronate receptor is associated with actin filaments, *J. Cell Biol.,* 105, 1395, 1987.
77. **Roughley, P. J., White, R. J., and Poole, A. R.,** Identification of a hyaluronic acid-binding protein that interferes with the preparation of high-buoyant-density proteoglycan aggregates from adult human articular cartilage, *Biochem. J.,* 231, 129, 1985.
78. **Mollenhauer, J. and von der Mark, K.,** Isolation and characterization of a collagen-binding glycoprotein from chondrocyte membranes, *EMBO J.,* 2, 45, 1983.
79. **Von der Mark, K., Mollenhauer, J., Kühl, U., Bee, J., and Lesot, H.,** Anchorins: a new class of membrane proteins involved in cell-matrix interactions, in *The Role of Extracellular Matrix in Development,* Trelstad, R. L., Ed., Alan R. Liss, New York, 1984, 67.
80. **Hewitt, A. T., Kleinman, H. K., Pennypacker, J. P., and Martin, G. R.,** Identification of an attachment factor for chondrocytes, *Proc. Natl. Acad. Sci. U.S.A.,* 77, 385, 1980.
81. **Hewitt, A. T., Varner, H. H., Silver, M. H., Dessau, W., Wilkes, C. M., and Martin, G. R.,** The isolation and partial characterization of chondronectin, an attachment factor for chondrocytes, *J. Biol. Chem.,* 257, 2330, 1982.
82. **Weiss, C., Rosenberg, L., and Helfet, A. J.,** An ultrastructural study of normal young adult human articular cartilage, *J. Bone Joint Surg. (Am.),* 50, 663, 1968.
83. **Clarke, I. C.,** Articular cartilage: a review and scanning electron microscopy study. 1. The interterritorial fibrillar architecture, *J. Bone Joint Surg. (Br.),* 53, 732, 1971.
84. **Clarke, I. C.,** Articular cartilage: a review and scanning electron microscopic study. II. The territorial fibrillar architecture, *J. Anat.,* 118, 261, 1974.
85. **Clark, J. M.,** The organization of collagen in cryofractured rabbit articular cartilage: a scanning electron microscopic study, *J. Orthop. Res.,* 3, 17, 1985.
86. **Benninghoff, A.,** Form und Bau der Gelenkknorpel in ihren Beziehungen zur Funktion 1. Die modellierenden und formerhaltenden Faktoren des Knorpelreliefs, *Z. Anat. Entwickl,-Gesch.,* 76, 43, 1925.
87. **Benninghoff, A.,** Form und Bau der Gelenkknorpel in ihren Beziehungen zur Funktion. II. Der Aufbau des Gelenkknorpels in seinen Beziehungen zur Funktion, *Z. Zellforsch.,* 2, 783, 1925.
88. **Bucher, O.,** Beitrag zum funktionellen Bau des hyalinen Knorpels (auf Grund von Untersuchungen im polarisierten Lichte), *Z. Zellforsch.,* 32, 281, 1942.
89. **Dallek, M.,** *Polarisationsoptische und Rasterelektronenmikroskopische Untersuchungen über den Kollagenfaserverlauf in kindlichen Epiphysenfugen,* Dissertation, Fachbereich Medizin der Universitat Hamburg, Hamburg, 1981.

90. **Dallek, M., Jungbluth, K. H., and Holstein, A. F.**, Studies on the arrangement of the collagenous fibers in the infant epipyseal plates using polarized light and the scanning electron microscope, *Arch. Orthop. Trauma. Surg.*, 101, 239, 1983.
91. **Yarker, Y. E., Aspden, R. M., and Hukins, D. W. L.**, Birefringence of articular cartilage and the distribution of collagen fibril orientations, *Connect. Tissue Res.*, 11, 207, 1983.
92. **Yarker, Y. E., Hukins, D. W. L., and Nave, C.**, X-ray diffraction studies of tibial plateau cartilage using synchrotron radiation, *Connect. Tissue Res.*, 12, 337, 1984.
93. **Vidal, B. C. and Vilarta, R.**, Articular cartilage: collagen type II-proteoglycan interactions. Availability of reactive groups. Variation in birefringence and differences as compared to collagen I, *Acta Histochem. (Jena)*, 83, 189, 1988.
94. **Módis, L., Módis-Süveges, I., and Conti, G.**, Quantitative polarisationsoptische Untersuchungen an Mucopolysacchariden, *Acta Histochem. (Jena)*, Suppl., 12, 169, 1972.
95. **Módis, L.**, *Topo-Optical Investigation of Mucopolysaccharides (Acid Glycosaminoglycans)*, Handbuch der Histochemie, Vol. 2, Part 4, Graumann, W. and Neumann, K., Eds., G. Fischer Verlag, Stuttgart, 1974.
96. **Kern, M., Módis, L., Antal, M., and Földes, I.**, Polarisationsmikroskopische Beobachtungen an pathologischer Knorpelgrundsubstanz, *Acta Histochem. (Jena)*, Suppl., 15, 361, 1975.
97. **Heuft, G. and Böhm, N.**, Altersabhangigkeit des Toluidinblau-Dichroismus und der Anisotropie der Fasern in der Wachstumszone menschlicher Rippen, *Histochemistry*, 59, 45, 1978.
98. **Vidal, B. C.**, Acid glycosaminoglycans and endochondral ossification: microspectrophotometric evaluation and macromolecular orientation, *Cell. Mol. Biol.*, 22, 45, 1977.
99. **Módis, L., Ádány, R., and Lakatos, I.**, Polarisationsoptische Analyse der menschlichen embryonalen Knorpelmatrix, *Acta Histochem. (Jena)*, Suppl., 26, 305, 1982.
100. **Dunham, J., Shackleton, D. R., Nahir, A. M., Billingham, M. E. J., Bitensky, L., Chayen, J., and Muir, H.**, Altered orientation of glycosaminoglycans and cellular changes in the tibial cartilage in the first two weeks of experimental canine osteoarthritis, *J. Orthop. Res.*, 3, 258, 1985.
101. **Dunham, J., Shackleton, D. R., Billingham, M. E. J., Bitensky, L., Chayen, J., and Muir, H. I.**, A reappraisal of the structure of normal canine articular cartilage, *J. Anat.*, 157, 89, 1988.
102, **Zambrano, N. Z., Montes, G. S., Shigihara, K. M., Sanchez, E. M., and Junqueira, L. C. U.**, Collagen arrangement in cartilages, *Acta Anat.*, 113, 26, 1982.
103. **Toole, B. P.**, Hyaluronate turnover during chondrogenesis in the developing chick limb and axial skeleton, *Dev. Biol.*, 29, 321, 1972.
104. **Földes, I., Módis, L., Antalffy, J., and Ádám, I.**, Ultrastructure of extracellular matrix of embryonic chick limb bud cartilage, *Acta Biol. Acad. Sci. Hung.*, 31, 81, 1980.
105. **Von der Mark, K.**, Differentiation, modulation and dedifferentiation of chondrocytes, in *Connective Tissue: Biological and Clinical Aspects*, Kühn, K. and Krieg, T., Eds., Karger, Basel, 1986, 272.
106. **Shinomura, T., Kimata, K., Oike, Y., Noro, A., Hirose, N., Tanabe, K., and Suzuki, S.**, The occurence of three different proteoglycan species in chick embryo cartilage, *J. Biol. Chem.*, 258, 9314, 1983.
107. **Shinomura, T., Kimata, K., Oike, Y., Maeda, N., Yano, S., and Suzuki, S.**, Appearance of distinct types of proteoglycan in a well-defined temporal and spatial pattern during early cartilage formation in the chick limb, *Dev. Biol.*, 103, 211, 1984.
108. **Solursh, M., Ahrens, P. B., and Reiter, R. S.**, Tissue culture and analysis of the steps in limb morphogenesis, *In vitro*, 14, 51, 1978.
109. **Hadházy, Cs., László, M. B., and Kostenszky, K. S.**, Cartilage differentiation in micro-mass cultures of chicken limb buds, *Acta Morphol. Acad. Sci. Hung.*, 30, 65, 1982.
110. **Sommarin, Y. and Heinegard, D.**, Four classes of cell-associated proteoglycans in suspension cultures of articular-cartilage chondrocytes, *Biochem. J.*, 233, 809, 1986.
111. **Módis, L., Hadházy, Cs., László, M. B., Kostenszky, K. S., and Földes, I.**, Proteoglycan biosynthesis is stimulated by D-penicillamine in chondrifying high density cell cultures, *Exp. Pathol.*, 35, 159, 1988.
112. **Szirmai, J. A.**, Structure of cartilage, in *Aging of Connective and Skeletal Tissue*, Engel, A. and Larsson, T., Eds., Nordiska Bokhandeln Förlag, Stockholm, 1969, 163.
113. **Weiss, P.**, From cell to molecule, in *The Molecular Control of Cellular Activity*, Allen, J. M., Ed., McGraw-Hill, New York, 1962, 1.
114. **Weiss, A. P.**, The compounding of complex macromolecular and cellular units into tissue fabrics, *Proc. Natl. Acad. Sci. U.S.A.*, 42, 819, 1956.
115. **Meyer, K.**, Reflections on "mucopolysaccharides" and their protein complexes, in *Chemistry and Molecular Biology of the Intercellular Matrix*. Vol. 1, Balazs, E. A., Ed., Academic Press, London, 1970, 5.

Chapter 5

THE ORGANIC BONE MATRIX

TABLE OF CONTENTS

I. CONSTITUENTS OF THE ORGANIC BONE MATRIX

Type I collagen is the major constituent of the organic matrix of bone tissue. Collagen makes up as much as 70% and 90% (w/w) of the organic matrix in cancellous bone and compact bone, respectively.[1] Depending upon the age of the specimen, collagen type I molecules form differently sized fibers. In humans, cross striated collagen fibers of infant bone have a diameter ranging from about 20 to 60 nm, while fibers with a diameter up to 150 nm are characteristic of the bones of mature individuals.[2]

Collagen fibers exhibit different microscopic arrangements in different bones, ages, and species. A great variety of different bone types are formed according to the combinations of size and orientation of fiber bundles, the ratio of volume of fiber bundles to volume of ground substance, and the number, orientation and spacing of cells[3] in higher animals, based on the arrangement of the collagen fibers, two major varieties of bone can be distinguished: woven (coarse-fibered or primary) and lamellar (finely fibered of secondary) bone. Both of them may form compact and spongy bone. The woven bone is permeated by large vascular spaces, and the coarse, interweaving fibers run in all directions. Osteocytes are scattered more or less at random in the matrix without any preferential orientation to the vascular channels. Adult skeleton contains very little woven bone. Generally, bone is deposited initially as woven bone in the embryo and during fracture healing and then it is gradually replaced by lamellar bone. Woven bone persists in traces in the osseous labyrinth, tooth sockets, bony sutures, and regions of tendon-bone attachment in the adult skeleton.

Considering histochemical findings, it is thought that woven and lamellar bone matrix have different compositions, e.g., the woven bone matrix can be stained more strongly with cationic dyes than can the lamellar bone matrix.[4] The structure of the lamellar bone will be discussed separately in this Chapter.

Other noncollagenous proteins of bone tissue form the minor but more specific constituents of the matrix. A 23 kDa glycoprotein with high sialic acid content was isolated from bovine and human bone.[1,5] This was named bone sialoprotein (BSP). It comprises 8 to 12% (w/w) of the total noncollagenous protein of bovine bone. Carbohydrate residues make up about 40% by weight of the molecule and sialic acid accounts for about half of this value.[1,6] It is rich in aspartate and glutamate, and it contains 2 to 4 mol of phosphate/mol of protein. These chemical features explain the polyanionic character of this molecule. It is supposed to be involved in the binding of calcium, but its function is not yet understood.

Two γ-carboxyglutamic acid (Gla)-containing proteins have been isolated from bone tissue: osteocalcin or bone Gla-protein (BGP) and matrix Gla-protein (MGP). Both of them are supposed to play essential roles in mineralization because of the characteristic interaction of calcium ions with Gla residues. Osteocalcin (BGP) is a small (6 kDa) protein, and is found in mineralized bone and in the dentine matrix of all vertebrates examined.[6-8] It comprises up to 15% (w/w) of the noncollagenous protein content of compact bone in most species but this value is only 1% in man.[6] It is produced by osteoblasts and osteocytes,[9,10] and it also occurs in blood serum.[11]

MGP has been isolated from bovine bone matrix. This 10 kDa molecule accumulates in the matrix parallel with mineralization.[12,13]

Proteoglycan molecules are also present in the bone matrix. They comprise about 4% (w/w) of the total noncollagenous protein of bovine bone.[6] Chondroitin-4-sulfate was identified as the predominating GAG of human bone, and its concentration was inversely related to age in growing subjects, but no change was noted later in life.[14] The size of bone PGs is much smaller compared to the cartilage PGs.[15] Three populations of PGs having different compositions have been isolated from the calcified matrix of mature bovine bone.[16,17] The core protein of the major PG fraction has a molecular mass of 30 kDa, and one or two 34 kDa chondroitin sulfate chains is attached to it.[17] Two classes of chondroitin sulfate PGs have been found in fetal bovine bone.[18,19] Both of them had a protein core of 38 kDa to which one of two 40 kDa chondroitin sulfate side-chains is attached. Supposedly the bone PGs interact with collagen fibrils hereby affecting fibrillogenesis. It is not known whether PGs have the same inhibitory function on mineralization processes in bone matrix which has been shown in cartilage matrix.[20]

An interesting functional aspect of bone PGs has been suggested by the experiments of Skerry et al.[21] Turkey ulnae were subjected to external dynamic loading in vivo over a short period. Using quantitative polarization microscopic analysis, these authors demonstrated a marked and transient reorientation of PGs within compact bone tissue after loading. In the animals killed immediately after loading, the retardation values of birefringence attributed to the orientation of bone PGs was increased by 36% compared to nonloaded controls. In sections taken 24 h after loading, the birefringence in loaded bones still exceeded that of the controls by 25%. By 48 h after loading, there was no significant difference between sections from loaded and contralateral controls. According to these authors, this slow re-orientation of PGs provides a persistent record of the tissue's recent history. They believe that such a record has the ability not only to "capture" strain transients but also to "update" and "average" them. They regard the orientation pattern of PGs as strain memory in bone tissue. These interesting speculations need, however, more experimental support. It is not sure that the staining conditions used in this work allow a specific demonstration of the orientation of PGs. The authors used mineralized sections stained with alcian blue in the presence of 0.025 mol/l $MgCl_2$ and at pH 5.8. It is not certain that hydroxyapatite crystals cannot bind alcian blue under these conditions. These experiments should be repeated with demineralized sections. Furthermore, the 0.025 mol/l $MgCl_2$ concentration is too low to allow a specific staining of chondroitin sulfate molecules. Using this low $MgCl_2$ concentration, each carboxyl group of different matrix constituents can also bind the cationic alcian blue. The use of alcian blue in the presence of 0.2 to 0.3 mol/l $MgCl_2$ and in combination with antecedent enzymatic extraction of GAGs would have made the interpretation of these experimental data more reliable.

Phosphoproteins are also present in bone matrix. The phosphate groups covalently bound to some of their serine and threonine residues can engage in strong ionic interactions with calcium ions. These molecules, therefore, may influence the mineralization processes.[22,23] There are 15 phosphoprotein components with molecular weights ranging from 5 kDa to 150 kDa.[24] The best characterized phosphoprotein of bone matrix is osteonectin. This 30 kDa glycoprotein has affinity both for collagen and hydroxyapatite, and therefore, it causes mineralization of type I collagen *in vitro*.[25,26] Osteonectin is the major non collagenous constituent of bovine and porcine bone and dentine, and it makes up as much as 23% (w/w) of the total noncollagenous protein.[6] It is produced also by different fibroblasts in culture[27] and human platelets.[28] A 44 kDa bone phosphoprotein has been isolated and characterized[29] which proved to be identical to osteopontin.[30] This protein has been immunohistochemically detected not only in the bone matrix, but in osteoblasts before the onset of mineralization.[31] Furthermore, it has also been found at ectopic sites in newborn rats, namely in proximal convoluted tubules of the kidney, neurons, sensory, and secretory cells of the internal ear.[32]

Different plasma proteins have also been found in bone matrix. Albumin makes up about 3% (w/w) of the non collagenous matrix of bovine bone[6] and it probably accumulates by absorption to the mineral phase during the formation of calcified tissue.[33] Plasma α_2HS 50 kDa glycoprotein is synthesized by the liver and subsequently accumulates in bone and dentine.[34] Apart from its possible effect on mineralization processes, α_2HS glycoprotein may be implicated in bone resorption. It has been shown that this glycoprotein is chemotactic for monocytes[35] and it enhances macrophage phagocytic function.[36]

Using immunohistochemistry and antibodies raised against collagen-free extract of bovine cortical bone, the noncollagenous proteins could predominantly be detected in the osteocyte capsules, in the wall of Haversian and Volkmann canals, and only a moderate reaction was shown in the bone lamellae.[37]

Although much progress has been made in the isolation and characterization of numerous bone matrix constituents we still have no exact data on their ultrastructure, their interactions with each other and with the mineral phase, and their physiological functions. From a biomechanical point of view, collagen fibers resist tension, while the primary function of hydroxyapatite attached to collagen and/or to different noncollagenous proteins is to resist compression. It is obvious that the bone matrix proteins, especially the non collagenous proteins, have a number of other possible functions. They may play roles in different phases of mineralization, in resorption, and they may be involved in the regulation of cell activity, matrix organization and morphogenesis under normal and pathological conditions.

II. STRUCTURE OF THE LAMELLAR BONE

The matrix of lamellar bone is a highly organized system. In this type of bone tissue collagen fibers form layers. Three different lamellae are distinguished in compact bone. The outer and inner circumferential lamellae are in the vicinity of peri- and endosteum. Secondly, the concentric lamellae, which are arranged around the vascular (Haversian) canals, form the Haversian system of osteons generally running along the long axis of the bone. Finally, the remnant of the concentric lamellae after resorption of the Haversian system are seen between the osteons and are called interstitial lamellae.

Osteons are generally regarded as the mechanical units of lamellar bone. The collagen fibers run in approximately in the same direction within one concentric layer of the osteon, but the orientation axis of the fibers generally differs by about 90° in adjacent lamellae in the most common type of osteons. These orientation patterns can easily be revealed by polarization microscopy and electron microscopy[38-40] (Figures 1 and 2). The internal orientation pattern of the lamellae sometimes exhibit a high post-mortem stability, which can be detected in some bones even after several hundreds or thousands of years (Figure 3).

The osteocytes are also arranged in a regular pattern. They are located within their lacunae and concentrically disposed between the lamellae. There are more concentric fiber lamellae than osteocyte layers within one Haversian system. The tiny processes of the osteocytes are found in the canaliculi which exhibit a radial arrangement within the osteon. Canaliculi of osteocytes found in adjacent layers communicate with each other within one Haversian system, and many of the osteocyte processes are in contact with each other by communicating (gap) and tight junctions. Osteocytes and their processes are separated from the wall of mineralized lacunae by a thin basophilic layer called the osteocyte capsule. Its ultrastructure will be discussed separately.

Considering their chemical composition — including mineralization — collagen orientation pattern, and biomechanical behavior, osteons are not uniform structures. Bone matrix, therefore, shows structural and functional microheterogeneity. This "patchwork" structure of bone[3] could be revealed by — among other techniques — investigating mechanically isolated osteons. Using polarization microscopic techniques, three major types of osteons have been described based on fiber bundle orientation in successive lamellae: transverse (light), alternate and longitudinal (dark) types[41-44] (Figure 4). These types can easily be recognized in cross sections of compact bone under the polarization microscope using either linearly or circularly polarized light. In the transversal type of osteon the fibers of adjacent lamellae show a distinctly transverse spiral course appearing homogeneously bright in polarized light. In the alternate type, however, the fiber bundles in one lamella exhibit a longitudinally spiral course, alternating in the next lamella with fibers following an almost transversally spiral course. In this case, the collagen fibers in two successive lamellae form an angle of nearly 90°. In polarized light, this osteon type contains alternate

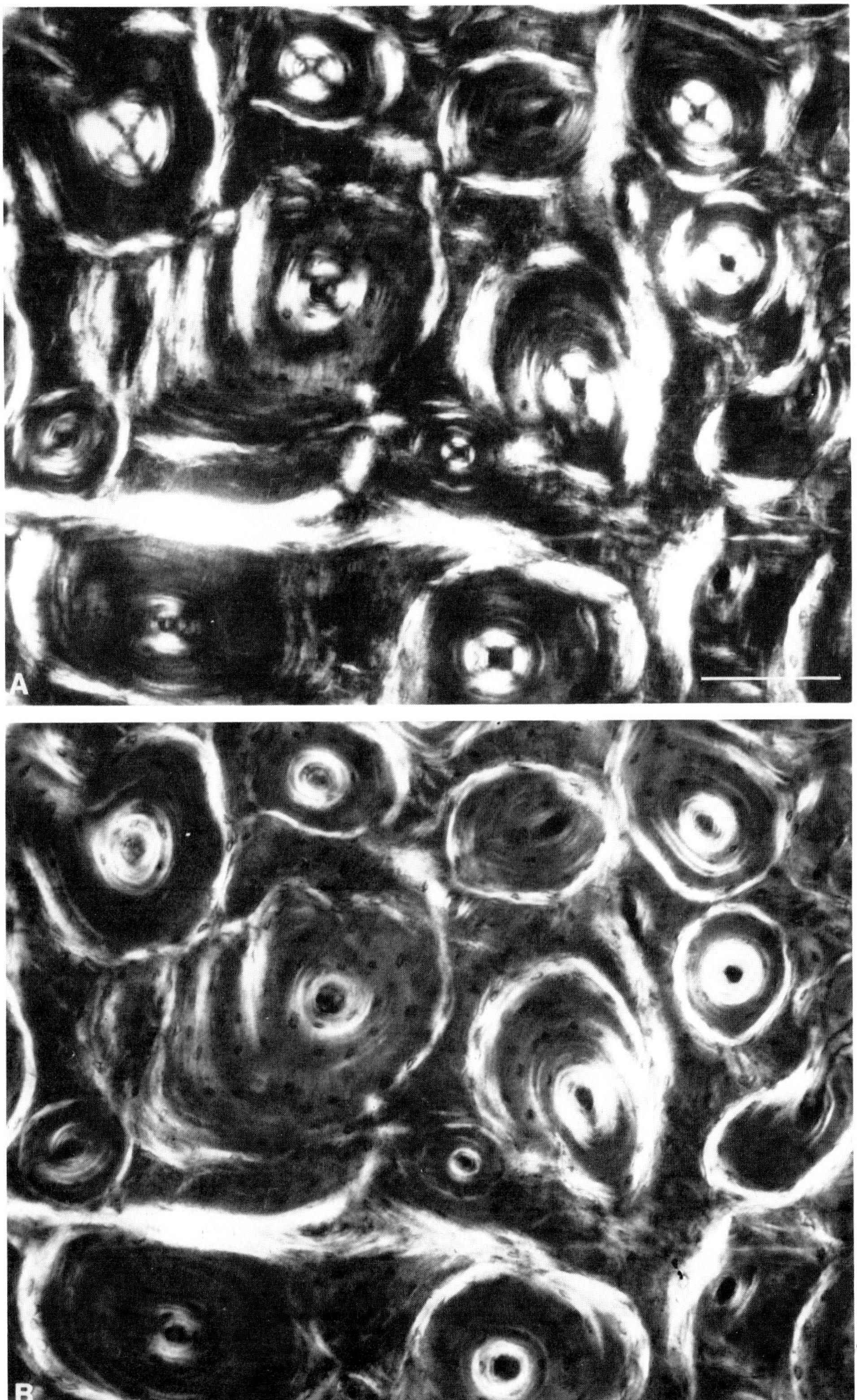

FIGURE 1. Unstained cross sections of adult canine compact bone taken in linearly (A) and circularly polarized light (B). Note the differently sized osteons with differently birefringent concentric lamellae. Scale: 100 μm.

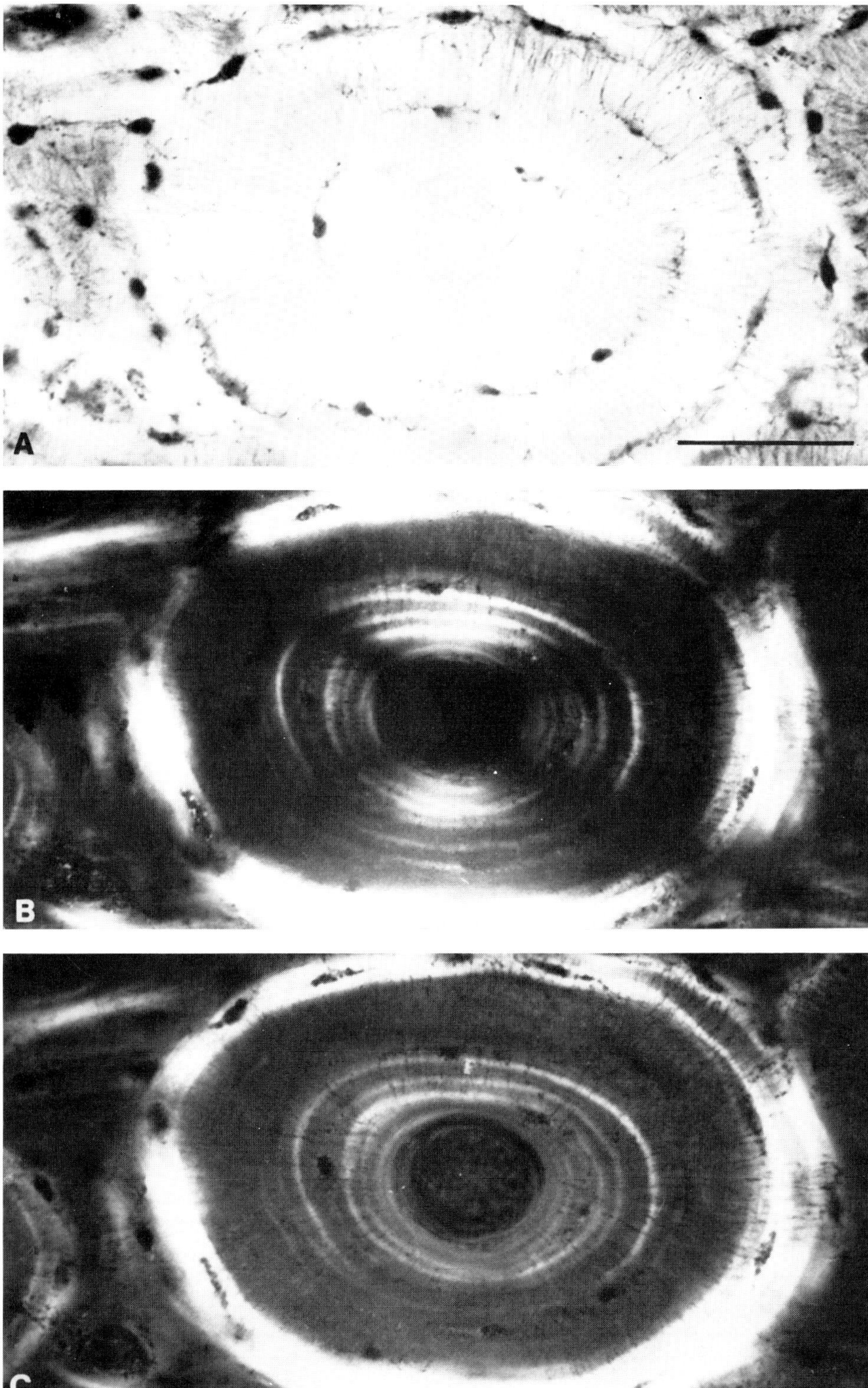

FIGURE 2. Cross section of an osteon of adult canine compact bone after Schmorl-staining taken in normal light (A), in linearly (B) and circularly polarized light (C). In the outermost part of this osteon, lamellae with circular organization are seen. In the internal part, the orientation of collagen fibers is alternating in successive (bright and dark) lamellae. Between these and the peripheral lamellae, longitudinally organized (dark) lamellae predominate. Scale: 50 μm.

FIGURE 3. Cross sections of decalcified human compact bones after phenol reaction taken in polarized light. (A) approx. 600-year-old bone, (B) approx. 1000-year-old bone. Collagen lamellae are partially preserved in A, while they are almost normal in (B) Scale: 100 μm

dark and bright lamellae. In the longitudinal type, the collagen fibers have a predominantly longitudinal spiral course. The angle of the fiber bundles in one lamella is almost the same as that of the fibers of the next lamella. In cross sections this type of osteon appears dark in polarization microscopy. According to other investigations, osteons can be classified into five different categories according to the proportion of dark (longitudinally oriented) and light (transversally oriented) lamellae: two light, one intermediate and two dark types are distinguished in human bones.[45] It is likely that different transitional forms exist between the classical osteon types.

Furthermore, a significant correlation was found between age and formation of certain osteon types: more light osteons and fewer dark osteons are formed in older individuals than in younger ones.[45] Different osteon types and differently oriented lamellae are not homogeneously distributed in compact bone.[38,45] In human tibial compact bone, dark osteons

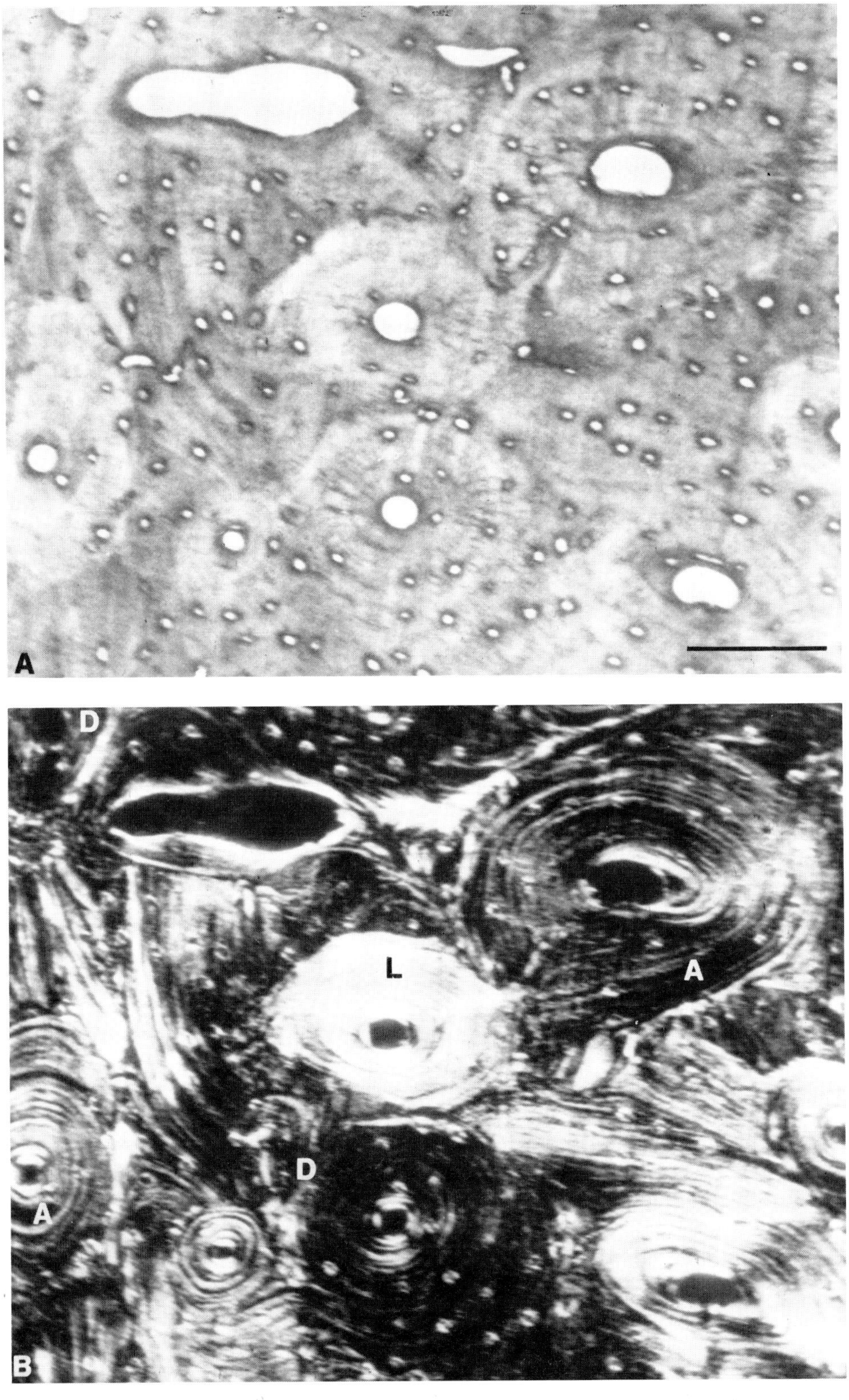

FIGURE 4. Cross section of decalcified adult canine compact bone after picrosirius red staining reaction taken in normal (A) and linearly polarized light (B). The light (L), alternate (A) and dark (D) osteons are labeled in (B). Scale: 100 μm.

predominate in the anterior quadrant while more light osteons are found posteriorly.[45] Using circularly polarized light combined with computerized image analysis,[38] the quantitative distribution pattern of different osteons and lamellae could be described within one bone type. From such data it may be inferred that in the femoral shaft the distribution of transverse lamellae undergoes a rotation from the medial toward the posterior wall going from the proximal to the distal end, whereas longitudinal lamellae undergo a rotation in the opposite direction from the lateral toward the anterior wall.[44] All these data suggest that compact bone exhibit a structural and functional heterogeneity even within one bone, e.g., in the femoral shaft the medial wall is resistant to compression, and the lateral wall to tension.[44]

Analyzing biomechanical features of isolated osteons with different structures, it became evident that these different anatomically structured osteons have different mechanical properties with regard to tension and compression. Osteons with longitudinally running collagen fibers are more capable of withstanding loading by tension, and those with transversally oriented fibers are better able to withstand loading by compression.[41-44]

The bone lamellae are, of course, complex systems. In addition to their collagen fibers, they contain hydroxyapatite crystals and noncollagenous proteins. Small-angle X-ray scattering studies suggest that the apatite crystal rods show a preferential orientation of their principal axis parallel to the longitudinal axis of the bone, and are located within the collagen fiber bundles.[46] X-ray diffraction studies carried out on bovine femur shafts indicate that the dominant orientation of apatite crystals is parallel to the osteon axis.[47] Quantitative polarization microscopic analyses reveal that the apatite crystals are perpendicularly oriented to the collagen fiber bundles.[48] X-ray diffraction method applied to single osteon samples indicate that in all three major types of osteons there are fiber bundles and crystals running parallel to the osteon axis and they increase progressively in number from the transversal to the alternate and longitudinal types of osteons.[49] The crystal orientation in an axial direction increases in fully calcified osteons.[49] This observation is in agreement with the electron-microscopic finding that the long needle-shaped crystals increase in number as calcification proceeds.[43,49] Radially arranged crystals have also been demonstrated by polarization microscopy after a staining with congo red in osteons made artificially collagen free.[50] They may be connected with the osteocyte canaliculi and probably with the radially arranged fibrils which link successive lamellae.[49]

Very little is known about the localization of noncollagenous proteins within the osteon. Scanning electron microscopy reveals collagen fiber-related spherical structures of unknown chemical composition and interlamellar ground substance. Furthermore, based on X-ray microanalysis and the results of enzymatic digestion, this ground substance is thought to be chondroitin sulfate containing noncollagenous proteins, most probably, proteoglycans.[51]

Using the copper-containing phthalocyanin dye, cupromeronic blue, proteoglycan particles of rat bone matrix have been demonstrated by electron microscopy.[52] PGs appeared as interfibrillar filaments oriented parallel to the fibril axis. No intrafibrillar PGs were found in bone matrix. The gap zone of the collagen fibers — which is occupied by orthogonally arranged small dematan sulfate — PGs in several noncalcifying connective tissues such as tendon, skin, cornea, sclera — is free of PGs in the bone.[53-56] The absence of gap-zone-associated PGs in bone and their presence in nonmineralizing tissue suggests their inhibitory role during calcification[52,53,56] because the gap zone has been implicated as the site of nucleation of calcification.[57] Besides the type of PGs, their arrangement and relationship to collagen also seems to be involved in the regulation of the calcification process. This hypothesis is supported by the observation that in calcifying turkey tendon the orientation of gap-zone-associated PG filaments changes from the orthogonal to axially parallel, as the calcification front moves along the tissue.[56]

The light microscopic demonstration of PGs in the bone matrix is more ambiguous because of their low local concentration. In addition to this, the generally used decalcification procedures lead to partial dissolution of PGs from the tissue which further decreases the chance to show GAGs or PGs morphologically. The use of an alcoholic solution for decalcification can minimize the loss of these molecules during demineralization.[52]

Using antibodies against noncollagenous proteins of the bone matrix, immunohistochemistry reveals a moderate reaction in the lamellae and a fairly strong reaction in the osteocyte.[37] It is not known, however, which antigens are responsible for these reactions.

Polarization microscopy has also been used to try to identify the spatial orientation pattern of GAG in the bone matrix.[21,58,59] Kent et al.[58] stained undecalcified bone sections with alcian blue in the presence of either 0.025 mol/1 or 0.5 mol/1 $MgC1_2$. Samples were taken from fractured necks of femurs, from osteoarthrotic femoral heads, and from the equivalent sites of otherwise normal subjects at necropsy. Measuring of the natural birefringence of the collagen showed no difference in the orientation of the collagen in all three types of specimens. However, the orientation of glycosaminoglycans, measured by the optical anisotropy of alcian blue bound to these molecules, was 45% lower in the specimens from the fractured necks of femurs than in the other specimens, even though the total content of GAGs was unchanged. Although the decreased orientation was most marked close to the site of fracture, it was still apparently 15 mm from that site. In the same location, relatively large apatite crystals were found by the same authors. They speculate that the decreased orientation of GAGs allows the formation of relatively large apatite crystals which are considered to cause the increased fragility of the bone. Unfortunately, these authors did not determine the sign of birefringence after alcian blue staining. Therefore, the orientation pattern of GAGs remained unknown. Furthermore, the staining conditions they used are not specific for chondroitin sulfate which is the characteristic GAG molecule in the bone matrix: the $MgCl_2$ concentration of their staining solutions is either too low (0.025 mol/1) or too high (0.5 mol/1) to ensure the required specificity. It is also questionable whether undecalcified sections can properly be used for polarization microscopic analysis of the GAGs. A simple enzymatic digestion carried out prior to the staining to remove chondroitin sulfate would have been a useful control in the above experiments. One encounters the same technical problems in another interesting paper of the same group where they described a decreased orientation of PGs in osteoporotic bones.[59]

In our experience decalcified lamellar bone exhibits the following staining pattern using cationic dyes under conditions suitable for polarization microscopic analysis of sulfated GAGs. The lamellae generally remain unstained or they are moderately colored (Figure 5). Using some dyes (e.g., thiazine dyes), although no staining effect can be observed in normal light, the dye-uptake of GAGs can be demonstrated at the same sites under the polarization microscope between crossed polaroids by the fact that the birefringence of the collagen lamellae is extinguished by the dye molecules bound perpendicularly to the lamellae. Consequently, the bone lamellae are isotropic after such stainings (Figure 6). The explanation of this phenomenon is simple. If the selected dye (e.g., toluidine blue, dimethylemethylene blue, alcian blue, Janus red) binds to the GAGs in an axiperpendicular manner, an oriented dye bond is formed on the GAGs associated with the collagen fibrils. Due to the oriented pattern of the dye molecules, the dye bond has an extraordinary refractive index (n_e) perpendicular to the n_e of the collagen structure. Thus, the dye molecules extinguish (compensate) the birefringence of the collagen. This can be observed even if the number of dye molecules is too low to observe their light absorption in normal light. We can conclude that the topo-optical reactions available at present for polarization optical analysis of sulfated GAGs can be applied for the investigation of lamellar bone matrix only under strong limitations. Contrary to this, osteocyte capsules are generally strongly basophilic and birefringent when using topo-optical reactions specific for sulfated GAGs (Figure 6).

Summing up, we can conclude that while many data are available on the chemical constituents and the collagen morphology of the bone matrix, little is known about the

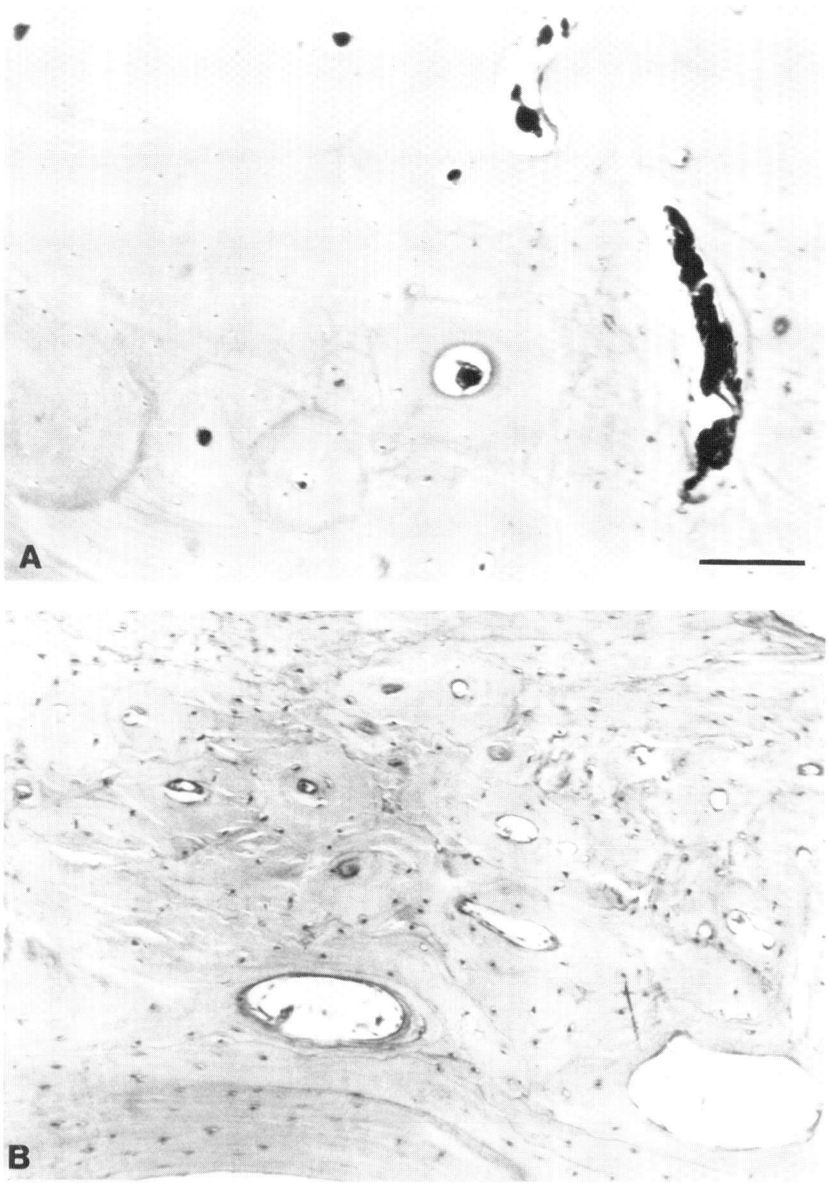

FIGURE 5. Cross sections of adult canine compact bone decalcified with 1 *N* HCl after staining with toluidine blue at pH 2.5(A) and 5.2 (B) taken in normal light. The vessels and connective tissues of the Haversian canals are stained, but the bone matrix remains unstained at low pH and it exhibits a moderate basophilic staining at pH 5.2 where all of the anionic groups are dissociated in the matrix. Scale: 100 μm.

ultrastructure of noncollagenous proteins. Furthermore, the mechanism of the formation of the highly oriented collagenous structure is not understood. We do not know how the differently organized osteons are formed. It seems to be probable that mechanical signals play a role in this process. The question, however, how the mechanical signals are "translated" by the bone cells cannot be answered yet. In searching for some kind of explanation for the formation of lamellae and osteons of different orientation patterns, it is tempting to suppose that the cell orientation pattern — like that of other dense connective tissues such as tendon and cornea — determines the collagen orientation pattern of the lamellae formed

FIGURE 6. Longitudinal section of compact bone of an adult albino rat after decalcification in a 1:1 mixture of formic acid and 70% ethanol. The microphotographs were taken after a staining with Janus red at pH 2.5 in normal light (A) and polarized light (B). Note the stained and birefringent osteocyte capsules. Due to the perpendicular binding of cationic dyes to the bone matrix fibers, their original birefringence of a positive sign is completely extinguished resulting in an optical isotropy of the bone matrix (B). Scale: 100 μm.

by a row of cells. No specific location of osteocyte lacunae was found, however, with respect to lamellae with scanning electron microscopic investigations.[60] Thus, this problem remains one of the secrets of bone research.

III. OSTEOCYTE CAPSULE

It has been known for a long time that osteocytes and their processes are surrounded by a thin, basophilic matrix layer called the osteocyte capsule (OC).[61-65] This unmineralized compartment of the bone matrix fills the pericellular space within the bone lacunae and canaliculi, and is interconnected with the perivascular matrix of the Haversian canals.

Histochemical analyses reveal a heterogeneous chemical composition of the OC. Its predominant constituents are PGs with their chondroitin sulfate side-chains.[66-70] Presence of PAS-positive glycoproteins has also been show.[68, 71] Noncollagenous proteins could be detected immunohistochemically.[37] Lipids[66, 71] and alkaline phosphatase[66, 72] have also been found. Some fine collagen fibrils have been revealed by electron microscopy in the OCs.[73-78] A thin osmiophilic layer called lamina limitans separates the OC from the mineralized bone matrix which disappears during bone resorption.[79]

It is supposed that the OC is involved in the nutrition of the bone. It forms a fairly large surface of about 250 mm^2/mm^3 bone tissue.[73] It has been suggested that this is the structure where diffusion processes take place between pericapillary spaces and bone cells.[67, 75] Studying the flow of tissue fluid within bone by colloidal thorium dioxide, it was found that the tissue fluid moves from the vessels of the Haversian canals towards the peripheral lacunae of the osteons through the osteocyte capsules.[80] Using ^{125}I-labeled albumin and polyvinylpyrrolidone, Owen et al.[81] demonstrated the movement of tissue fluid through the cortical bone from the endosteal to the periosteal surface in the OCs. It has been suggested that the OCs are involved in the transport of calcium between the vessels and bone cells as well as between the osteocytes and mineralized matrix.[82, 83] The "pumping" function of the osteocytes may be at least one of the factors which drive the tissue fluid movement within the OCs.[84]

The structure and composition of the OC depend on the functional state of the osteocyte. Three different types of osteocytes have been distinguished: formative (osteoblastic), resorptive (osteoclastic) and degenerative.[85-87] According to these authors, the OC of formative (young) osteocytes contains more collagen fibrils that that of the two other types, while GAGs predominate in the OC of the resorptive cell type.

There are some data indicating that the structure of the OCs is changed under pathological conditions. Parathormone treatment and hypervitaminosis D causes a thickening of the OC in experimental animals.[88-90]

The interfibrillar substance of OCs is generally considered as an amorphous material. Based on polarization microscopic analysis, however, we have shown that GAGs are arranged in an oriented manner in OCs of different species (chick, rat, human).[70, 90-92]

Using metachromatic cationic dyes as topo-optical reagents, the OCs exhibited a strong staining and birefringence both in spongy and compact bone. The sign of birefringence was dependent on the type of dye used. *N,N'*-diethylpseudoisocyanine induced a birefringence with positive sign related to the tangent of the cell membranes. This staining resulted in the strongest optical anisotrophy of the OCs even around the cell processes (Figures 7, 8). In OCs stained with toluidine blue and dimethyl-methylene blue, however, we observed a birefringence of a negative sign (Figure 9). Both staining and birefringence were abolished by testicular hyaluronidase digestion prior to stainings. It follows that chondroitin sulfate molecules are responsible for these topo-optical reactions. Considering the orientation pattern of the dye molecules bound to GAGs (parallel binding of *N,N'*-diethylpseudoisocyanine, and perpendicular orientation of thiazine dyes related to the long axis of chromatropes), all of these mean that the chondroitin sulfate molecules of the OCs are deposited parallel to the tangent of the cell membrane. The perpendicular binding of the GAGs to the core protein of PGs implies that at least the great majority of the PGs of the OC are presumably in perpendicular orientation related to the cell membranes.

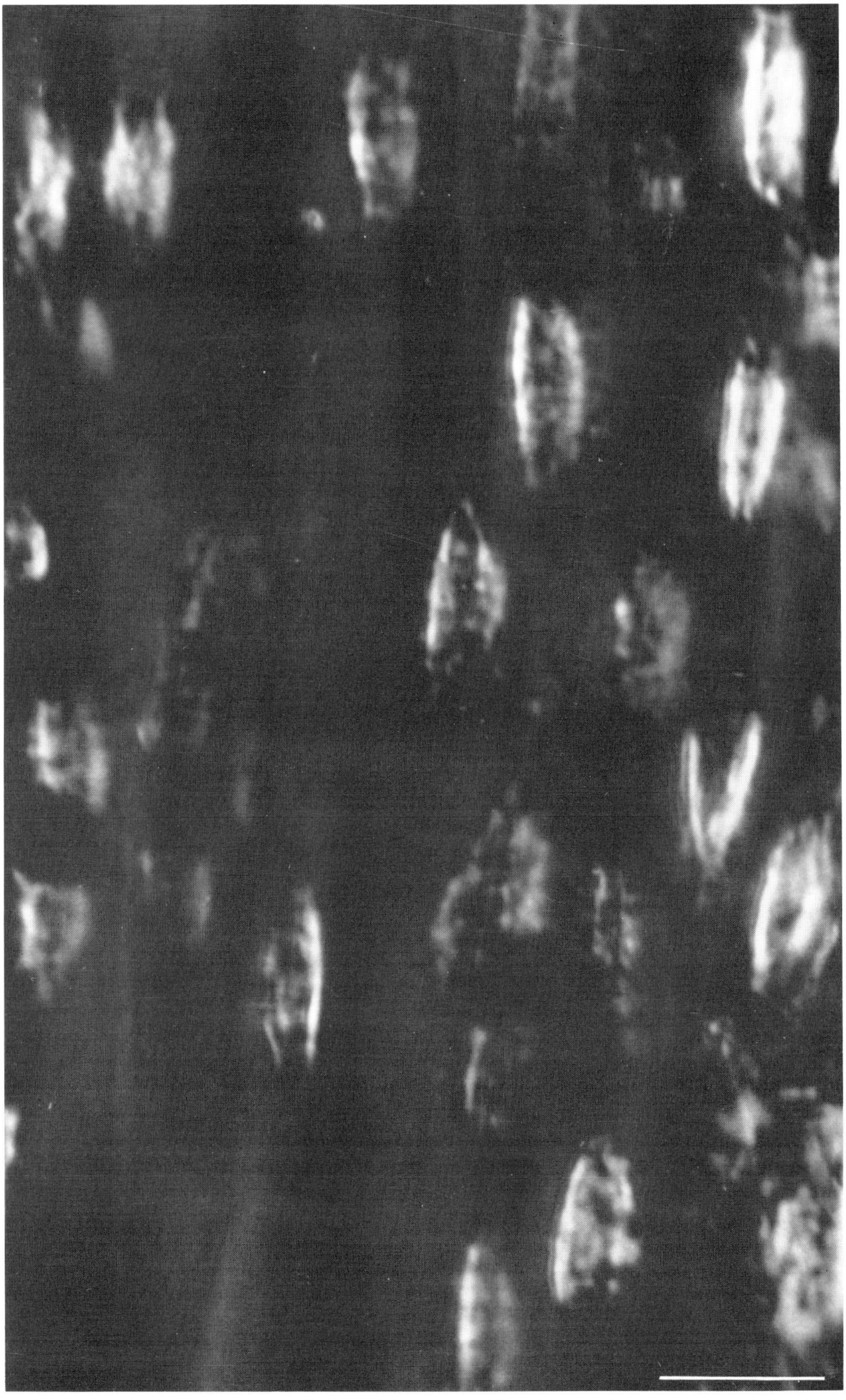

FIGURE 7. Decalcified compact bone section of an adult albino rat after staining with *N,N'*-diethylpseudoisocyanine and taken in polarized light. The osteocyte capsules are strongly birefringent. Scale: 20 μm.

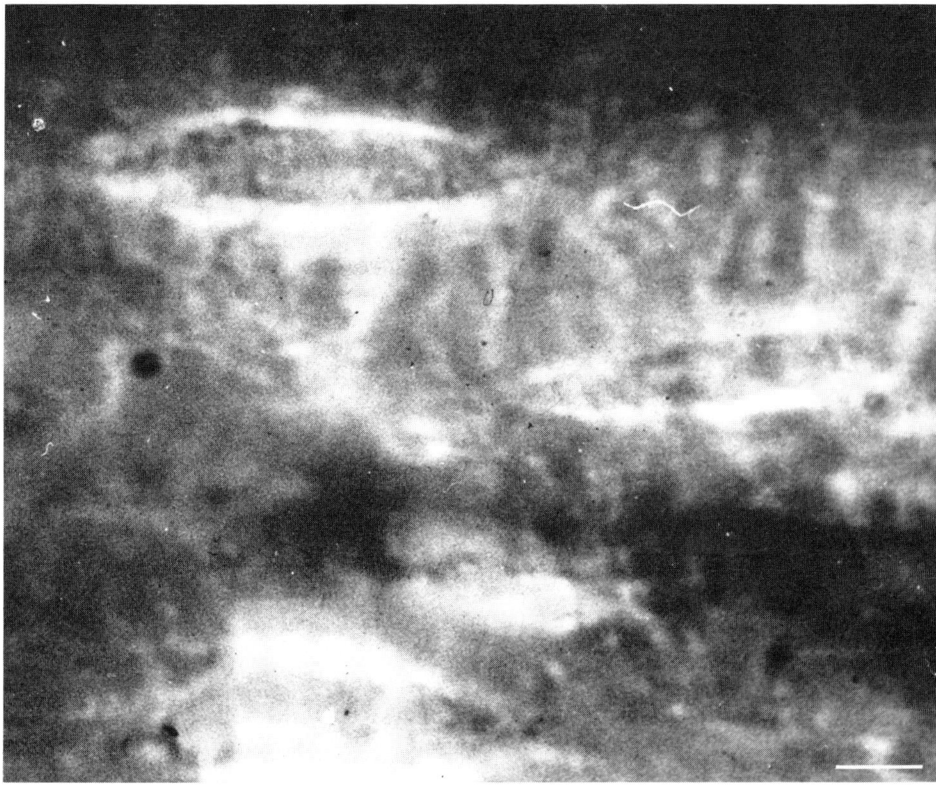

FIGURE 8. Another portion of the same specimen shown in Figure 7 at higher magnification. The polarized light micrograph demonstrates the birefringent osteocyte capsule surrounding the osteocytes and their processes. Scale: 10 μm.

OCs with oriented GAG structure can be detected even at early stages of embryonic bone differentiation. Furthermore, the formation of this oriented structure can be followed during the differentiation of osteoblasts and osteocytes in spongy bone (Figure 9). Differentiating osteocytes being embedded into osteoid seams on the surface of bone trabeculae are already surrounded by birefringent capsules.

Quantitative polarization microscopic investigations were carried out to study the degree of orientation of the chondroitin sulfate molecules in the OCs. For this purpose, anomalous dispersion curves of birefringence were taken in embryonic bone trabeculae after different topo-optical staining reactions, i.e., retardation values were measured at different wavelengths of the transluminating light. The retardation values were plotted against the wavelengths. Such anomalous dispersion curves of OCs, cartilage matrix trabecules in the core of bone trabeculae and osteoid seams were compared. One example is shown in the Figure 10. As this figure demonstrates, the qualitative orientation pattern of the GAG molecules is identical but the degree of submicroscopic orientation of chondroitin sulfate molecules is highly different in the three structures investigated. The highest degree of orientation was found in the osteoid seams, while GAGs of the cartilage matrix residues exhibited moderate values of birefringence, and curves taken in the OCs suggested the lowest level of orientation of the chondroitin sulfate molecules in this structure. As for the measurements carried out in the same section, these differences could not be attributed to the variation of the section thickness. To try to explain these differences, the qualitative orientation patterns of GAGs and collagen were compared in these three structures. For polarization microscopic analysis of the collagen, picrosirius red F3B topo-optical reaction was chosen (Figure 11). It turned

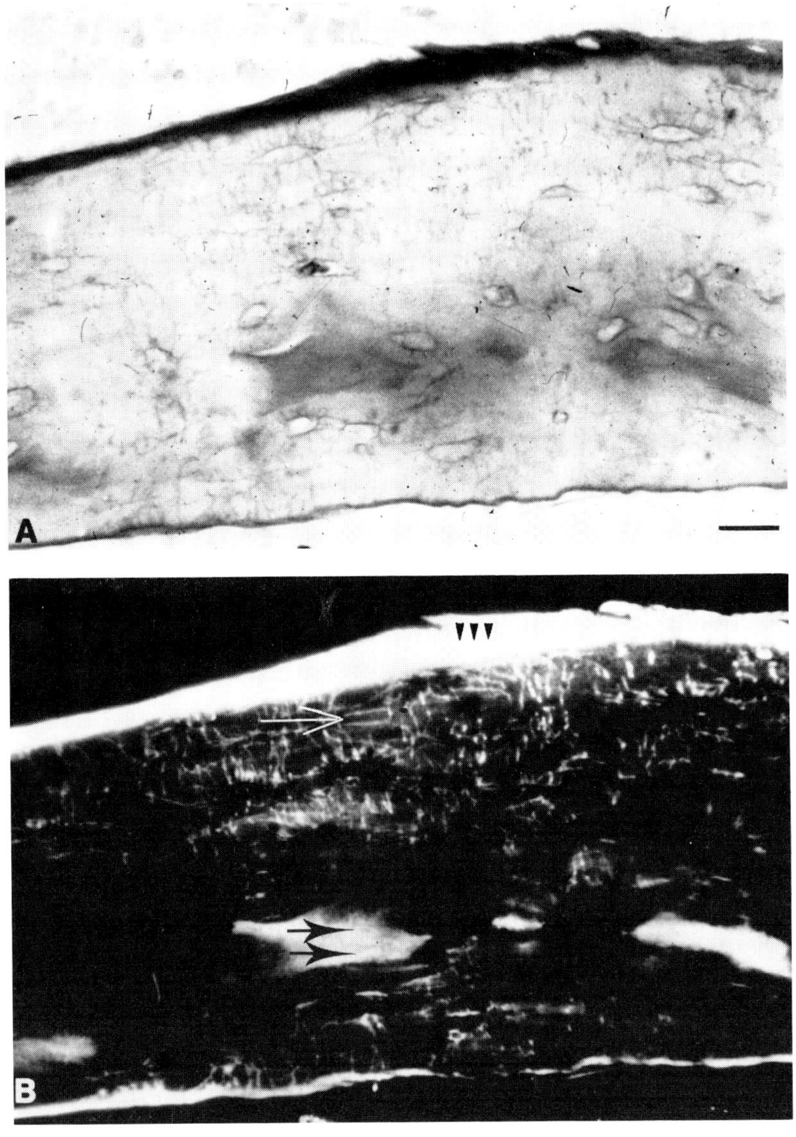

FIGURE 9. Bone trabecula of decalcified tibia metaphysis of a 6-month-old human fetus stained with toluidine blue at pH 2.5 and taken in normal (A) and polarized light (B). Note the staining and birefringence of the osteocyte capsules (arrow), the cartilage matrix residue (double arrows) and the osteoid border (arrowheads). Scale: 20 μm.

out that both chondroitin sulfate molecules and collagen fibrils are similarly oriented in the structures investigated. They run parallel with respect to the long axis of bone trabeculae in the osteoid seams and cartilage matrix residues, while they are arranged parallel to the tangent of the cell membrane in the OCs. It is known that the osteoid seams are very rich in densely packed collagen fibrils. Cartilage matrix trabeculae also contain longitudinally oriented collagen type II fibrils (see Chapter 4). In the OCs, however, only a few collagen fibrils are found electron microscopically which form a loosely oriented system embedded in abundant ground substance (Figure 12). These data suggest that the level of orientation of the GAG molecules is influenced by the degree of orientation of the collagen fibrils. This and the fact that these two constituents are oriented parallel to each other also suggest that

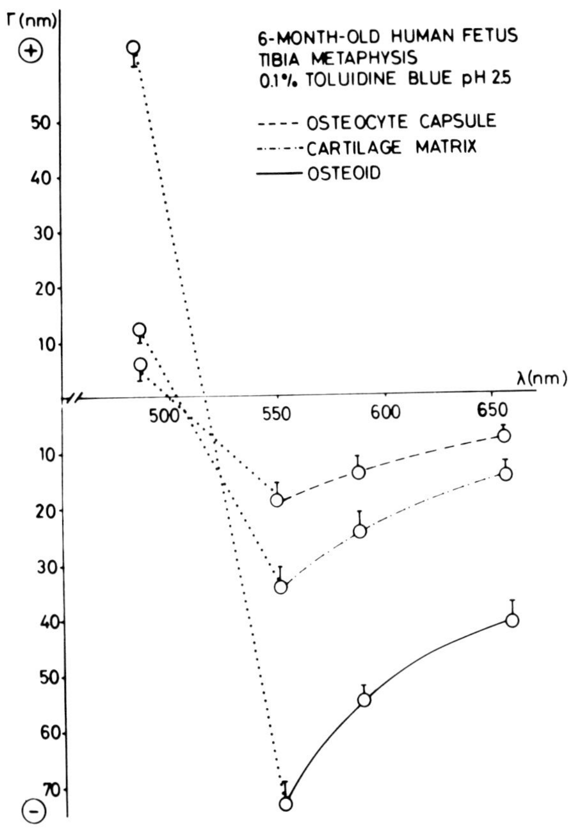

FIGURE 10. Anomalous dispersion curves taken after a metachromatic staining for sulfated GAGs in three different structures of metaphyseal bone trabeculae of a 6-month-old human fetus. The retardation values measured at different wavelengths clearly indicate that the degree of orientation of the GAG molecules decreases in the following sequence: osteoid seam — cartilage matrix — osteocyte capsule.

their electrostatic interaction is one of the factors which control the spatial orientation of the PG moleules and their GAG side chains in the bone matrix. Similar conclusions could be drawn after analysis of other connective tissues. The assumption that collagen orientation regulates the orientation of GAGs in the tissues has also been experimentally proved (see Chapter 8).

It has been found that the oriented GAG structure of the OCs is severely altered under several pathological conditions.[70, 90, 91] Bones of albino rats subjected to different treatments leading to functional and structural damages of different degrees have been used for such studies. It has been demonstrated that hypervitaminosis A, hypervitaminosis D, hypo and hyperthyroidism, treatments with papain, cortisone, ACTH, and also experimentally induced rickets result in a significant decrease in the number of birefringent osteocyte capsules. The number of optically anisotropic OCs was decreased even if the OCs became thicker and their basophilic staining was stronger when comparing to the controls. This was the case for instance in the experimental hypervitaminosis D.[90] Although some diminution in number of the stained OCs has also been registered in the different experimental groups as compared to the relevant control groups, the decrease in birefringence was always more dramatic.[90-92] Thus, the decrease of the GAG content in OCs of the experimentally damaged bones cannot be the only explanation for their decreased orientation level reflected by the decreased induced

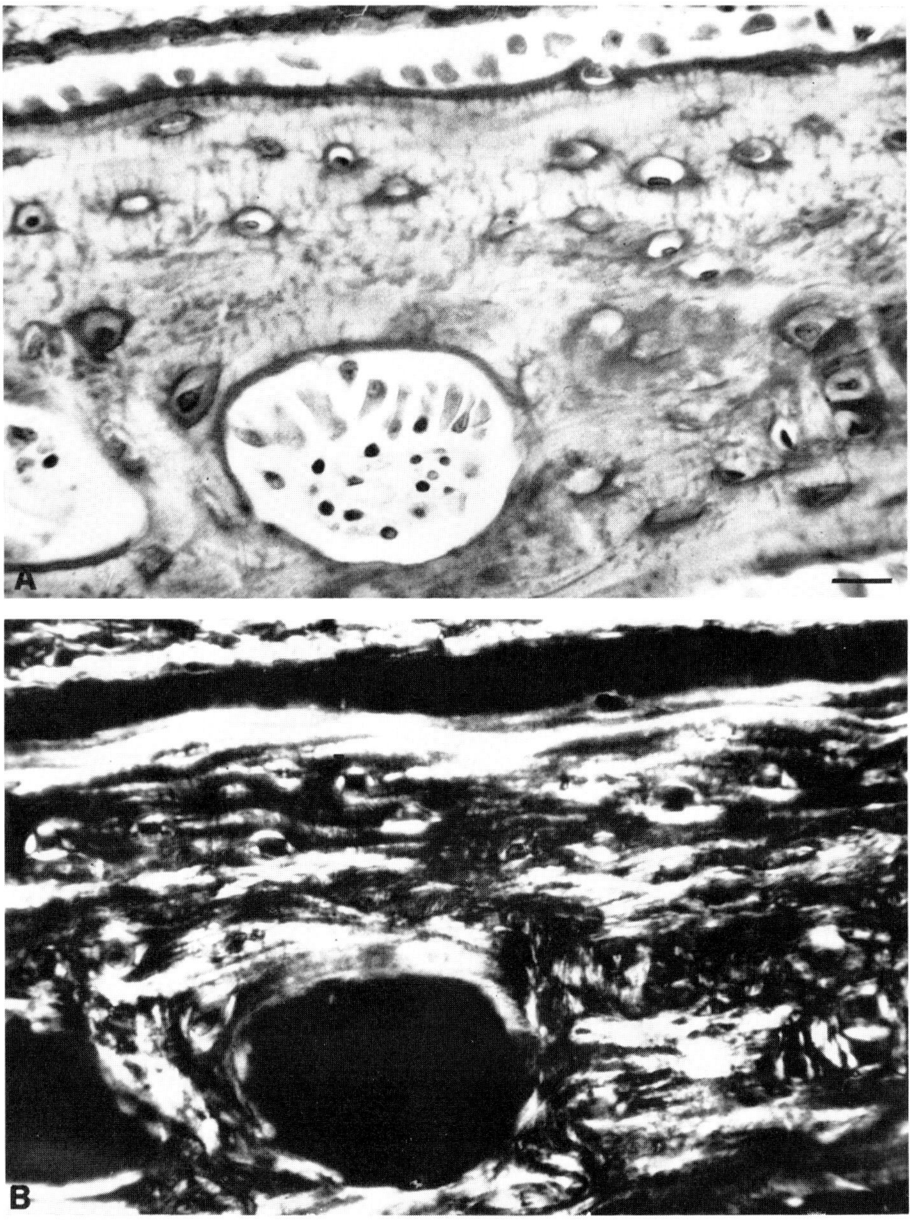

FIGURE 11. Decalcified spongy bone section after picrosirius red F3B staining taken in normal (A) and polarized light (B). Note the stained and birefringent osteocyte capsules. Scale: 20 μm.

birefringence. We do not know at this point what happens with the proteoglycans and the intricate organization of the OCs in the experimentally induced bone alterations mentioned above. In any case, we believe that the polarization microscopical analysis of birefringence of the OCs after a GAG-specific topo-optical reaction is a simple and sensitive tool for analyzing pathological bones. It must be added, unfortunately, that demineralization procedures with strong acids which are generally used for adult human compact bones do not allow for proper polarization microscopic studies of the OCs. This is probably due to the deterioration of the spatially oriented GAG structure by strong acids (e.g., HCl).

Based on the above data we suggest that the submicroscopically oriented GAG structure

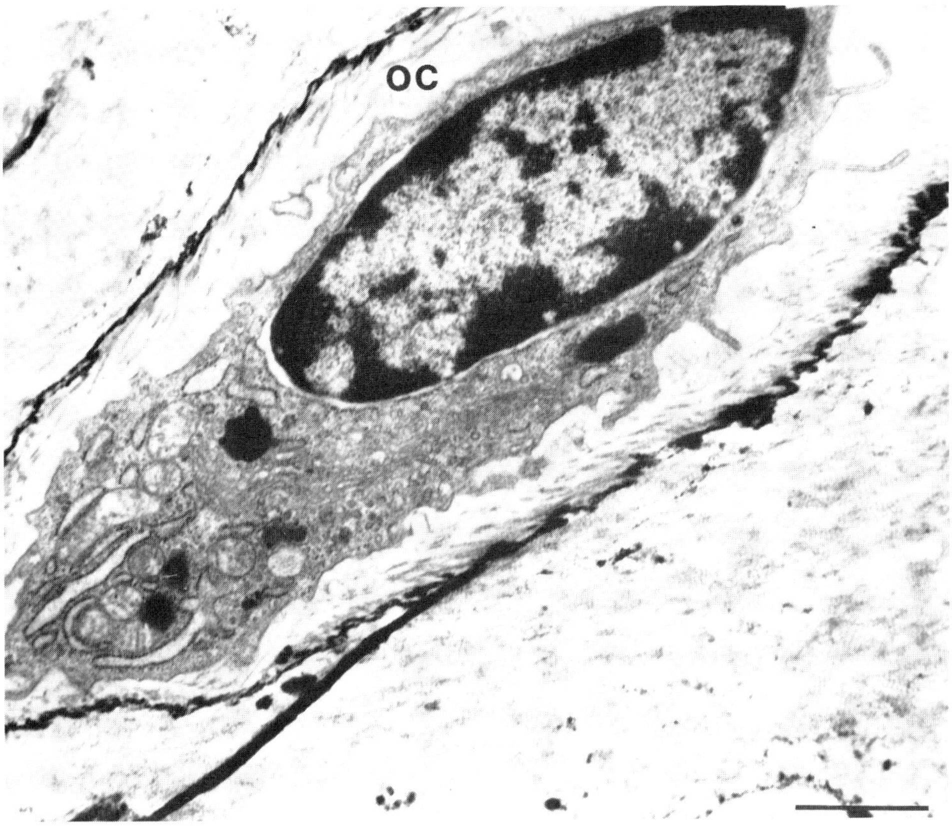

FIGURE 12. Electro micrograph taken from decalcified compact bone of an adult albino rat. Between the osmiophilic border of the lacuna and the cell membrane the osteocyte capsule (OC) is visible with its fine collagen fibrils. Scale: 2 μm.

of the osteocyte capsule may play a role in the different anabolic and catabolic transport processes of the bone. Its involvement in Ca^{2+} transport should be especially considered.

ACKNOWLEDGMENTS

The first observations on the ultrastructure of osteocyte capsules were made by the author during his stay of two years in the laboratory of Professor G. Conti (Department of Histology, University of Fribourg, Fribourg, Switzerland). The stimulative and amicable atmosphere of that Department is highly appreciated. I wish to thank to Drs. Clara Matesz (Department of Anatomy, University Medical School, Debrecen) and B. Telek (2nd Department of Medicine, University Medical School, Debrecen), who carried out some parts of the investigations on osteocyte capsules.

REFERENCES

1. **Herring, G. M.,** The organic matrix of bone, in *The Biochemistry and Physiology of Bone,* Vol. 1., Bourne, G. H., Ed., Academic Press New York, 1972, chap. 5.
2. **Gay, S. and Miller, E. J.,** *Collagen in the Physiology and Pathology of Connective Tissue,* G. Fischer Verlag, Stuttgart, 1978.
3. **Ascenzi, A.,** Microscopic dissection and isolation of bone constituents, in *Skeletal Research: An Experimental Approach,* Vol. 2., Kunin, A. S. and Simmons, D. J., Eds., Academic Press, New York, 1983, chap. 7.
4. **Hancox, N. M.,** *Biology of Bone,* University Press, Cambridge, 1972.
5. **Sheatlar, M. R., Hern, D., and Chien, S. F.,** Isolation and characterization of human bone sialoprotein, *Tex. Rep. Biol. and Med.* 30, 339, 1972.
6. **Triffitt, J. T.,** The special proteins of bone tissue, *Clin. Sci.,* 72, 399, 1987.
7. **Gundberg, C., M., Hauschka, P. V., Lian, J. B., and Gallop, P. M.,** Osteocalcin: isolation, characterization and detection. *Methods Enzymol.,* 107, 516, 1984.
8. **Price, P. A.,** Vitamin K-dependent formation of bone Gla-protein (osteocalcin) and its function, *Vitam. Horm.,* 42, 65, 1985.
9. **Bronckers, A. L. J. J., Gay, S., Dimuzio, M. T., and Butler, W. T.,** Immunolocalization of γ-carboxyglutamic acid containing proteins in developing rat bones, *Coll. Rel. Res.,* 5, 273, 1985.
10. **Bianco, P., Hayashi, V., Silvestrini, G., Termine, J. D., and Bonucci, E.,** Osteonectin and Gla-protein in calf bone: ultrastructural immunohistochemical localization using the protein A-gold method, *Calcif. Tissue Int.,* 37, 684, 1985.
11. **Price, P. A. and Nishimoto, S. K.,** Radioimmunoassay for the vitamin K-dependent protein of bone and its discovery in plasma, *Proc. Natl. Acad. Sci. U.S.A.,* 77, 2234, 1980.
12. **Price, P. A., Urist, M. R., and Otawara, Y.,** Matrix Gla-protein; a new γ-carboxyglutamic acid-containing protein which is associated with the organic matrix of bone, *Biochem. Biophys. Res. Commun.,* 117, 765, 1983.
13. **Price, P. A. and Williamson, M. K.,** Primary structure of bovine matrix Gla-protein, a new vitamin K-dependent bone protein, *J. Biol. Chem.,* 260, 14971, 1985.
14. **Vejlens, L.,** Glycosaminoglycans of human bone tissue. 1. Pattern of compact bone in relation to age, *Calcif. Tissue Res.,* 7, 175, 1971.
15. **Engfeldt, B. and Hjerpe, A.,** Glycosaminoglycans and proteoglycans of human bone tissue at different stages of mineralization, *Acta Pathol. Microbiol. Scand.,* 84, 95, 1976.
16. **Franzén, A. and Heinegard, D.,** Extraction and purification of proteoglycans from mature bovine bone, *Biochem. J.,* 224, 47, 1984.
17. **Franzén, A. and Heinegard, D.,** Characterization of proteoglycans from the calcified matrix of bovine bone, *Biochem. J.,* 224, 59, 1984.
18. **Fisher, L. W., Termine, J. D., Dejter, S. W., Whitson, S. W., Yanagishita, M., Kimura, J. H., Hascall, V. C., Kleinman, H. K., Hassell, J. R., and Nilsson, B.,** Proteoglycans of developing bone, *J. Biol. Chem.,* 258, 6588, 1983.
19. **Fisher, L. W. and Termine, J. D.,** Noncollagenous proteins influencing the local mechanisms of calcification,. *Clin. Orthop.,* 200, 362, 1985.
20. **Howell, D. S. and Pita, J. C.,** Calcification of growth plate cartilage with special reference to studies on micropuncture fluids, *Clin. Orthop.,* 118, 208, 1976.
21. **Skerry, T. M., Bitensky, L., Chayen, J., and Lanyon, L. E.,** Loading-related reorientation of bone proteoglycan *in vivo.* Strain memory in bone tissue?, *J. Orthop. Res.,* 6, 547, 1988.
22. **Spector, A. R. and Glimcher, M. J.,** The extraction and characterization of soluble anionic phosphoproteins from bone, *Biochim. Biophys. Acta.* 263, 593, 1972.
23. **Glimcher, M. J.,** Recent studies of the mineral phase in bone and its possible linkage to the organic matrix by protein-bound phosphate bonds, *Philos. Trans. R. Soc. Lond. (Biol.),* 304, 479, 1984.
24. **Uchiyama, A., Suzuki, M., Leftrion, B., and Glimcher, M. J.,** Phosphoproteins of chicken bone matrix: heterogeneity in molecular weight and composition, *Trans. Orthop. Res. Socl,* 11, 222, 1986.
25. **Termine, J. D., Kleinman, H. K., Whitson, S. W., Conn, K. M., McGarvey, M. L., and Martin, G. R.,** Osteonectin; a bone-specific protein linking mineral to collagen, *Cell,* 26, 99, 1981.
26. **Romberg, R. W., Werness, P. G., Lollar, P., Riggs, B. L., and Mann, K. G.,** Isolation and characterization of native adult osteonectin, *J. Biol. Chem.,* 260, 2728, 1985.
27. **Gehron-Robey, P., Denholm, L. J., Drum, M. A., Fisher, L. W. Termine, J. D., and Boyce, R.,** Biosynthesis of bone matrix protein by osteogenesis imperfecta connective tissue cells, *J. Bone Mineral Res.,* 1, 90A, 1986.
28. **Stenner, D., Tracy, R. P., Riggs, B. L., and Mann, K. G.,** Human platelets contain and secrete osteonectin, *J. Bone Mineral Res.,* 1, 287A, 1986.

29. **Prince, C. W., Oosawa, T., Butler, W. T., Tomana, M., Bhown, A. S., Bhown, M., and Schrohenloher, R. E.,** Isolation, characterization and biosynthesis of a phosphorylated protein from rat bone, *J. Biol. Chem.*, 262, 2900, 1987.
30. **Oldberg, A., Franzén, A., and Heinegard, D.,** Cloning and sequence analysis of rat bone sialoprotein (osteopontin). cDNA reveals an Arg-Gly-Asp cell binding sequence, *Proc. Natl. Acad. Sci. U.S.A.*, 83, 8819, 1986.
31. **Mark, M. P., Prince, C. W., Oosawa, T., Gay, S., Bronckers, A. L. J. J., and Butler, W. T.,** Immunohistochemical demonstration of a 44-kDa phosphoprotein in developing rat bones, *J. Histochem. Cytochem.*, 35, 707, 1987.
32. **Mark, M. P., Prince, C. W., Gay, S., Austin, R. L., and Butler, W. T.,** 44-kDa bone phosphoprotein (osteopontin) antigenicity at ectopic sites in newborn rats; kidney and nervous tissues, *Cell Tissue Res.*, 251, 23, 1988.
33. **Triffitt, J. T. and Owen, M. E.,** Preliminary studies on the binding of plasma albumin to bone tissue, *Calcif. Tissue Res.*, 23, 303, 1977.
34. **Triffitt, J. T., Gebauer, U., Ashton, B. A., Owen, M. E., and Reynolds, J. J.,** Origin of plasma α_2HS-glycoprotein and its accumulation in bone, *Nature*, 262, 226, 1976.
35. **Lewis, J. G. and Andre, C. M.,** Binding of α_2HS-glycoprotein to peripheral blood monocytes, *Immunol. Commun.*, 10, 541, 1981.
36. **Lewis, J. G. and Andrew, C. M.,** Enhancement of human monocyte phagocytic function by α_2HS-glycoprotein, *Immunology*, 42, 481, 1981.
37. **Glant, T.., Hadházy, Cs., Bordán, L., and Harmati, S.,** Antigenicity of bone tissue. 1. Immunological and immunohistochemical study of noncollagenous proteins of the bovine cortical bone, *Acta Morphol. Acad. Sci. Hung.*, 23, 111, 1975.
38. **Boyde, A., Bianco, P., Portigliatti Barbos, M., and Ascenzi, A.,** Collagen orientation in compact bone: 1. A new method for the determination of the proportion of collagen parallel to the plane of compact bone sections, *Metab. Bone Dis. Rel. Res.*, 5, 299, 1984.
39. **Ascenzi, A. and Benvenuti, A.,** Orientation of collagen fibers at the boundary between two successive osteonic lamellae and its mechanical interpretation, *J. Biomech.*, 19, 455, 1986.
40. **Giraud-Guille, M. M.,** Twisted plywood architecture of collagen fibrils in human compact bone osteons, *Calcif. Tissue Int.*, 42, 167, 1988.
41. **Ascenzi, A. and Bonucci, E.,** The tensile properties of single osteons, *Anat. Rec.*, 158, 375, 1967.
42. **Ascenzi, A. and Bonucci, E.,** The compressive properties of single osteons, *Anat. Rec.*, 161, 377, 1968.
43. **Ascenzi, A. and Bonucci, E.,** Relationship between ultrastructure and "pin test" in osteons, *Clin. Orthop.*, 121, 275, 1976.
44. **Ascenzi, A., Boyde, A., Bianco, P., and Portigliatti Barbos, M.,** Relationship between mechanical properties and structure in secondary bone, *Connect. Tissue Res.*, 15, 73, 1986.
45. **Vincentelli R.,** Relation between collagen fiber orientation and age of osteon formation in human tibal compact bone, *Acta Anat. (Basel)*, 100, 120, 1978.
46. **Matsushima, N., Akiyama, M., and Terayama, Y.,** Quantitative analysis of small-angle x-ray scattering of bone: determination of sizes of its collagen and apatite components, *Jpn. J. Applied Phys.*, 20, 699, 1981.
47. **Nightingale, J. P. and Lewis, D.,** Pole figures of the orientation of apatite in bones, *Nature*, 232, 334, 1971.
48. **Ascenzi, A.,** On the existence of bonds between ossein and inorganic bone fraction, *Science*, 112, 84, 1950.
49. **Ascenzi, A., Bonucci, E., Generali, P., Ripamonti, A., and Roveri, N.,** Orientation of apatite in single osteon samples as studied by pole figures, *Calcif. Tissue Int.*, 29, 101, 1979.
50. **Schmidt, W. J.,** Grenzscheiden der Lakunen und Kittlinien des Knochengewebes. Polarisationsoptische Analyse kollagenfreier Kongorotgefärbter Schliffe, *Z. Zellforsch.*, 50, 275, 1959.
51. **Fransca, P., Harper, R. A., and Katz, J. L.,** Scanning electron microscopy studies of collagen, mineral and ground substance in human cortical bone, *Scan. Electron Microsc.*, 111, 339, 1981.
52. **Scott, J. E. and Haigh, M.,** Proteoglycan-type 1 collagen fibril interactions in bone and non-calcifying connective tissues, *Biosci. Rep.*, 5, 71, 1985.
53. **Scott, J. E. and Haigh, M.,** Proteoglycan-type 1 collagen fibril interactions in calcifying and non-calcifying connective tissues, *Biochem. Soc. Trans.*, 13, 933, 1985.
54. **Scott, J. E.,** Proteoglycan-collagen interactions, *CIBA Found. Symp.*, 124, 104, 1986.
55. **Scott, J. E.,** Proteoglycan-collagen interactions. Structure in the amorphous ground substance, *Collection Czechoslovak Chem. Commun.*, 53, 202, 1988.
56. **Scott, J. E.,** Proteoglycan-fibrillar collagen interactions, *Biochem. J.*, 252, 313, 1988.
57. **Fitton-Jackson, S.,** The fine structure of developing bone in the embryonic flow, *Proc. R. Soc. Lond. (Biol.)*, 146, 270, 1957.

58. **Kent, G. N., Dodds, R. A., Klenerman, L., Watts, R. W. E., Bitensky, L., and Chayen, J.,** Changes in crystal size and orientation of acidic glycosaminoglycans at the fracture site in fractured necks of femur, *J. Bone Joint Surg. (Br.)*,. 65, 189, 1983.
59. **Ferris, B. D., Klenerman, L., Dodds, R. A., Bitensky, L., and Chayen, J.,** Altered organization of non-collagenous bone matrix in osteoporosis, *Bone*, 8, 285, 1987.
60. **Boyde, A. and Hobdell, M.,** Scanning electron microscopy of lamellar bone, *Z. Zellforsch. Mikrosk. Anat.*, 93, 213, 1969.
61. **Virchow, R.,** Die Identität von Knochen, Knorpel und Bindegewebskörperchen, sowie über Schleimgewebe, *Verh. Physik. Med. Ges. Würzburg*, 2, 150, 1851.
62. **Neumann, G.,** *Ein Beitrag zur Kenntnis des normalen Zahnbein- und Knochengewebes*, Leipzig, 1863.
63. **Broesike, G.,** Über die feinere Struktur des normalen Knochengewebes, *Arch. Mikr. Anat.*, 21, 695, 1882.
64. **Schaffer, J.,** Die Stützgewebe, in *Handbuch der mikroskopischen Anatomie des Menschen*. Vol. 11/2, Möllendorf, W., Ed., Springer-Verlag, Berlin, 1930, 304.
65. **Weidenreich, F.,** Das Knochengewebe, in *Handbuch der mikroskopischen Anatomie des Menschen*, Vol. 11/2, Möllendorf, W., Ed., Springer-Verlag, Berlin, 1930, 391.
66. **Lorber, M.,** A study of the histochemical reactions of the dental cementum and alveolar bone, *Anat. Rec.*, 111, 129, 1951.
67. **Harris, W. R. and Ham, A. W.,** The mechanism of nutrition in bone and how it affects its structure, repair and fate of transplantation, in *CIBA Foundation Symposium on Bone Structure and Metabolism*. Wolstenholme, G. E. W. and O'Connor, C. M., Eds., Churchill, London, 1956, 135.
68. **Baud, C. A. and Dupont, D.,** Démonstration histochimique des mucopolysaccarides dans le tissu osseux non decalcifié, *Bull. Micr. Appl.*, 11, 121, 1961.
69. **Pawlicki, R.,** Methacrylate section technique as a convenient method of localization of mucopolysaccharides in bone tissue. *Folia Histochem. Cytochem.*, 10, 193, 1972.
70. **Módis, L.,** Topo-optical investigation of mucopolysaccharides (acid glycosaminoglycans) in, *Handuch der Histochemie*, Vol. II. Part 4, Graumann, W. and Neumann, K., Eds., G. Fischer Verlag, Stuttgart, 1974.
71. **Lipp, W.,** Neuuntersuchungen des Knochengewebes (Morphologie, Histochemie und Beeinflussung durch das periphere, vegetative Nervensystem, durch Fermente und Hormone), *Acta Anat. (Basel)*, 20, 162, 1954.
72. **Hekkelman, J. W.,** Studies on the alkaline phosphatase activity of the surface of living bone cells, *Calcif. Tissue Res.*, Suppl. 4, 73, 1970.
73. **Baud, C. A.,** Morphologie et structure inframicroscopique des osteocytes, *Acta Anat. (Basel)*, 51, 209, 1962.
74. **Baud, C. A. and Dupont, D.,** The fine structure of the osteocytes in the adult compact bone, in *Fifth International Congress for Electronmicroscopy* (Philadelphia, 29th Aug.-5th Sept. 1962.) Vol. 2, Breese, S. S. Jr., Ed., Academic Press, New York, 1962, QQ/10.
75. **Baud, C. A. and Morgenthaler, P. W.,** Structure submicroscopique du rebord lacunocanaliculaire osseux, *Morph. Jb.*, 104, 476, 1963.
76. **Wassermann, F. and Yaeger, J. A.,** fine structure of the osteocyte capsule and of the wall of the lacunae in bone, *Z. Zellforsch*, 67, 636, 1965.
77. **Tonna, E. A.,** An electron microscopic study of skeletal cell aging. II. The osteocyte, *Exp. Gerontol.*, 8, 9, 1973.
78. **Pawlicki, R.,** Correlation between the structure of the wall of the bone lacuna and the localization of the osteocyte within this lacuna, *Acta Anat. (Basel)*. 95, 421, 1976.
79. **Scherft, J. P.,** The lamina limitans of the organic matrix of calcified cartilage and bone, *J. Ultrastruct. Res.*, 38, 318, 1972.
80. **Seliger, W. G.,** Tissue fluid movement in compact bone *Anat. Rec.*, 166, 247, 1970.
81. **Owen, M., Howlett, C. R., and Triffitt, J. T.,** Movement of ^{125}I albumin and ^{125}I polyvinylpyrrolidone through bone tissue fluid, *Calcif. Tissue Res.*, 23, 103, 1977.
82. **Cooper, R. R., Milgram, J. W., and Robinson, R. A.,** Morphology of the osteon. An electron microscopic study, *J. Bone Joint Surg. (Am.)*, 48, 1239, 1966.
83. **Talmage, R. V.,** Morphological and physiological considerations in a new concept of calcium transport in bone, *Am. J. Anat.*, 129, 467, 1970.
84. **Arnold, J. S., Frost, H. M., and Buss, R. O.,** The osteocyte as a bone pump, *Clin. Orthop.*, 78, 47, 1971.
85. **Jande, S. S.,** Fine structural study of osteocytes and their surrounding bone matrix with respect to their age in young chicks, *J. Ultrastruct. Res.*, 37, 279, 1971.
86. **Jande, S. S. and Bélanger, L. F.,** Ultrastructural changes associated with osteocytic osteolysis in normal trabecular bone, *Anat. Rec.*, 163, 204, 1969.
87. **Jande, S. S. and Bélanger, L. F.,** Electron microscopy of osteocytes and the pericellular matrix in rat trabecular bone, *Calcif. Tissue Res.*, 6, 280, 1971.
88. **Dupont, D. A.,** Contribution a l'étude de la croissance et du remainement osseux chez la souris, Thése, Albert Kundig, Geneva, 1964.

89. **Remagen, W., Höhling, H. J., Hall, T. A., and Caesar, R.,** Electron microscopical and microprobe observations on the cell sheets of stimulated osteocytes, *Calcif. Tissue Res.*, 4, 60, 1969.
90. **Matesz, K., Földes, I., and Módis, L.,** Structure of the osteocyte capsule in rats with hypervitaminosis-D and rickets, *Acta Morphol. Acad. Sci. Hung.*, 23, 217, 1975.
91. **Módis, L., Matesz, K., Szabó, L. E., Pápay, Á, and Telek, B.,** Functional fine structure of the osteocyte capsule in experimentally damaged bones, in *Modern Trends in Orthopaedic Surgery,* Bozdech, Z., Horn, V., Eds., Acta Fac. Med. Univ. Brunensis J. E., Purkyne Univ., Brno, 1975, 53.
92. **Módis, L., Matesz, C., Telek, B., and Ádány, R.,** Submicroscopic structure of glycosaminoglycans in osteocyte capsule of human embryonic bone as revealed by polarizaiton microscopy, *Arch. Ital. J. Anat. Embryol.*, 94, 283, 1989.

Chapter 6

STRUCTURE OF THE CORNEAL STROMA

TABLE OF CONTENTS

I. INTRODUCTION

The corneal stroma represents a unique connective tissue. Its collagenous structure is highly oriented and regular, and, due to this, it is a transparent tissue. In this short chapter we want to demonstrate polarization microscopic findings obtained in embryonic human and chick corneas. These observations strongly suggest that GAGs are also oriented in the corneal lamellae, and the embryonic differentiation of the stromal lamellae — which occurs in a posterior-anterior direction — includes the oriented axiparallel deposition of the GAGs. Furthermore, it will be shown that the acellular primary cornea matrix has also a spatially oriented microstructure.

II. MAJOR CONSTITUENTS OF THE CORNEAL STROMA

The differentiated cornea is a transparent tissue comprising a stratified squamous epithelium and its basement membrane anteriorly, an endothelial layer and its basement membrane on the posterior surface and an avascular stroma interposed between the epithelium and endothelium.

The transparency of the corneal stroma is unique among adult connective tissues and it is thought to depend on at least three related structural features: (1) highly ordered collagen structure; (2) proper proteoglycan composition; (3) proper water content.

Polarization microscopy revealed long ago that the collagen fibers are oriented in the lamellae of the stroma; they run perpendicularly to the corneal surface.[1,2] Subsequent electron-microscopic investigations showed that collagen fibrils have the same orientation within a lamella, but collagen fibrils of adjacent lamellae are oriented approximately orthogonally.[3-6] Within one lamella, the collagen fibrils are uniform in size (24 to 25nm) and the spacing between them is also regular within one species. The interfibrillar spacing varies from species to species over a wide range (31 to 77 nm).[7,8]

The dry collagen stroma contains about 70% collagen,[2] and its major constituent is of type I.[9] Type V is also present, and accounts for 10% or more.[10-12] It has been shown that type V collagen filaments are embedded in type I fibrils in avian cornea.[13] Type VI collagen was also found in the stroma of bovine, rabbit, and human corneas.[14-18] In human corneas, type VI comprises as much as one quarter of the dry weight of the cornea.[14] It has been suggested that type VI is located in the gap region of the collagen fibrils.[15] Type IV and VII collagens are shown to be present in epithelial basement membrane, and type IV is also present in endothelial basement membrane.[19] In these basement membranes, laminin, nidogen, and heparan sulfate molecules have also been identified.[20] Cartilage-specific collagen type II and type IX have been detected in the primary stroma of the cornea.[21-23] The presence of type III collagen in the adult cornea seems to be controversial. It is shown to be present in adult human corneas[24] and in embryonic bovine and rabbit corneas.[25, 26]

Electron microscopic observations suggest that the development of the highly oriented collagenous lamellar structure of the corneal stroma is under the control of corneal fibroblasts. Corneal fibroblasts are dispersed between the collagen lamellae, and they extend long processes between the lamellae. The processes of neighboring fibroblasts are interconnected by tight junctions[27] and gap junctions.[28] It is supposed that the rows of corneal fibroblasts might function synchronously through the gap junctions, for example, in the synthesis of collagen and proteoglycans.[28] High voltage electron microscopy revealed that the corneal fibroblasts and their processes had two major axes at approximately right angles to one another.[6] According to Birk and Trelstad[6] the corneal fibroblast's cell surface defines at least three hierarchical extracellular matrix compartments involved in the control of collagen fibrillogenesis, fibril bundle formation, and the morphogenesis of the corneal stroma. The first compartment is represented by small recesses associated with the cell surface, and they contain 1 to 12 collagen fibrils. The recesses are supposed to be formed by tandem fusion of secretory vacuoles with the cell surface and with one another. The next compartment is a large surface folding developed from the lateral fusion of several small recesses, and it contains 50 to 100 collagen fibrils. The largest compartments are surrounded by the cell and its processes, which are formed by the fusion of the surface foldings, and are the sites of formation of the fiber bundles and lamellae. Since the corneal fibroblasts and their processes are aligned along two distinct axes at about 90° to one another, the cells simultaneously form fiber bundles in two directions perpendicular to one another. Migration of the fibroblasts along one of the axes would result in one straight bundle and one bundle with a 90° bend. As Birk and Trelstad[6] suggest, this process would provide the necessary translocation and explain the 90° changes in bundle direction.

The orthogonal lamellae describe a gradual clockwise shift from epithelium to endothelium, and the clockwise spiral pattern has the same handedness in both eyes of the mature chick cornea stroma.[5,29]

The interfibrillar spaces of the collagen lamellae are occupied by non-collagenous proteins. Among these constituents, the proteoglycans are fairly well characterized. These molecules control the degree of hydration in the corneal stroma hereby maintaining its transparency.[30-34] Normally, the differentiated and transparent corneal stroma is in a relatively dehydrated state. It has been shown that corneal dehydration and increase in transparency are correlated events in the development of the cornea of the chick and the rabbit.[35-37] Epithelium abrasion and especially endothelial damage is followed by stromal hydration, swelling, and opacity in the differentiated cornea indicating that epithelial and endothelial cells form barriers to water movement into the corneal stroma.[38] The development of a water pumping mechanism in the epithelia is in correlation with the dehydration of the corneal stroma during embryonic differentiation.[39]

The composition of the corneal PGs varies from species to species. Except for mouse cornea, the predominating corneal PG is the keratan sulfate (KS) proteoglycan (for review see Poole,[40] and Scott and Haigh.[41]) In corneal KS-PG, one or two KS chains of M_r7,000 are linked to a protein core of M_r about 30,000 to 40,000.[42-44] The other class of corneal PGs is a chondroitin sulfate (CS)-dermatan sulfate (DS)-PG (some of the glucuronic acid residues of the CSs are epimerized to iduronic acid, and are therefore, regarded as DSs[45]). This is also a small PG containing 1 to 2 GAG chains of about 55,000 molecular weight attached to a core protein of about 80 kDa.[44, 46] The core protein of corneal CS-DS-PG is immunologically related to the small DS-PG of cartilage and sclera and has the same size.[40]

In normal adult corneal stroma these small PGs fill the interfibrillar spaces. In opaque corneal scars, however, the interfibrillar spaces are enlarged and contain an unusually large PG, with the KS-PG being absent.[47] When the scars heal, the KS-PG reappears and glucuronic acid-rich large PG returns to its normal size. These data indicate that PGs may regulate collagen fibril spacing in the corneal stroma which is one of the most important prerequisites of the transparency.[2]

The ultrastructure of the corneal PGs has been established by electron microscopic histochemical reactions. In mouse corneal stroma, which contains predominantly CS-DS-PGs, these molecules are located mainly at the d band of the collagen fibrils and in a predominantly orthogonal position.[41] In rats and rabbits, however, where KS-PGs are also present, PGs are found to be orthogonally oriented to the a, c, d, and e bands of the fibrils. It is suggested that a and c bands are specific binding sites for KS-PGs.[41, 48] Using synchrotron X-ray diffraction techniques, regularly spaced and oriented PGs have been found in the unstained fresh corneal stroma.[33, 49] In bovine corneal stroma stained with the cationic dye cupromeronic blue for GAGs, however, the synchrotron X-ray diffraction method revealed — in addition to the PGs associated periodically with the collagen — randomly distributed PG particles, too.[50] Electron microscopic immunohistochemistry — using monoclonal antibodies against KS-PGs and CS-DS-PGs — has also demonstrated that corneal PGs are associated with the collagen fibrils.[51] The anterior and posterior regions of the stroma were found to be richer in CS-DS-PG than the middle part, whereas KS-PG showed a homogeneous distribution throughout the stroma.[51]

The biosynthesis and deposition of KS-PGs is thought to be regulated by the O_2 tension.[41] During embryonic development of the eye, the atrophy of the hyaloid artery is accompanied by a thickening of the cornea and appearance of KS.[52] A gradient of KS content in the corneal stroma has been observed, the posterior regions being richer in KS-PG than the anterior lamellae.[53] This gradient seems to follow the declining O_2 tension across the cornea in anterior-posterior direction.[54]

III. PRIMARY MATRIX

The morphogenesis of chick corneal stroma has been studied in detail (for review see Hay and Revel[4]). At the beginning, the corneal epithelium deposits the primary corneal stroma beneath its basal surface. The primary acellular stroma appears as a layer about 1 μm thick at the end of stage 18 according to Hamburger and Hamilton,[55] and it is 5 μm thick in the center and 10 μm thick at the periphery by stage 22.[4] It contains striated fibrils which tend to be arranged in orthogonal layers. The primary corneal stroma contains collagen types I, II and IX.[21-23, 56]

Between the 5th and 12th day of development, presumptive corneal fibroblasts migrate from the limbus first to posterior layer of the primary stroma. These cells seem to originate from neuroectoderm.[4]

The beginning of cell migration coincides with increased hydration and swelling of the acellular stroma enriched in hyaluronate.[57, 58] Subsequently, when the stroma loses water, condenses, and the cell migration ceases, hyaluronidase activity increases within the cornea and the hyaluronate will be enzymatically depolymerized.[57] The hyaluronate is secreted by the endothelial cells.[57, 59] These cells, like the presumptive corneal fibroblasts, derive from the neuroectoderm (neural crest).[60, 61] After mesenchymal invasion, the hyaluronate is synthesized by both the mesenchymal and endothelial cells.[59] The epithelium synthesizes chondroitin sulfate PGs.[59, 62] The chondroitin sulfate is supposed to be in interaction with the collagen influencing the precise spacing of the fibrils.[5] Other experimental data, however, argue against the role of CS-PGs in the formation of the regular collagen structure.[61] CS-PGs seem to influence the metabolic activity of the epithelial cells.[63] Electron microscopy revealed a well ordered pattern of CS-PG stained with ruthenium red in both the epithelial basement membrane and the collagen fibrils of the primary stroma in embryonic chick corneas.[59]

The structure of the primary corneal matrix of chick embryos has been studied in our laboratory using topo-optical reactions. Picrosirius red F3B staining resulted in an optical anisotropy in the primary stroma, which can be observed from stage 22 (Figures 1, 2). The sign of birefringence was positive relating to the tangent of the cornea anlage. Since the binding of the sirius red molecules to the collagen results in an axiparallel orientation of the dye, the positive sign indicates that collagen is arranged tangentially in the primary matrix. This finding is in agreement with electron-microscopic observations showing oriented deposition of the collagen fibrils.[59, 61] Polarization microscopy, of course, cannot reveal whether orthogonal fibrils form lamellae[4, 59] or not.[61]

Topo-optical reactions used for polarization microscopic detection of GAGs resulted generally in a moderate metachromatic staining and weak birefringence. *N,N'*-diethylpseudoisocyanine staining induced a stronger optical anisotropy in the primary corneal matrix (Figure 2). This topo-optical reaction could be completely abolished by antendent digestions with testicular hyaluronidase and chondroitinase ABC. It follows that predominantly chondroitin

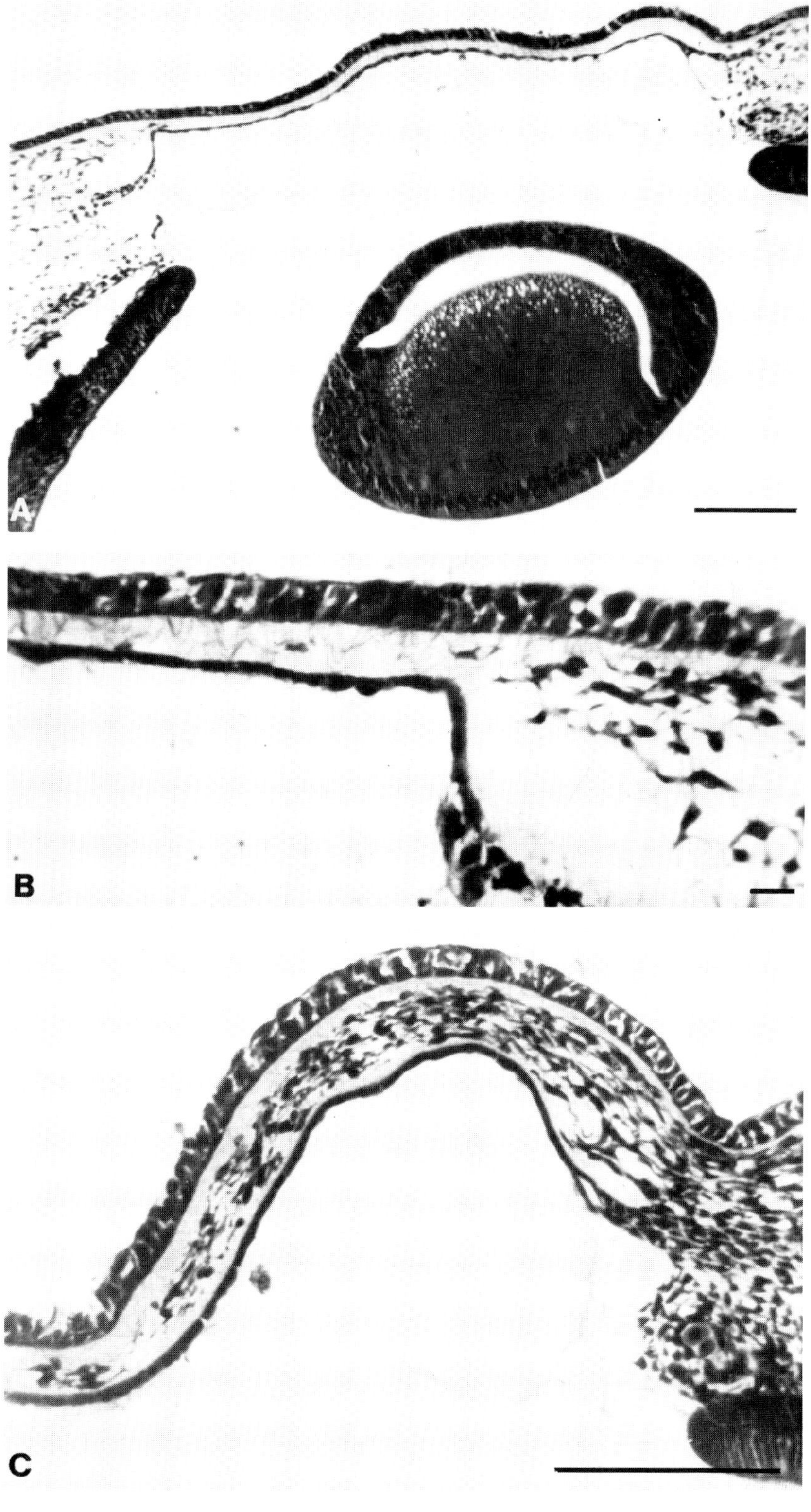

FIGURE 1. Primary corneal stroma of chick embryo after Goldner staining. (A, B): stage 26, (C) stage 29. The invasion of presumptive fibroblasts is visible in C. Scales: 100 μm (A, C), 10 μm (B).

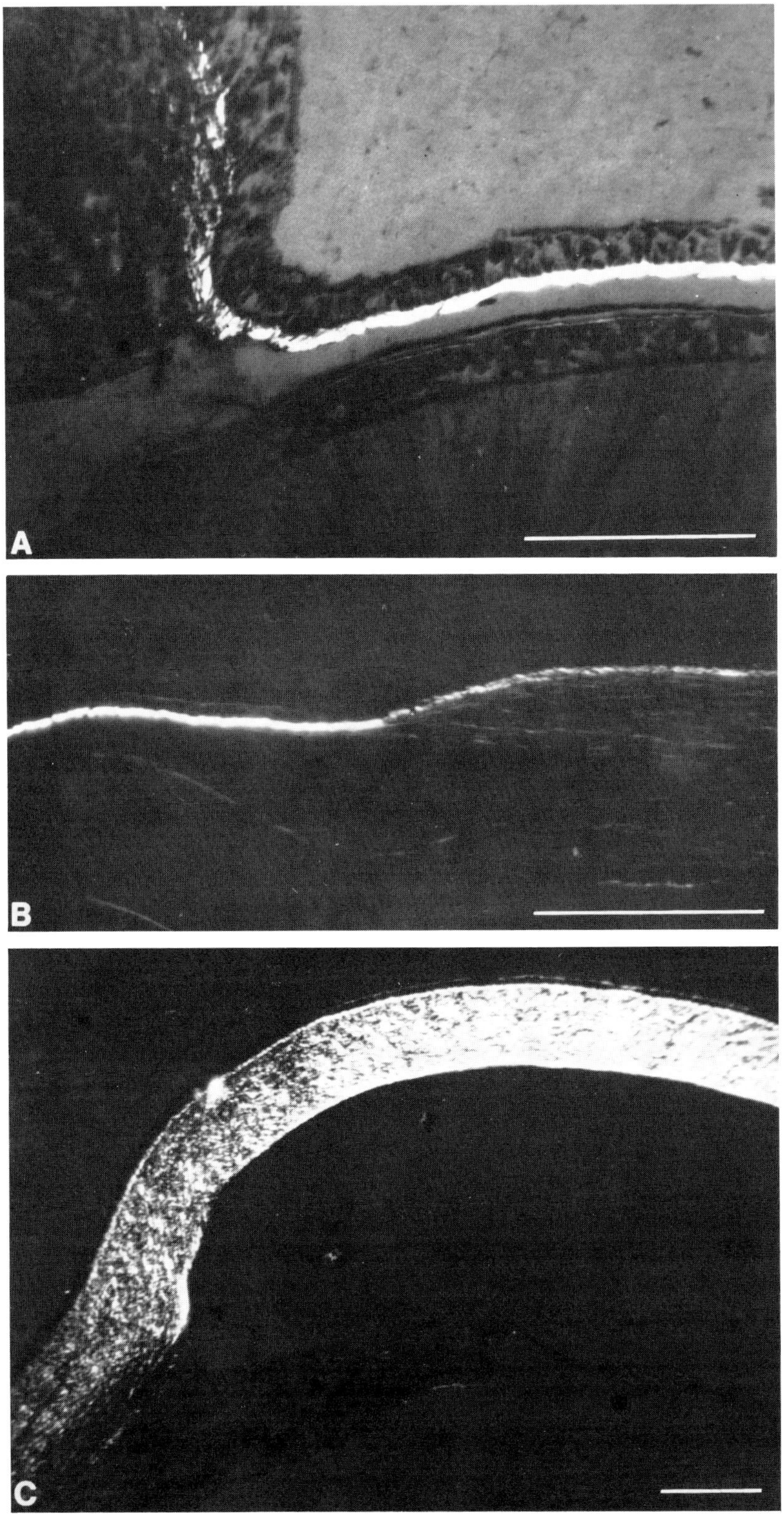

FIGURE 2. Birefringence of the primary (A, B) and secondary corneal matrix (C) after staining with picrosirius red (F3B for collagen (A, C) and *N,N'*-diethylpseudoisocyanine metachromatic reaction for GAGs (B). Scales: 100 μm.

sulfate molecules are responsible for the dye-binding. The sign of birefringence induced by *N,N'*-diethylpseudoisocyanine is positive related to the tangent of the cornea. We know that this dye binds to GAGs in an axiparallel manner (see Chapter 3). Our finding, therefore, suggests that the chondroitin sulfate molecules are tangentially arranged in the primary matrix. This conclusion fits in with the earlier electron-microscopic result showing an oriented arrangement of CS-PG particles.

We suppose that the spatially oriented primary matrix may facilitate the migration of the presumptive corneal fibroblasts which will invade the primary corneal stroma. Furthermore, this oriented microstructure may be involved in the regulation of the alignment of the cells and the formation and orientation of the first lamellae of the secondary matrix.

IV. SECONDARY MATRIX

After the primary matrix has been invaded by the presumptive corneal fibroblasts, these cells secrete the constituents of the secondary matrix (collagen type I, VI, KS-PG, CS-DS-PG), which will be deposited around the cells forming the definitive lamellae. The lamellae are deposited in sequence, beginning in the deepest layer just beneath the basement membrane of the endothelium.[36, 64-66] The layers of the secondary matrix exhibit a highly oriented microstructure (see Section II), and they gradually fill the corneal stroma providing the morphological basis of the transparency. The acellular primary matrix disappears from differentiated cornea of most species, but it persists as Bowman's membrane in primates.

According to biochemical investigations, the GAG pattern of the cornea changes during embryonic development. In the chicken cornea, a stepwise disappearance of the nonsulfated GAG and an increase of KS and CS content has been demonstrated.[67] After hatching KS becomes the predominant GAG of the corneal stroma.[67] In fetal human cornea the concentration of the sulfated GAGs is increased by 33% at seven months gestation and remained constant in the rest of the gestation period.[52] The concentration of sulfated GAGs was found to be the same in newborns and adults, but C/KS ratio varied with age. CS predominates in human fetal corneal stroma while KS is the predominant GAG in adults.[52, 68] The local concentration of the sulfated GAGs is the highest in the central cornea, while the peripheral cornea has 24% less GAGs.[69] In addition, the central corneal stroma is richer in KS than the peripheral parts. This finding coincides with the change of regularity in collagen fibers: the uniformity of fiber diameters and their arrangement decreases distally from the central cornea.[69] These data suggest that the KS-PGs may be predominantly responsible for the control of the development of the highly oriented collagen structure in the transparent secondary corneal stroma.

The posterior lamellae of the fetal human corneal stroma exhibit a stronger staining reaction of GAGs, and this pattern persists during the first decade of postnatal life.[70] Using monoclonal antibodies against KS-PGs, immunohistochemistry reveals the appearance of KS-PG inside of stromal cells deriving from the neural crest at stage 29 (day 6), and in the matrix of the posterior stroma at stage 35 (day 9) in embryonic chick cornea.[71] KS-PG content is then gradually increased during the embryonic life and also after the hatching.[71] Similar changes of KS-PGs have been demonstrated by immunohistochemical methods during the development of rabbit cornea.[72] $^{35}SO_4$-incorporation studies and different staining methods have shown that the differentiation of the lamellar structure rich in sulfated GAGs progresses in a posterior-anterior direction in secondary corneal stroma of rabbit embryos.[73] The hydration of the cornea decreases until transparency appears in chicks[4] and rabbits.[73]

It was believed earlier, based on the observation that embryonic tissues are more or less translucent, that corneal transparency does not develop, but it is an example of an embryonic condition retained in adult life. The aforementioned data, however, clearly indicate that the transparency results from the differentiation of the secondary stroma. Our polarization microscopic studies demonstrate that the spatially oriented microstructure of sulfated GAGs is gradually formed in the lamellae of the secondary corneal stroma during human embryonic development.

The ultrastructure of adult corneal stroma has been studied with polarization microscopical methods without stainings[1, 2, 74, 75] and after topo-optical reactions.[76-79] Topo-optical reactions have revealed an oriented microstructure of the sulfated GAGs in stromal lamellae of the differentiated corneas. It has been shown in our laboratory, that the oriented structure of sulfated GAGs appears already in the embryonic corneal stroma.[80]

We investigated the development of human secondary corneal matrix with metachromatic reactions and picrosirius red F3B staining. Despite the presence of sulfated GAGs in embryonic and adult human corneas, these specimens cannot be stained by metachromatic dyes below pH 4.2 to 4.5.[81] The reason of this phenomenon is not known. One possible explanation is the relatively low local concentration of GAGs in this structure. It is about 13 to 35% of the GAG-content of the cartilage matrix.[82] It is known that hyaline cartilage matrix cannot be stained by the metachromatic toluidine blue after having lost 42% of its hexuronic acid content indicating the relative insensitivity of such stainings.[83] This feature of the human corneal stroma causes some difficulties in specific histochemical detection of sulfated GAGs in human corneal stroma.

We observed a strong metachromasia in the subendothelial layers of the secondary corneal stroma of human embryos after dimethylmethylene blue and toluidine blue staining at pH 4.5. The metachromatic areas exhibited birefringence with positive sign (Figures 3 to 5). Since these dyes are perpendicularly oriented to the GAGs, this optical phenomenon indicates that GAGs are arranged parallel to the long axis of the collagen fibrils. Picrosirius red staining, on the other hand, revealed birefringent collagen in the entire secondary stroma, but the intensity of optical anisotropy was stronger in the posterior stroma where the formation of collagen lamellae occured (Figure 6). The anterior part of the embryonic corneal stroma contained loosely arranged birefringent collagen fibers but we failed to detect birefringence here after metachromatic staining in 9-week-old embryos (Figure 3). At older stages, birefringence appeared in the anterior part of the stroma after metachromatic staining, but it could be abolished by an antecedent digestion with testicular hyaluronidase indicating that CS-PGs are the predominant GAGs in that region (Figure 4). The staining and birefringence induced by cationic metachromatic dyes in the posterior stroma, however, persisted — although somewhat weakened — after the same digestion, suggesting the predominance of KS-PGs in the lamellae formed posteriorly (Figure 4). In 7-month-old fetuses, like the adult corneas, metachromasia and birefringence could be observed throughout the secondary stroma (Figure 7), and antecedent digestion with testicular hyaluronidase did not essentially modify these optical phenomena.

These results suggest that the collagen lamellae of the secondary corneal stroma also contain spatially oriented, sulfated GAGs, predominantly keratan sulfates. They are arranged parallel to the optical axis of the collagen. This orientation pattern is in agreement with electron microscopic findings showing a predominantly perpendicular arrangement of the corneal PGs to the collagen fibrils.[41, 48, 49] Our observations, however, cannot exclude the coexistence of randomly distributed PGs observed by other techniques.[50] We suppose that the presence of oriented GAGs in the stromal lamellae may be involved in the formation of their organization, and in this way, in the development and maintenance of the transparency.

ACKNOWLEDGMENTS

I wish to express my thanks to my former students, Drs. Erika Bihari, Judit Kiss, and I. Kovács for their help in the collection and topo-optical analysis of the embryonic human and chicken corneas. Human materials were kindly provided by Department of Pathology

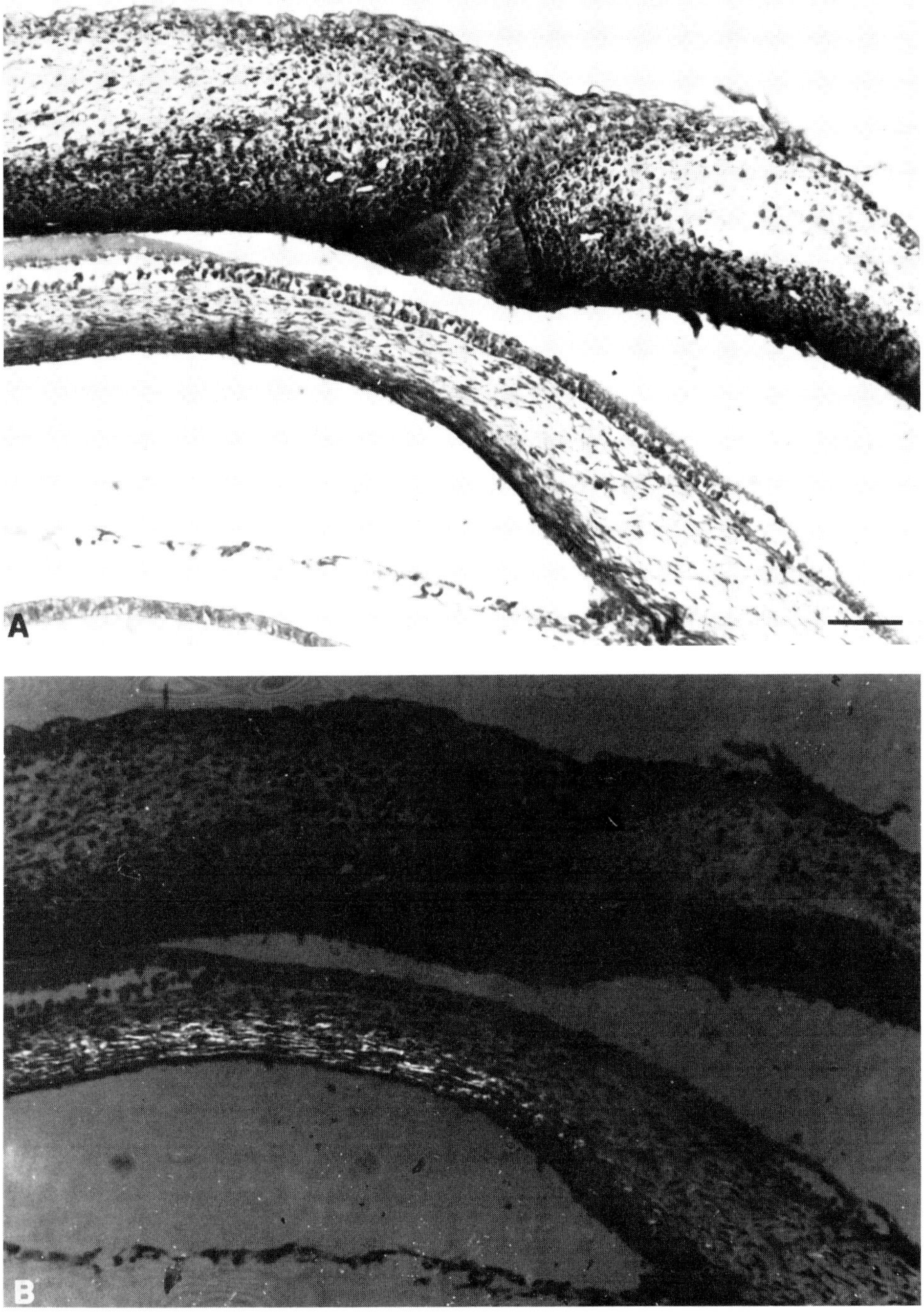

FIGURE 3. Nine-week-old human embryonic cornea and eyelids after dimethylmethylene blue staining at pH 4.5 taken in normal (A) and polarized light (B). Note the basophilic (metachromatic) staining and birefringence in the posterior stroma of the cornea. Scale: 100 μm.

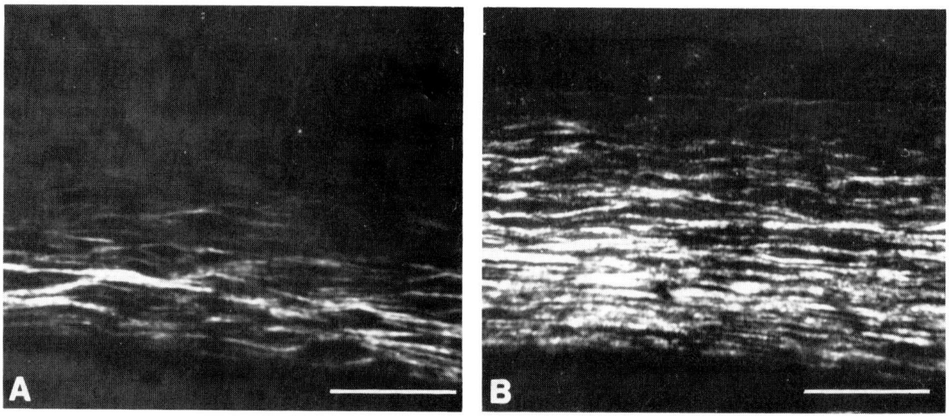

FIGURE 4. Twelve-week-old human embryonic corneas after dimethylmethylene blue staining at pH 4.5 taken in polarized light. The staining was performed after digestion with testicular hyaluronidase (A) and without any pretreatment (B). The birefringence seen in (A) is probably due to the oriented deposition of keratan sulfate molecules in the posterior lamellae. Scales: 50 μm.

FIGURE 5. Portion of an 18-week-old human embryonic eye after dimethylmethylene blue staining taken in normal (A) and polarized light (B). In the middle, the cornea is seen. Note the staining and induced birefringence of the posterior stroma. Eyelid (on the left side) and lens (on the right side) are also visible. Scale: 100 μm.

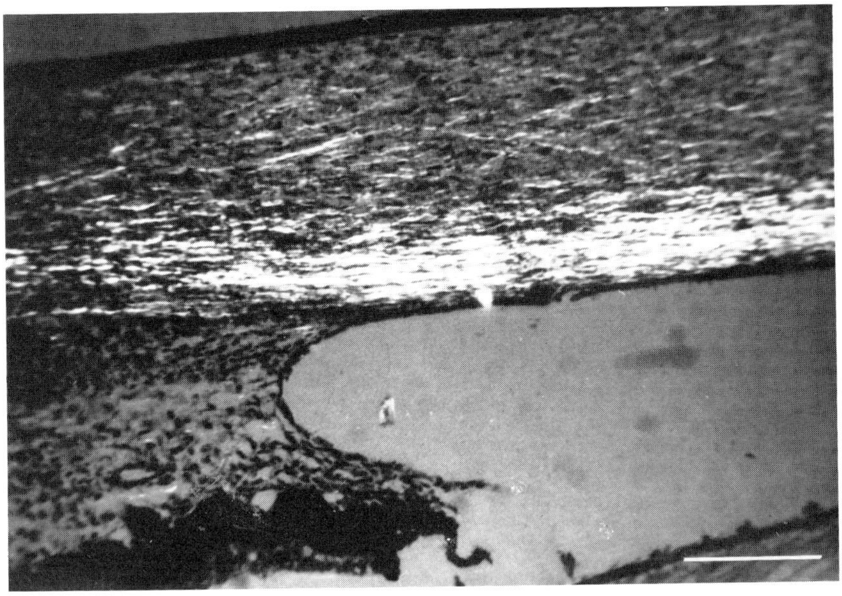

FIGURE 6. Eight-week-old human embryonic cornea after picrosirius red F3B taken in polarized light. Observe the strong birefringence of the compact posterior stroma. The loosely arranged collagen fibers of the anterior part are also birefringent. Scale: 100 μm.

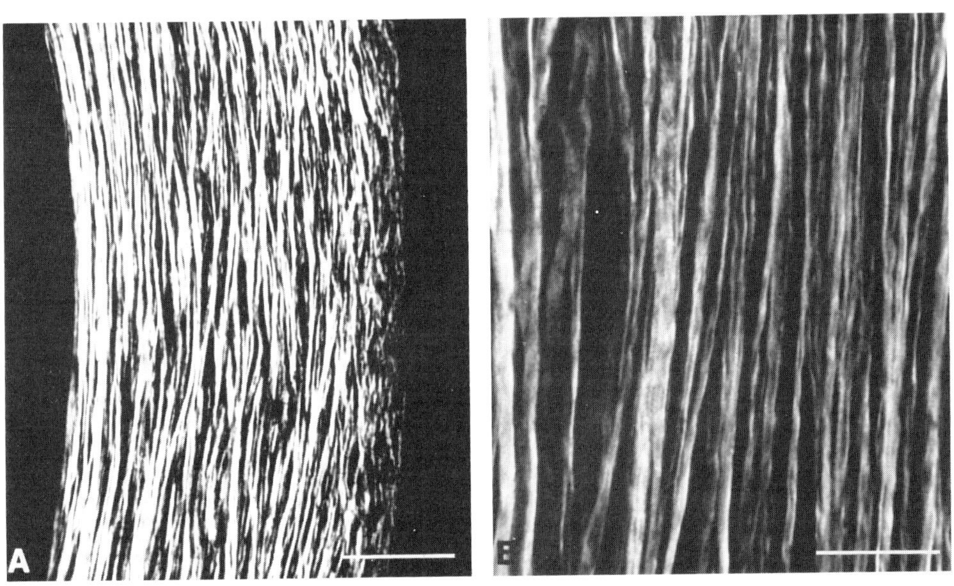

FIGURE 7. Seven-month-old human embryonic cornea after dimethylmethylene blue staining for sulfated GAGs taken in polarized light. Note that the birefringent lamellae fill the corneal stroma. Scales: 100 μm (A), 50 μm (B).

as well as Department of Obstetrics and Gynecology of this university in accordance with the laws of this country.

REFERENCES

1. **Naylor, E. J.,** Polarized light studies of corneal structure, *Br. J. Ophthalmol.,* 37, 77, 1953.
2. **Maurice, D. M.,** The structure and transparency of the cornea, *J. Physiol.,* 136, 263, 1957.
3. **Jakus, M., Ed.,** *The Structure of the Eye,* Academic Press, New York, 1961.
4. **Hay, E. D. and Revel, J. P.,** *Fine Structure of the Developing Avian Cornea,* S. Karger, Basel, New York, 1969.
5. **Treistad, R. L. and Coulombre, A. J.,** Morphogenesis of the collagenous stroma in the chick cornea, *J. Cell Biol.,* 50, 840, 1971.
6. **Birk, D. E. and Treistad, R. L.,** Extracellular compartments in matrix morphogenesis: collagen fibril, bundle, and lamellar formation by corneal fibroblasts, *J. Cell Biol.,* 99, 2024, 1984.
7. **Hamada, R., Giraud, J.-P., Graf, B., and Pouliquen, Y.,** Étude analytique et statistique des lamelles, des kératocytes, des fibrilles de collagéné de la région centrale de la cornée humaine normale, (Microscopie optique et électronique), *Arch. Ophthalmol.,* 32, 563, 1972.
8. **Gyi, T. J. and Meek, K. M.,** Structural role of proteoglycans in the corneal stroma: a species study, *Biochem. Soc. Trans.,* 15, 1086, 1987.
9. **Kay, E. P.,** Molecular anatomy of the vertebrate eye: distribution of collagen in ocular tissues, in *Collagen, Biochemistry,* Vol. I, Nimni, M. E., Ed., CRC Press, Boca Raton, FL, 1988, 207.
10. **Welsh, C., Gay, S., Rhodes, R. K., Pfister, R., and Miller, E. J.,** Collagen heterogeneity in normal rabbit cornea. I. Isolation and biochemical characterization of the genetically-distinct collagens, *Biochim. Biophys. Acta.,* 625, 78, 1980.
11. **Linsenmayer, T. F., Fitch, J. M., and Mayne, R.,** Extracellular matrices in the developing avian eye: type V collagen in corneal and noncorneal tissues, *Invest. Ophthalmol. Vis. Sci.,* 25, 41, 1984.
12. **Bailey, A. J.,** Structure, function and ageing of the collagens of the eye, *Eye,* 1, 175, 1987.
13. **Fitch, J. M., Birk, D. E., Mentzer, A., Hasty, K. A., Mainardi, C., and Linsenmayer, T. F.,** Corneal collagen fibrils: dissection with specific collagenases and monoclonal antibodies, *Invest. Ophthalmol. Vis. Sci.,* 29, 1125, 1988.
14. **Zimmermann, D. R., Trüeb, B., Winterhalter, K. H., Witmer, R., and Fischer, R. W.,** Type VI collagen is a major component of the human cornea, *FEBS Lett.,* 197, 55, 1986.
15. **Wall, R. S., Elliott, G. F., Gyi, T. J., Meek, K. M., and Branford-White, C. J.,** Bovine corneal stroma contains a structural glycoprotein located in the gap region of the collagen fibrils, *Biosci. Rep.,* 8, 77, 1988.
16. **Cintron, C. and Hong, B.-S.,** Heterogeneity of collagens in rabbit cornea: type VI collagen, *Invest. Ophthalmol. Vis. Sci.,* 29, 760, 1988.
17. **Alper, R.,** The bovine corneal SGP-complex is related to the tissue form of type VI collagen, *Curr. Eye Res.,* 7, 31, 1988.
18. **Murata, Y., Yoshioka, H., Iyama, K., and Usuku, G.,** Distribution of type VI collagen in the bovine cornea, *Ophthalmic Res.,* 21, 67, 1989.
19. **Zimmermann, D. R., Fischer, R. W., Winterhalter, K. H., Witmer, R., and Vaughan, L.,** Comparative studies of collagens in normal and keratoconus corneas, *Exp. Eye Res.,* 46, 431, 1988.
20. **Schittny, J. C., Timpl, R., and Engel, J.,** High resolution immunoelectron microscopic localization of functional domains of laminin, nidogen, and heparan sulfate proteoglycan in epithelial basement membrane of mouse cornea reveals different topological orientations, *J. Cell Biol.,* 107, 1599, 1988.
21. **Linsenmayer, T. F., Smith, G. N., Jr., and Hay, E. D.,** Synthesis of two collagen types by embryonic chick corneal epithelium *in vitro,* Proc. Natl. Acad. Sci. U.S.A., 74, 39, 1977.
22. **Von der Mark, K., Von der Mark, H., Timpl, R., and Trelstad, R. L.,** Immunofluorescent localization of collagen types I, II, and III in the embryonic chick eye, *Dev. Biol.,* 59, 75, 1977.
23. **Fitch, J. M., Mentzer, A., Mayne, R., and Linsenmayer, T. F.,** Acquisition of type IX collagen by the developing avian primary corneal stroma and vitreous, *Dev. Biol.,* 128, 396, 1988.
24. **Newsome, D. A., Gross, J., and Hassell, J. R.,** Human corneal stroma contains three distinct collagens, *Invest. Ophthalmol. Vis. Sci.,* 22, 376, 1982.
25. **Lee, R. E. and Davison, P. F.,** Collagen composition and turnover in ocular tissues of the rabbit, *Exp. Eye Res.,* 32, 737, 1981.
26. **Tseng, S. C. G., Smuckler, D., and Stern, R.,** Comparison of collagen types in adult and fetal bovine corneas, *J. Biol. Chem.,* 257, 2627, 1982.

27. **Hogan, M. J., Alvarado, J. A., and Weddell, J. E.,** The cornea, in *Histology of the Human Eye: An Atlas and Textbook,* Hogan, M. J., Ed., W. B. Saunders, Philadelphia, 1971, 55.
28. **Ueda, A., Nishida, T., Otori, T., and Fujita, H.,** Electron-microscopic studies on the presence of gap junctions between corneal fibroblasts in rabbits, *Cell Tissue Res.,* 249, 473, 1987.
29. **Trelstad, R. L.,** The bilaterally asymmetrical architecture of the submammalian corneal stroma resembles a cholesteric liquid crystal, *Dev. Biol.,* 92, 133, 1982.
30. **Hedbys, B. O.,** The role of polysaccharides in corneal swelling, *Exp. Eye Res.,* 1, 81, 1961.
31. **Hodson, S.,** Why the cornea swells, *J. Theor. Biol.,* 33, 419, 1971.
32. **Bettelheim, F. A. and Plessy, B.,** The hydration of proteoglycans of bovine cornea, *Biochim. Biophys. Acta.,* 381, 203, 1975.
33. **Meek, K. M., Elliott, G. F., Sayers, Z., and Whitburn, S. B.,** The distribution of electron density in corneal collagen fibrils, *Curr. Eye Res.,* 1, 281, 1981.
34. **Plessy, B. and Bettelheim, F. A.,** Water vapor sorption of keratan sulfate, *Mol. Cell. Biochem.,* 6, 85, 1975.
35. **Coulombre, A. and Coulombre, J.,** Corneal development. I. Corneal transparency, *Cell. Comp. Physiol.,* 51, 1, 1958.
36. **Coulombre, A. and Coulombre, J.,** The development of the structural and optical properties of the cornea, in *The Structure of the Eye,* Smelser, G., Ed., Academic Press, New York, 1961, 405.
37. **Smelser, G. K.,** Corneal hydration: comparative physiology of fish and mammals, *Invest. Ophthalmol.,* 1, 11, 1962.
38. **Stanley, J. A.,** Water permeability of the human cornea, *Arch. Ophthalmol.,* 87, 568, 1972.
39. **Coulombre, A. J.,** Problems in corneal morphogenesis, *Adv. Morphogenes.,* 4, 81, 109, 1965.
40. **Poole, A. R.,** Proteoglycans in health and disease: structures and functions, *Biochem. J.,* 236, 1, 1986.
41. **Scott, J. E. and Haigh, M.,** Keratan sulphate and the ultrastructure of cornea and cartilage: a "stand-in" for chondroitin sulphate in conditions of oxygen lack?, *J. Anat.,* 158, 95, 1988.
42. **Axelsson, I. and Heinegard, D. K.,** Characterization of the keratan sulfate proteoglycans from bovine corneal stroma, *Biochem. J.,* 169, 517, 1978.
43. **Hassel, J. R., Newsome, D. A., and Hascall, V. C.,** Characterization and biosynthesis of proteoglycans of corneal stroma from rhesus monkey, *J. Biol. Chem.,* 254, 12346, 1979.
44. **Gregory, J. D., Cöster, L., and Damle, S. P.,** Proteoglycans of rabbit corneal stroma. Isolation and partial characterization, *J. Biol. Chem.,* 257, 6965, 1982.
45. **Stuhlsatz, H. W., Muthiah, P. L., and Greiling, H.,** Occurrence of dermatan sulphate in calf cornea, *Scand. J. Clin. Lab. Invest.,* 29, Suppl., 123, 31, 1972.
46. **Axelsson, I. and Heinegard, D.,** Characterization of chondroitin sulphate rich proteoglycans from bovine corneal stroma, *Exp. Eye Res.,* 30, 57, 1980.
47. **Hassel, J. R., Cintron, C., Kublin, C., and Newsome, D. A.,** Proteoglycan changes during restoration of transparency in corneal scars, *Arch. Biochem. Biophys.,* 222, 362, 1983.
48. **Scott, J. E. and Haigh, M.,** "Small"-proteoglycan: collagen interactions: keratan sulphate proteoglycan associates with rabbit corneal collagen fibrils at the "a" and "c" bands, *Biosci. Rep.,* 5, 765, 1985.
49. **Meek, K. M., Elliott, G. F., Hughes, R. A., and Nave, C.,** The axial electron density in collagen fibrils from human corneal stroma, *Curr. Eye Res.,* 2, 471, 1982/1983.
50. **Meek, K. M., Elliott, G. F., and Nave, C.,** A synchrotron X-ray diffraction study of bovine cornea stained with cupromeronic blue, *Coll. Relat. Res.,* 6, 203, 1986.
51. **Valesco, A. and Hidalgo, J.,** Ultrastructural demonstration of proteoglycans in adult rat cornea, *Tissue Cell,* 20, 567, 1988.
52. **Breen, M., Johnson, R. L., Sittig, R. A., Weinstein, H. G., and Veis, A.,** The acid glycosaminoglycans in human fetal development and adult life: cornea, sclera and skin, *Connect. Tissue Res.,* 1, 291, 1972.
53. **Bettelheim, F. and Goetz, D.,** Distribution of hexosamines in bovine cornea, *Invest. Ophthalmol.,* 15, 301, 1976.
54. **Fatt, I.,** *Physiology of the Eye,* Butterworths, London, 1978, chap. 6.
55. **Hamburger, V. and Hamilton, L.,** A series of normal stages in the development of the chick embryo, *J. Morphol.,* 88, 49, 1951.
56. **Hendrix, M. J., Hay, E. D., von der Mark, K., and Linsenmayer, T. F.,** Immunohistochemical localization of collagen types I and II in the developing chick cornea and tibia by electron microscopy. *Invest. Ophthalmol. Vis. Sci.,* 22, 359, 1982.
57. **Toole, B. P. and Trelstad, R. L.,** Hyaluronate production and removal during corneal development in the chick, *Dev. Biol.,* 26, 28, 1971.
58. **Toole, B. P.,** Glycosaminoglycans in morphogenesis, in *Cell Biology of Extracellular Matrix,* Hay, E. D., Ed., Plenum Press, New York, 1981, 259.
59. **Trelstad, R. L., Hayashi, K, and Toole, B. P.,** Epithelial collagens and glycosaminoglycans in the embryonic cornea. Macromolecular order and morphogenesis in the basement membrane, *J. Cell Biol.,* 62, 815, 1974.

60. **Johnston, M. C., Noden, D. M., Hazelton, R. D., Coulombre, J. L., and Coulombre, A. J.,** Origins of avian ocular and periocular tissues, *Exp. Eye Res.*, 29, 27, 1979.
61. **Bard, J. B. L., Bansal, M. K., and Ross, A. S. A.,** The extracellular matrix of the developing cornea: diversity, deposition and function, *Development* Suppl., 103, 195, 1988.
62. **Meier, S. and Hay, E. D.,** Synthesis of sulfated glycosaminoglycans by embryonic epithelium, *Dev. Biol.*, 35, 318, 1973.
63. **Meier, S. and Hay, E. D.,** Stimulation of extracellular matrix synthesis in the developing cornea by glycosaminoglycans, *Proc. Natl. Acad. Sci. U.S.A.*, 71, 2310, 1974.
64. **Smelser, G. K., Ed.,** *The Structure of the Eye,* Academic Press, New York, 1961.
65. **Meyer, D. B. and O'Rahilly, R.,** The development of the cornea in the chick, *J. Embryol. Exp Morphol.*, 7, 303, 1959.
66. **Coulombre, A. J.,** The eye, in *Organogenesis,* Haan, R. L., and Urprung H., Eds., Holt, Rinehart & Winston, New York, 1965, 219.
67. **Praus, R. and Brettschneider, I.,** Glycosaminoglycans in the developing chicken cornea, *Ophthalmol. Res.*, 2, 367, 1971.
68. **Praus, R. and Brettschneider, I.,** Glycosaminoglycans in embryonic and postnatal human cornea, *Ophthalmic Res.*, 7, 452, 1975.
69. **Borcherding, M. S., Blacik, L. J., Sitting, R. A., Bizzell, J. W., Breen, M., and Weinstein, R.,** Proteoglycans and collagen fiber organization in human corneoscleral tissue, *Exp. Eye Res.*, 21, 59, 1975.
70. **Gemelotto, G. and Patrone, C.,** Rilievi isto-chimici sulla distribuzione del mucopolisaccaride nella cornea e nella sclera del'occhio umano nelle diverse eta, *Giorn. Ital. Oftalmol.*, 8, 42, 1955.
71. **Funderburgh, J. L., Caterson, B., and Conrad, G. W.,** Keratan sulfate proteoglycan during embryonic development of the chicken cornea, *Dev. Biol.*, 116, 267, 1986.
72. **Sundarraj, N., Chao, J., Gregory, J. D., and Damle, S. P.,** Ocular distribution of keratan sulfates during pre- and postnatal development in rabbits, *J. Histochem. Cytochem.*, 34, 971, 1986.
73. **Smelser, G. K. and Ozanics, V.,** Morphological and functional development of the cornea, in *The Transparency of the Cornea.* Duke-Elder, S. and Perkins, E. S., Eds., Blackwell Scientific, Oxford, 1960, 23.
74. **Stanworth, A. and Naylor, E. J.,** The polarization optics of the isolated cornea, *Br. J. Ophthalmol.*, 34, 201, 1950.
75. **Stanworth, A. and Naylor, E. J.,** Polarized light studies of the cornea. I. The isolated cornea, *J. Exp. Biol.*, 30, 160, 1953.
76. **Romhányi, G.,** Über die submikroskopische strukturelle Grundlage der metachromatischen Reaktion, *Acta Histochem. (Jena)* 15, 201, 1963.
77. **Módis, L.,** Topo-Optical Investigation of Mucopolysaccarides (Acid Glycosaminoglycans), *Handbuch der Histochemie.* Vol. II, Part, 4, Graumann, W. and Neumann, K., Eds., G. Fischer Verlag, Stuttgart, 1974.
78. **Süveges, I.,** Ultrastrukturelle Veränderung bei Keratoconus, *Klin. Monatsbl. Augenheilkd.*, 174, 462, 1979.
79. **Süveges, I.,** Histological and ultrastructural studies of the cornea in Maroteaux-Lamy Syndrome, *Graefes Arch. Clin. Exp. Ophthalmol.* 212, 29, 1979.
80. **Módis, L.,** Polarization microscopical studies on the submicroscopic structure of glycosaminoglycans in embryonic connective tissues, *Uppsala J. Med. Sci.*, 82, 144, 1977.
81. **Marré, M.,** Die Stärke der Metachromotropie der Cornea in Abhängigkeit vom Lebensalter, *Graefes Arch. Klin. Exp. Ophthalmol.*, 174, 177, 1967.
82. **Buddecke, E.,** *Grundriss der Biochemie,* 3rd. ed., Walter de Gruyter, Berlin, 1973.
83. **Poole, A. R.,** The relationship between toluidine blue staining and hexuronic acid content of cartilage matrix, *Histochem. J.*, 2, 425, 1970.

Chapter 7

STRUCTURE OF KIDNEY BASEMENT MEMBRANES

TABLE OF CONTENTS

I. INTRODUCTION

Basement membranes (BM) are specialized extracellular matrix structures deposited in the close vicinity of epithelial, muscle, fat, and nerve cells. The best known example is the BM seen at the interface between the basal surface of epithelial cells and the underlying connective tissue.

The term BM is used to describe different structures creating some confusion in the literature. The term basement membrane has been used for a long time by light microscopists to denote a well visible argirophilic and PAS-positive connective tissue condensation separating epithelial tissue and stromal connective tissue. Electron microscopy and biochemistry revealed that the "classical" BM can be separated into two different structures. Just beneath the epithelial and muscle cells there is a 20 to 200-nm-wide zone called basal (or basement) lamina. Its constituents are produced by the epithelial or muscle cells. The basal lamina can be subdivided into lamina rara and lamina densa. The lamina rara or lucida is rich in PGs and it is associated with the cell coat. The lamina densa contains predominantly collagen and in some places it is connected to underlying reticular (type III collagen-containing) fibrils.[1] The condensation of reticular fibrils represent the outer and thicker part of the classical BM. The classical BM term refers to the basal lamina plus the reticular fibrils and associated macromolecules produced by connective tissue cells of the environment (Figure 1). Today the term BM is generally used as a synonym of the basal lamina in electron microscopic descriptions and by biochemists. Light microscopists, however, frequently use this nomenclature in the classical sense.

BMs are considered to control a number of functions such as expression of cell phenotype, organogenesis, invasion of cells, and transport of macromolecules under normal and pathological conditions. The importance of BM explains the rapid accumulation of data about its composition, ultrastructure, and functions.[3-5]

II. A SHORT OVERVIEW OF MAJOR CONSTITUENTS OF THE BASEMENT MEMBRANES

Collagen type IV, laminin, nidogen (entactin), and heparan sulfate PGs are the most common and characteristic constituents of the BM.[6-11] They are in specific interaction with each other, forming a three dimensional supramolecular structure. Some of them also interact with the cell membrane and the underlying stroma (in places where it exists) (for review see Timpl[11]). If purified BM constituents are mixed under certain experimental conditions, owing to their interactions, precipitation occurs and formation of basement membrane-like structures are observed electron microscopically.[12, 13]

The characteristic collagen of BM is designated as type IV. Collagen type IV molecule has a heterotrimer composition: $[\alpha_1(IV)]_2$ $[\alpha_2(IV)]$. As revealed by rotary shadowing, these chains are arranged into 390-nm-long triple-helical rods, containing a globule of 12 nm at the C-terminus (NC1 domain) and a 30-nm-long segment at the N-terminus called the 7S domain.[14, 15] Collagen type IV molecules have a high capacity to interact with each other. Tetramers are formed by crosslinking four N-terminal ends (7S domains) in an antiparallel fashion. The overlapped N-terminal ends are stabilized by inter- and intramolecular disulfide bonds and are relatively resistant to proteolytic degradation. On the other hand, dimers are formed by interactions between the globular domains (NC1). By alternating use of 7S and NC1 domains, a network of collagen type IV molecules is formed. In addition, lateral interactions may also occur, and the laterally associated collagen type IV molecules form a supertwisted helix.[11, 16] The head-to-head and side-by-side aggregation of the tetramers results in formation of huge supramolecular and linear aggregates.[10]

Laminin is the major noncollagenous glycoprotein in BM.[17] It has been shown to be a

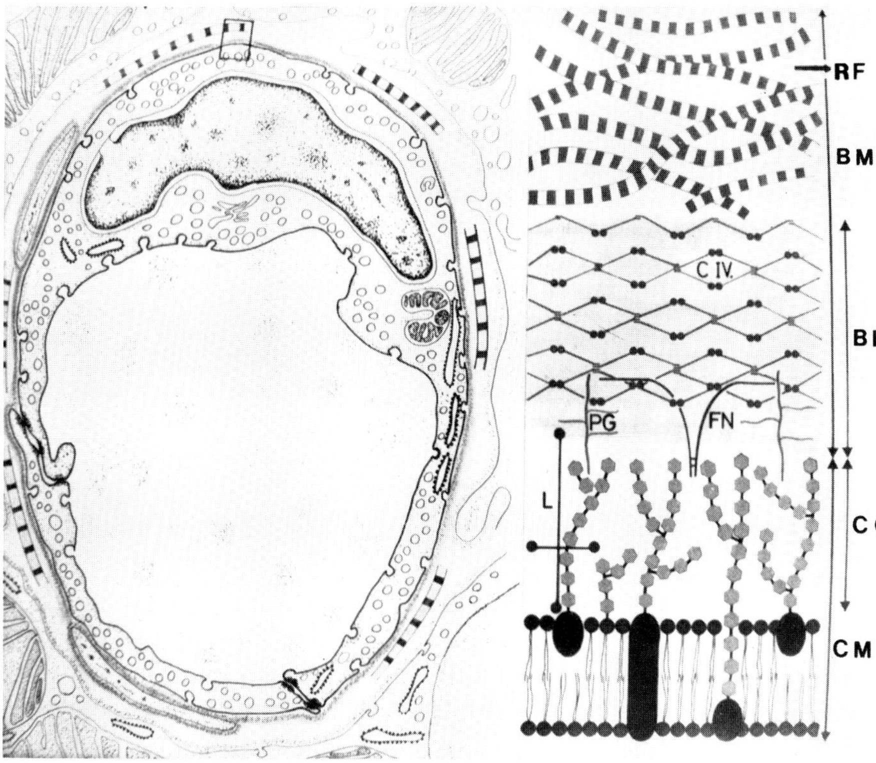

FIGURE 1. The diagram illustrates the structural differences between basement lamina (BL) and "classical" basement membrane (BM). On the left side, a blood capillary is drawn modified after Kristic.[2] The framed area (on the top) is enlarged on the right side. CM: cell membrane; CC: cell coat containing branching oligosaccharides of different membrane glycoproteins. The basal lamina (BL) contains, among other things, collagen type IV (CIV) and proteoglycans (PG). Collagen type IV is bound to the cell membrane by laminin (L) and in some cases by fibronectin (FN). Collagen fibrils (generally reticular fibers (RF) containing type III collagen molecules) are attached to the outer surface of the BL. The use of the term basement membrane (BM) in the classical sense denotes BL + RF.

multifunctional macromolecule, which can influence cell adhesion and migration, cell differentiation (e.g., neutrite outgrowth), and tissue development.[18] Laminin is a cross-shaped 900-kDa protein containing three polypeptide chains linked to each other by disulfide bonds.[11, 16, 19] As revealed by rotary shadowing technique, the elongated cross-form of laminin has three similar 37-nm short arms and a 77-nm-long arm.[20] The short arms consist of two globular domains, while the long arm terminates in a globular end. The globular regions of the short arms can be bound to collagen type IV, the major cell-binding domain is present near the intersection of the short arms. The globular domain of the long arm is the heparin- and heparan sulfate-binding site (this domain is a potent promoter of axonal growth.[19-22] Laminin preferentially binds to collagen type IV compared to other collagen types, and the laminin-binding sequence of the collagen type IV molecule is locate about 140 nm from the NC1 domain.[23, 24] The cell surface receptors for laminin belong to the family of integrins, but other laminin-binding proteins have also been shown.[11, 25, 26] Electron microscopic investigations show that laminin is distributed throughout the lamina rara and lamina densa of the BM.[27-29]

Nidogen is a 150-kDa dumbbell-shaped protein.[30, 31] It is suggested that nidogen is structurally similar to entactin, a sulfated glycoprotein of basement membranes.[7, 32, 33] Nidogen was shown to form nestlike self-aggregates,[30] and complexes with laminin.[7, 34] Laminin has the ability to self-aggregate in the presence of Ca^{2+}.[35] The hugh laminin-nidogen complexes interact with collagen type IV and heparan sulfate proteoglycans,[36-38] forming an intricate supramolecular structure in the BM.

The basophilic staining pattern of BMs proved to be due to the presence of heparan sulfate (HS) proteoglycans. Two distinct HS-PGs have been identified: small and large forms.[39, 40] The small HS-PG has an M_r of 130,000, and its core protein is substituted by heparan sulfate chains (M_r 29,000) of 32 nm length.[39] Similar PG molecules have been isolated from glomerular basement membranes of the kidney.[41] Small HS-PGs have been shown to bind to laminin, fibronectin, and collagen type IV.[39] The interactions are mediated by the HS chains. The large HS-PG has a 500-kDa protein core made up of a single polypeptide folded into six globular domains.[42] Three HS chains are emerging from one of the terminal globules, and their length varies between 100 and 170 nm.[42] The large HS-PG is supposed to have a number of additional interactions compared to the small HS-PGs. These additional interactions seem to be mediated by the protein core.[42] The protein core of the large HS-PG includes EGF (epidermal growth factor)-like repeats alternating with globular domains and repeats of disulfide-bonded loops similar to the cell adhesion molecule, N-CAM, comprising peptide loops typical for the immunoglobulin-gene superfamily.[11, 42, 43] Like the small HS-PGs, large HS-PGs are also detected in the glomerular basement membrane.[44]

Other proteoglycans and glycoproteins have also been identified in different BMs (for review see Paulsson[7] and Timpl[11]). Among these, a 40-kDa protein is mentioned, named BM-40 or SPARC (secreted protein which is acidic and rich in cysteine). It is a calcium-binding protein, and it is produced in high amounts by cells which synthesize BM components but is also secreted by other cells.[45, 46] The sequence of SPARC proved to be highly homologous to that of bovine osteonectin.[47] Osteonectin is supposed to be a specific bone protein with binding activities for calcium, hydroxyapatite, and collagen.[48]

The kidney glomeral basement membrane (GBM) is a peculiar structure. It is formed during differentiation by fusion of two basement membranes. In the kidney glomerulus of mammals, the capillary wall has the following layers: fenestrated endothelium; basement membrane comprising lamina rara interna, lamina densa, and lamina rara externa; and epithelium (foot processes of podocytes). The endothelial and epithelial basement membranes were fused together forming the trilaminated structure of the GBM in differentiated kidney. Therefore, GBM is thicker than other BMs, and at the periphery of the glomerular capillary loops where filtration is assumed to occur reticular fibers are never attached to its outer surface. Thus, GBM is a real basement lamina.

The GBM is specialized for the filtration of blood producing protein-free filtrate of the blood plasma. The glomerular filter has been shown to have both size-selective and charge-selective properties, i.e., an increase in size and negative charge density of macromolecules results in their increased retention in the blood plasma.[49] The filtration barrier function of the GBM is due to the presence of several molecular constituents. The cell coats both of endothelial and epithelial cells with their fixed negative charges (sialic acid residues of different membrane glycoproteins) and the cell membrane of podocytes, including the so-called filtration slit diaphragms which interconnect neighboring foot processes, are definitely involved in the barrier function.[50, 51] HS-PGs are localized electronmicroscopically mainly in the laminae rarae.[52-57] The permeability of GBM is considerably increased if heparan sulfate is removed enzymatically.[55, 58] It has also been shown that cationic stains revealed a reduced binding to anionic charges in the laminae rarae in experimental animals with glomerular disease associated with proteinuria.[59] These data suggest that HS-PGs may play a role in the barrier funciton of the GBM.

Collagen type IV is located in the lamina densa.[60] The presence of collagen type V, which was formerly supposed[61] to be in GBM is not justified.[62] Laminin appears localized in laminae rarae,[27-29, 63, 64] and entactin is found predominantly in the lamina rara interna.[63] Fibronectin has also been identified by immunoelectron microscopical methods showing either diffuse localization[60] or preferential localization in the laminae rarae.[65] An interconnection between the cytoskeleton of podocytes and GBM matrix via laminin/fibronectin receptor has been proposed.[66]

Very little is known about the spatial orientation pattern of the constituents of GBM. Electron micrographs taken after staining with cationic probes have suggested a regular distribution of PGs along the cell membranes.[52, 53, 55] Polarization microscopic observations made after picrosirius red staining indicate an oriented arrangement of the collagen in GBM.[67] In contrast, no birefringence of GBM could be demonstrated after the same staining technique.[68, 69] We decided, therefore, to undertake qualitative and quantitative polarization microscopical investigations after different topo-optical reactions. Topo-optical reactions of GBM are compared to those of peritubular (TBM) and Bowman's capsule basement membranes (BBM). Reticular fibers are attached to TBM and BBM but not to the GBM.

III. TOPO-OPTICAL INVESTIGATIONS OF KIDNEY BASEMENT MEMBRANES

CFY rats weighing 100 to 150 g were used for the experiments. Under narcosis with Nembutal (SERVA, Heidelberg), a laparatomy was performed, the aorta was isolated, and a polyethylene catheter was introduced below the renal arteries. A ligature was tied above the renal arteries. Open-circuit perfusion was carried out by making an opening in the renal veins. The renal vasculature was cleared of blood by 3- to 5-min perfusion of Ringer's solution. Following the washing out of blood from the vasculature, 15 ml Karnovsky's fixative was perfused. Kidneys were removed, and 2 to 3 mm transversal slices were cut and further fixed in Karnovsky's fixative overnight. After washing in 0.1 mol/l cacodylate buffer (pH 7.2), tissue samples were dehydrated, embedded in wax, and 6- to 8-μm sections were cut.

Other groups of animals were operated on under narcosis, and hypertension was produced by partially constricting the left renal artery with a Goldblatt's silver wire clip. The systolic blood pressure was determined according to Byrom[70] in every second day both in operated and sham-operated control animals. The animals were sacrificed 3 weeks after the operation when the blood pressure of the operated animals varied between 180 and 210 mmHg (blood pressure of the control animals was between 80 and 110 mmHg). The tissue samples were processed as mentioned before. Left and right kidneys of the operated animals were designated by G+ (ischemic kidney with Goldblatts clip) and G− (nonischemic kidney), respectively.

The following topo-optical reactions were carried out: picrosirius red F3B staining without and after methylation for collagen, toluidine blue metachromatic reaction at pH 4.5 without and after carboxymethylation with methyliodide for GAGs, and aldehyde-bisulfite-toluidine blue (ABT) without and after a mild alkaline hydrolysis (0.5% KOH in 70% ethanol at room temperature for 30 min). The reactions are described in Chapter 12.

To specify polyanionic structures stained by toluidine blue, sections were digested with one of the following three enzymes prior to staining: α-neuraminidase (SERVA, Heidelberg) (2 U enzyme in 1 ml 0.05 mol/l acetate buffer containing 0.04 mol/l $CaCl_2$, pH 5.5, 37°C, 1 and 2 h); heparitinase (Seikagaku Kogyo, Tokyo) (5 U in 1 ml 0.5 mol/l Tris-HCl buffer pH 7, 37°C, 2h); testicular hyaluronidase (Reanal, Budapest), (0.5 mg enzyme in 1 ml 0.1 mol/l phosphate buffer pH 6, 37°C, 18h).

Stained sections were evaluated under a polarization microscope (Zetopan-pol, Reichert, Vienna) equipped with λ/10 Brace-Köhler compensator and metal interference filters. Retardation values of birefringence were measured at $\lambda = 589$ nm, and they were plotted in

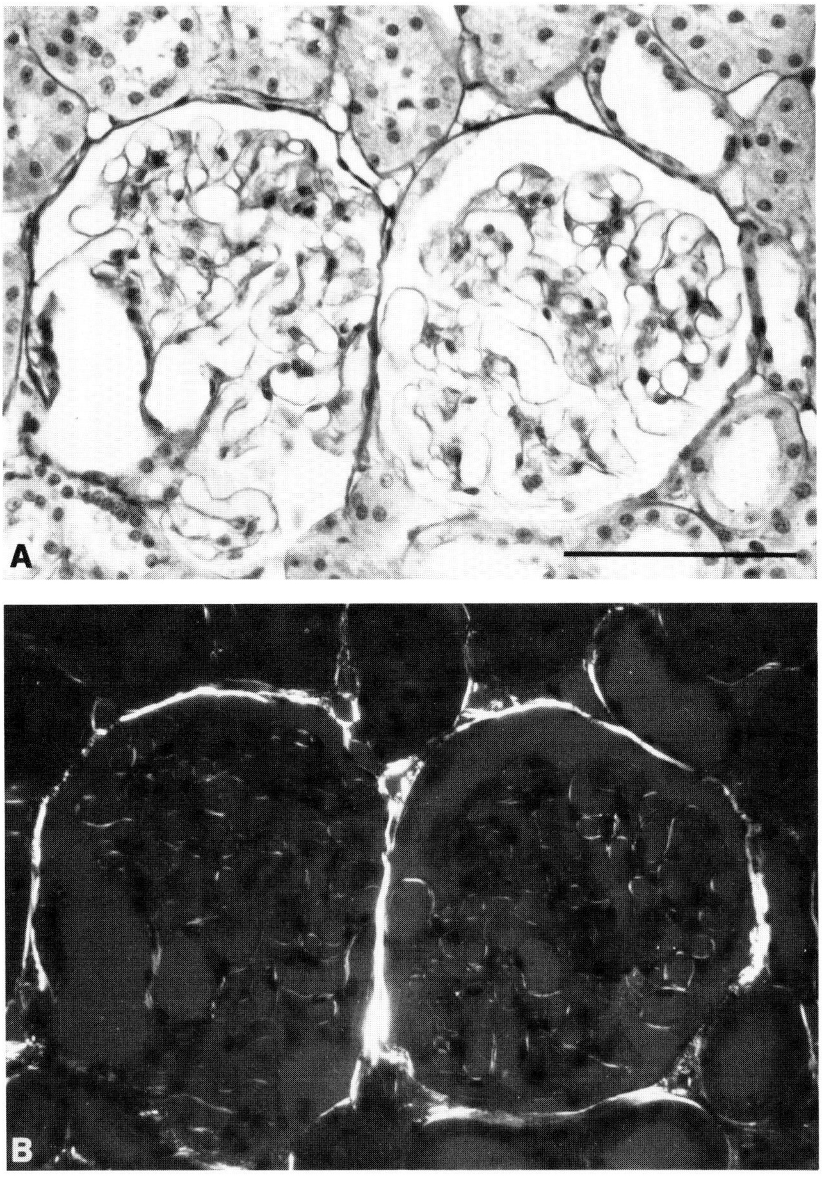

FIGURE 2. Kidney section of CFY rat after picrosirius red F3B staining taken in normal (A) and polarized light (B). Observe the moderately birefringent GBMs. Bowman's capsule BMs (BBMs) and tubular BMs (TBMs) are strongly birefringent. Scale: 100 μm.

histograms. Retardation values were determined in GBM, BBM, and TBM of kidneys of the untreated animals, and in GBM of operated animals. Histograms were compared with the ''ambiguity'' (A)-test (see Chapter 12).

A. COLLAGEN

Kidney basement membranes are birefringent after picrosirius red F3B staining. Intense optical anisotropy can be observed in BBM and TBM, while moderate birefringence is seen in GBM (Figure 2). The sign of birefringence is positive, relating to the tangents in each BMs studied (Figure 3). Considering the axiparallel binding of the sirius red molecules to collagen, it follows that collagen is tangentially oriented in these basement membranes.

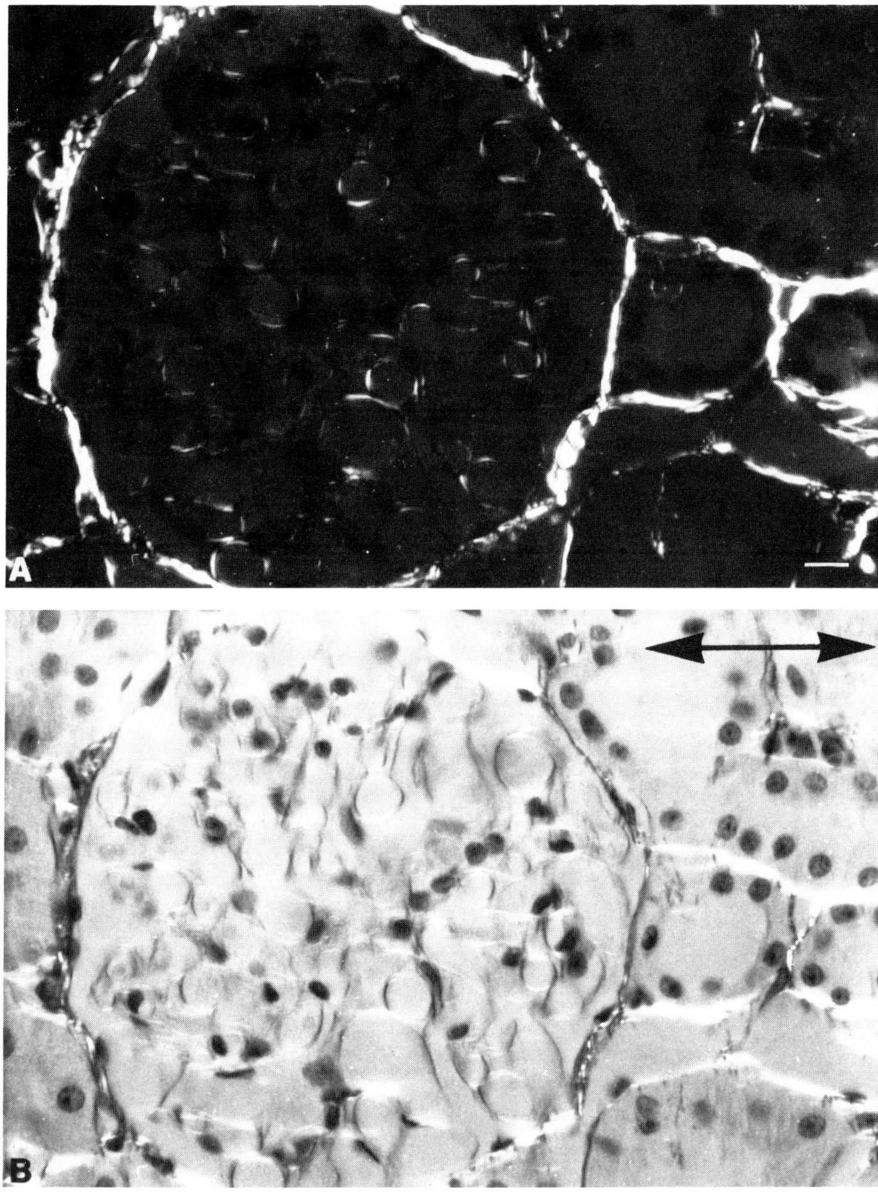

FIGURE 3. Kidney of CFY-rat after staining with picrosirius red F3B taken in polarized light without (A) and after compensation (B). The arrow indicates the direction of slow axis of the compensator. Note that several glomerular capillary walls cut transversely appear as Maltese crosses in A suggesting a possible circular orientation of the collagen. The appearance of dark segments of GBM, BBM, and TBM after compensation in the same (north-south) direction indicates that collagen is tangentially arranged in all of these basement membranes. Scale: 10 μm.

Using the same technique in their investigations, Junqueira et al.[67] have suggested that collagen is oriented in the GBM. These authors, however, did not describe the preferential direction of the orientation.

The retardation values measured in different basement membranes of the kidney significantly increase in the following sequence: GBM-TBM-BBM (Figure 4). This means that the most compact collagen structure characterizes the BBM, while GBM contains a looser collagen structure compared to TBM and BBM. It is known that collagen fibrils are attached

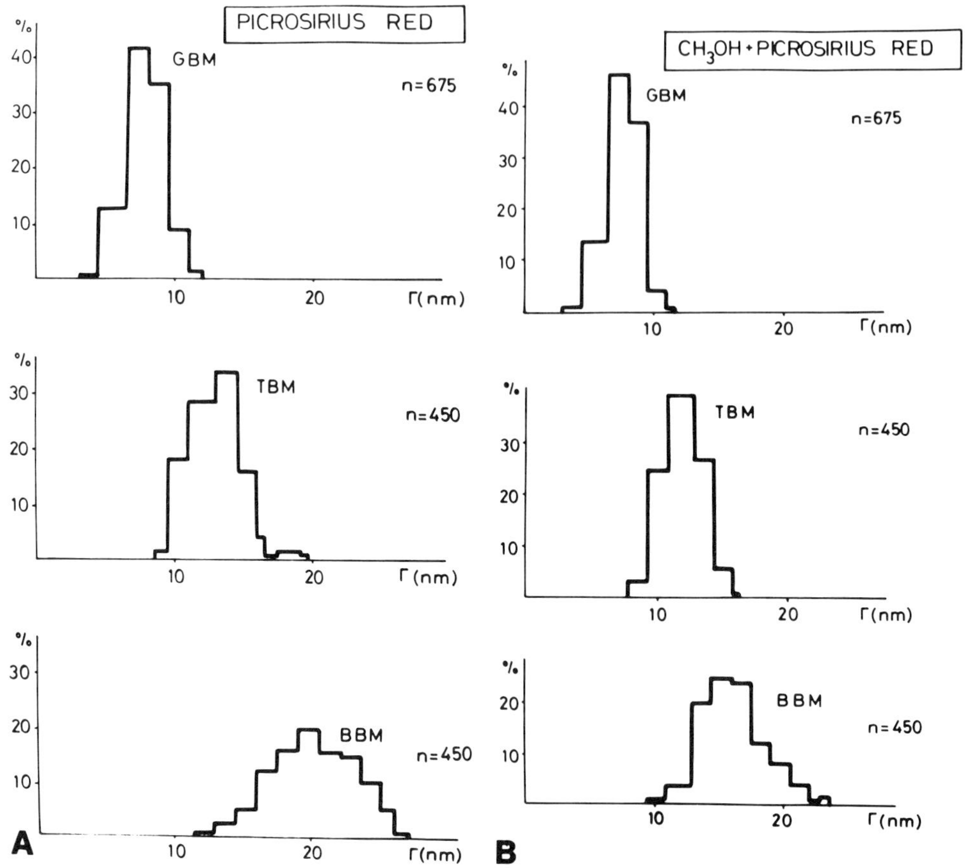

FIGURE 4. Retardation values determined in GBM, TBM, and BBM stained with picrosirius red without (A) and after methylation (B). Antecendent methylation does not result in an increase in retardation values. Lowest values were measured in the GBM and highest retardation values were recorded from the BBM.

to the TBM and BBM, but GBM contains only aggregated collagen type IV molecules and not highly oriented fibrils.

The fact that GBM is birefringent strongly suggests that aggregated collagen type IV molecules are not randomly distributed throughout the GBM. Taking into account the facts that these molecules form tetramers by interactions in the 7S domain and aggregates of neighboring tetramers are also formed, we must suppose, based on polarization microscopic obervations, that the long arms between the 7S and NC1 domains form an angle with the tangent of GBM of less than 45°. If these angles were higher, we could never observe a positive sign of birefringence after picrosirius red staining; in this case the sign would be negative relating to the tangent. In the case of a random distribution of collagen type IV molecules-tetramers, no optical anisotropy could be observed because nonoriented collagen fails to exhibit any birefringence.[67] It is reasonable to suppose that the end-to-end interactions of the collagen type IV molecule-tetramers explain why these collagen micellae are oriented along the plane of the GBM and not transversely. If glomerular capillaries are cross-sectioned, the optical anisotropy of their walls frequently appears in the form of Maltese crosses (Figure 3). This optical phemenon suggests that the collagen micellae may assume a predominantly circular orientation around the capillary lumen in the GBM.

No significant differences are found in retardation values between non methylated and methylated sections (Figure 4). It has been shown in Chapter 4 that the intensity of staining and birefringence is considerably increased if picrosirius red staining is carried out after a

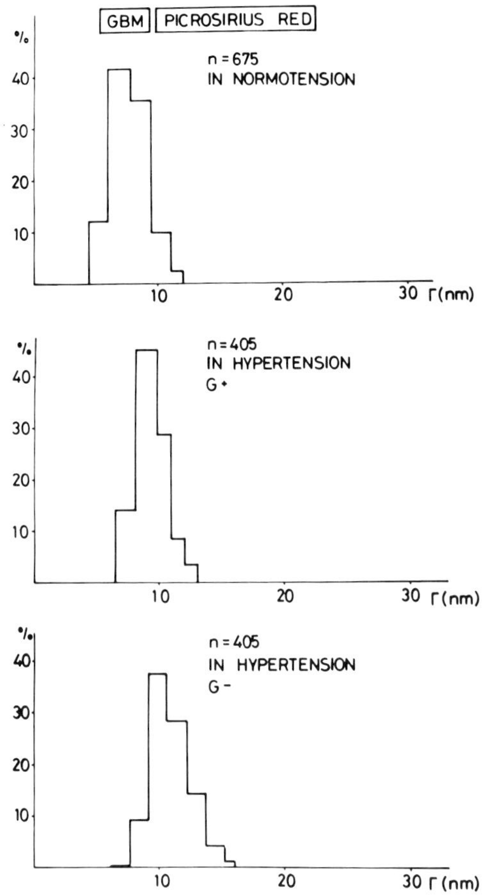

FIGURE 5. Retardation values of birefringence induced by picrosirius red in GBM of control and hypertensive (G + , G −) rats. The retardation values are higher in non-ischemic kidney (G −) (where the renal artery was not constricted by Goldblatt's clip) than in controls.

methylation or a digestion with testicular hyaluronidase in hyaline cartilage. The explanation of this phenomenon is as follows. Methylation blocks the carboxyl groups and removes the sulfate groups of different constituents (see Chapter 12), hereby decreasing the repulsive forces of tissue anionic groups against the anionic sirius red molecules. Testicular hyaluronidase has the same effect. PGs and their GAGs are in close contact with collagen fibrils in hyaline cartilage. Seemingly it is not the case in the GBM.

Similar retardation values are obtained after picrosirius red staining in GBMs of kidneys of sham-operated animals and in GBMs of ischemic kidneys (where the renal artery was partially constricted) (Figure 5). In contrast, a significant increase is detected (A = 0.4356) in anisotropy of GBMs in the non ischemic kidneys of the hypertensive rats compared to the controls (Figure 5). At this moment, we do not know the reason for this alteration. An augmented deposition of collagen type IV and/or molecular structural changes leading to an increased degree of collagen orientation may result in an increased birefringence induced by picrosirius red staining. Collagen type IV has been shown to accumulate in thickened glomerular capillary walls of patients with hypertensive syndromes of pregnancy.[71]

B. GLYCOSAMINOGLYCANS

Cationic dyes such as toluidine blue stain GBM at a pH lower than 4.5, but the optical anisotropy is rather weak if the staining is carried out at pH 2.5 to 3.5. Polarization microscopic measurements could not be carried out after these stainings. Therefore, we used our staining solution at a pH value (pH 4.5) which does not guarantee a specific reaction of sulfated GAGs, except when it is combined with an antecedent carboxymethylation. Using this combination, GBM, TBM, and BBM stain metachromatically. Strong birefringence is observed in BBM, and moderate optical anisotropy is found in GBM and TBM. Both staining and birefringence are abolished by a digestion with heparitinase carried out prior to the topo-optical reaction. Antecedent digestions with neuraminidase and testicular hyaluronidase, on the other hand, do not modify essentially the staining reaction, suggesting that predominantly heparan sulfate molecules are responsible for the dye uptake. If the toluidine blue staining is carried out at pH 4.5 without antecedent carboxymethylation, the staining and birefringence of GBM is somewhat decreased by a digestion with neuraminidase, but the reactions of TBM and BBM remain practically unaffected. It suggests a higher neuraminic acid content in GBM compared to the other two basement membranes.

The toluidine blue staining induces a birefringence with a negative sign in the kidney basement membranes (Figure 6). Toluidine blue molecules are perpendicularly bound to GAGs (see Chapter 3). The relative negative sign of birefringence related to the tangent of the basement membranes, therefore, indicates a tangential arrangement of heparan sulfate molecules in the basement membranes. A periodically distributed PG structure has been shown electron microscopically after labeling with cationized ferritin in laminae rarae of the GBM.[52, 53, 55] Using other cationic probes (safranin O and alcian blue), however, a rather random distribution of the PG particles has been demonstrated with electron-microscopic investigations of the GBM.[57] Our optical findings argue for a spatially oriented deposition of the PGs.

According to our quantitative data, the most compact heparan sulfate structure seems to be present in the BBM as revealed by the high retardation values measured in this structure (Figure 7). An alternative possibility for explaining this result is that the BBM contains a highly oriented collagen structure (Figure 4). This is probably the reticular fiber system, the reaction of which cannot be distinguished light microscopically from that of the basal lamina. The proteoglycans and their GAG side chains bound to these oriented reticular fibers may assume a higher degree of orientation compared to GAGs in GBM where no fibrillar collagen is present.

There is no significant difference in retardation values between GBM and TBM, but retardation values of BBM are significantly higher than those measured in TBM after toluidine blue staining (Figure 7). This result suggests that TBM is either poorer in GAGs than BBM or the GAGs of TBM form a more moderately oriented system.

The intensity of birefringence is significantly higher in the GBMs of the hypertensive rats than in the GBMs of the control animals (Figure 8A). We conclude that a more compact PG structure is formed in the GBM of the hypertensive rats.

C. PERIODATE-REACTIVE CARBOHYDRATES

ABT and KOH + ABT reactions induce optical anisotropy in kidney basement membranes and brush borders of the proximal tubules (Figure 6). The relative sign of birefringence is negative relating to the tangents of the basement membranes. It means that oligo- and polysaccharides containing periodate-reactive monosaccharides are tangentially arranged in the basement membranes. The retardation values of birefringence increase in the kidney basement membranes in the following sequence: GBM-TBM-BBM (Figure 9). No differences are found in the intensity of birefringence of GBM between normotensive and hypertensive rats (Figure 8B).

ABT reaction, like the classical PAS reaction, detects carbohydrates in which vicinal −OH groups are present[72, 73] (see Chapter 12). Antecendent, mild alkaline hydrolysis transforms −O-acyl substituted groups of carbohydrates to −OH groups.[74] If these newly formed

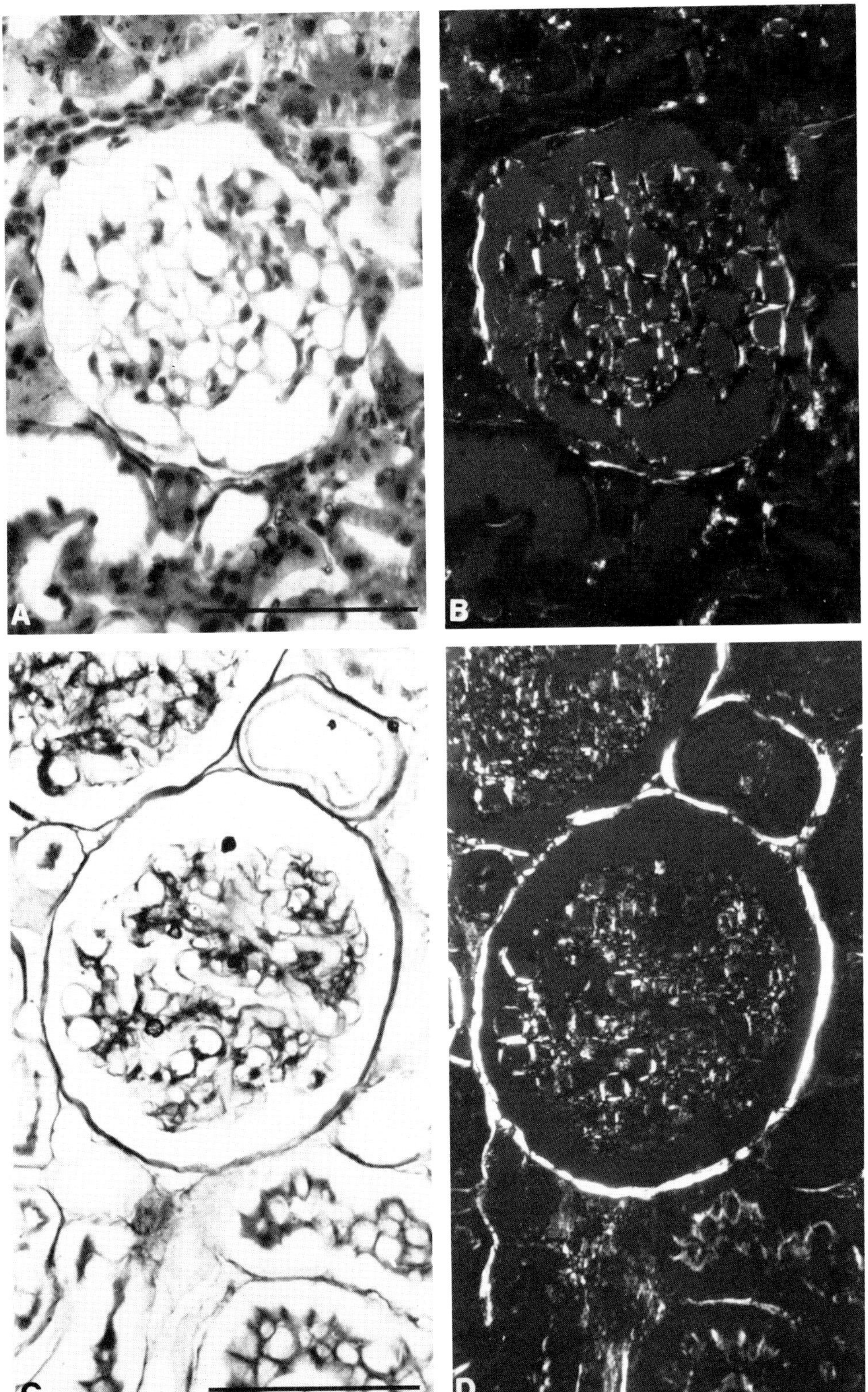

FIGURE 6. Kidney sections of CFY-rats after toluidine blue staining at pH 4.5 (A, B) and ABT reaction (C, D) taken in normal (A, C) and polarized light (B, D). Note the birefringence of the GBMs in (B) and (D). The optical anisotropy of BBM and TBM is weaker after toludine blue staining than after ABT reaction. The brush border of the proximal tubules in strongly stained and birefringent after ABT reaction (C, D). Scales: 100 μm.

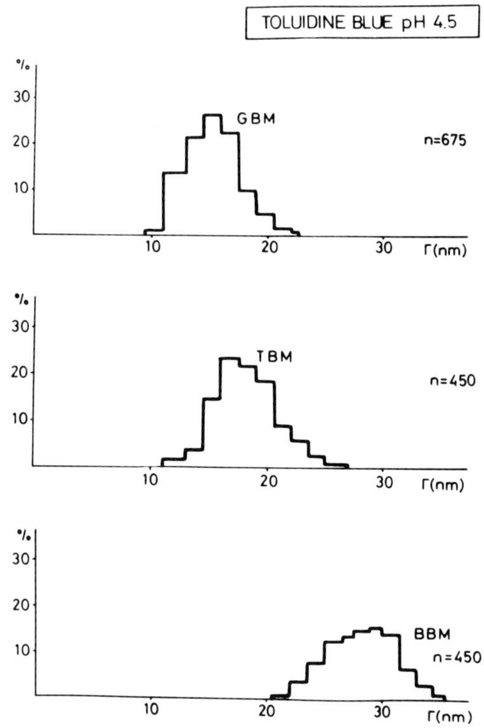

FIGURE 7. Retardation values measured after toluidine blue staining. No significant difference is found between GBM and TBM, but the birefringence of the BBM is significantly stronger than that of the TBM and GBM.

−OH groups are in the vicinity of other alcoholic hydroxyl groups, the intensity of the ABT reaction is increased because of the increased number of aldehydes generated by the periodic acid.

Although the mechanism of the ABT reaction is well understood, the interpretation of this topo-optical reaction is difficult. The same holds true for the PAS reaction. This difficulty is due to the simple fact that monosaccharides containing vicinal −OH groups may occur in a number of different glycoproteins. In the glomerular capillary wall, for example, different carbohydrate constituents of the cell coat, laminin, nidogen, and collagen type IV can be oxidized by periodic acid, i.e., can give the ABT reaction. Therefore, the ABT reaction cannot be used to specify an extracellular macromolecule.

IV. SUPRAMOLECULAR ORGANIZATION: SPECULATIVE MODELS

Our analyses in polarized light resulted in two points concerning the ultrastructure of the GBM: (1) the collagen is tangentially oriented; (2) the heparan sulfate molecules are arranged parallel to the tangent of the cell membranes. Based on these findings, a speculative model of the GBM structure can be constructed if we accept the following premises: (1) collagen type IV molecules form aggregates and they are located in the lamina densa; (2) heparan sulfate PGs are located in laminae rarae; (3) laminin interconnects collagen type IV molecules with cell membranes (Figure 10).

According to this model, the heparan sulfate side chains form an axiparallel orientation to the cell membranes both of endothelial and epithelial cells (podocytes). The core proteins

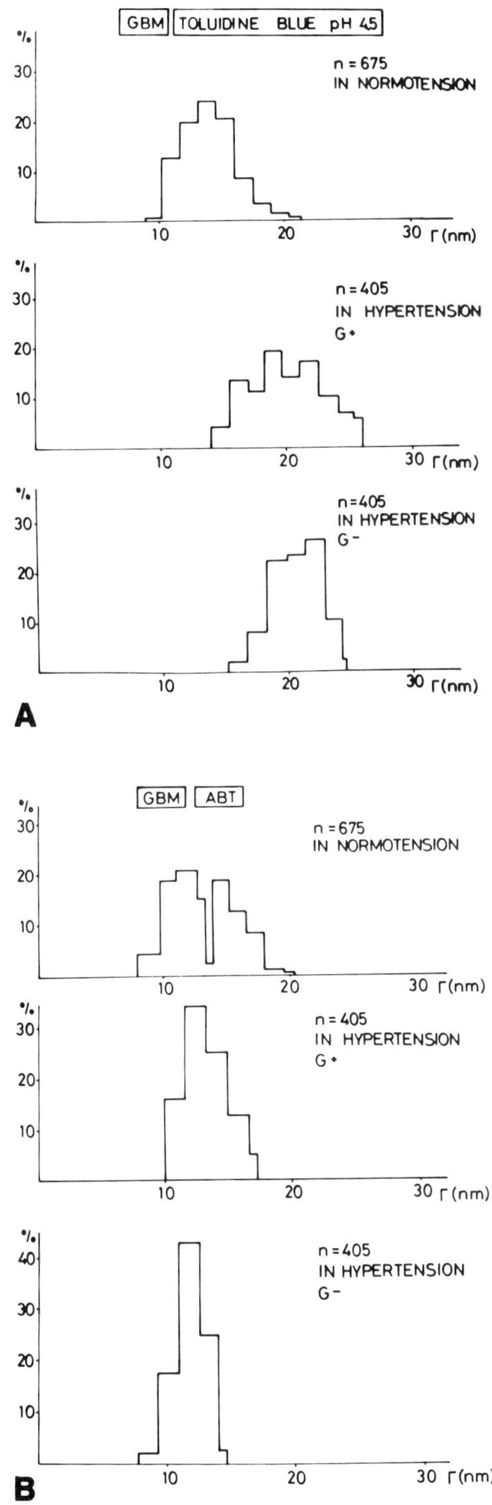

FIGURE 8. Retardation values of birefringence induced by toluidine blue (A) and ABT reactions (B) in glomerular basement membranes of normotensive and hypertensive rats. Significantly higher values were measured in GBMs of hypertensive rats than in those of the controls after toluidine blue staining. A-values: 0.5253 (between normotensive and G+ hypertensive groups) and 0.3411 (between normotensive and G− hypertensive groups). G+ ischemic kidneys with Goldblatt's clip, G−: non-ischemic kidneys. ABT reaction did not reveal any significant difference between control and experimental groups.

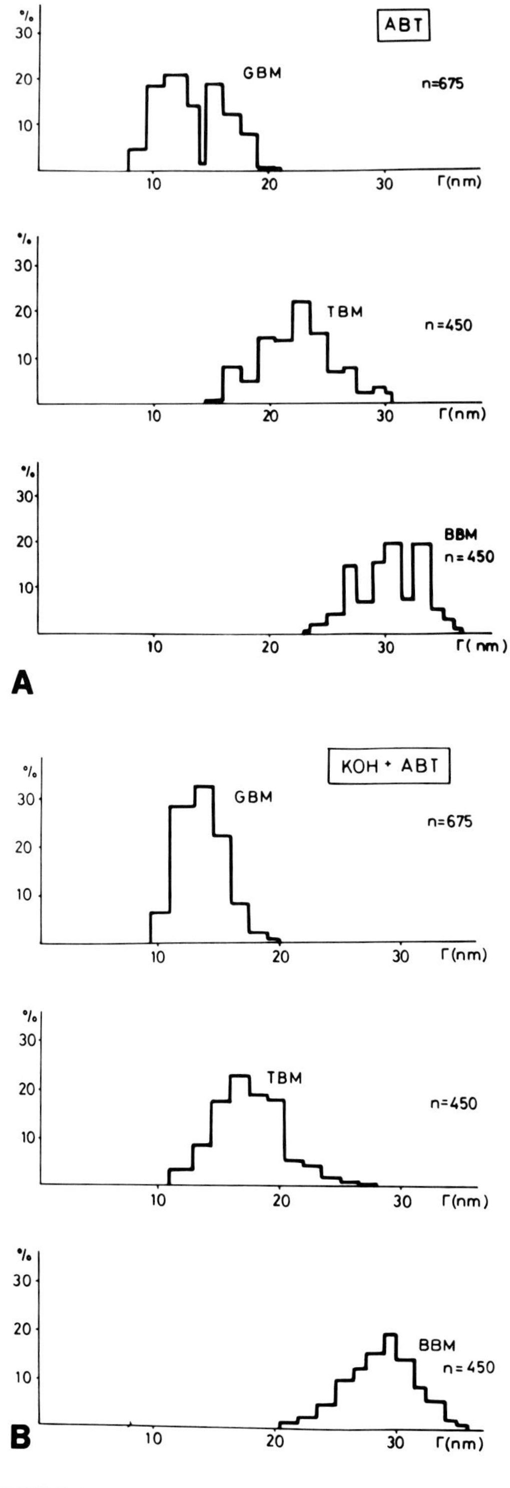

FIGURE 9. Retardation values determined in three different kidney basement membranes after ABT (A) and KOH + ABT reactions (B). An almost normal distribution of retardation values is seen after KOH + ABT. Lowest intensity of birefringence is seen in GBM. Both TBM and BBM exhibit significantly stronger birefringence than GBM. Retardation values recorded from the BBM are significantly higher than those of the GBM.

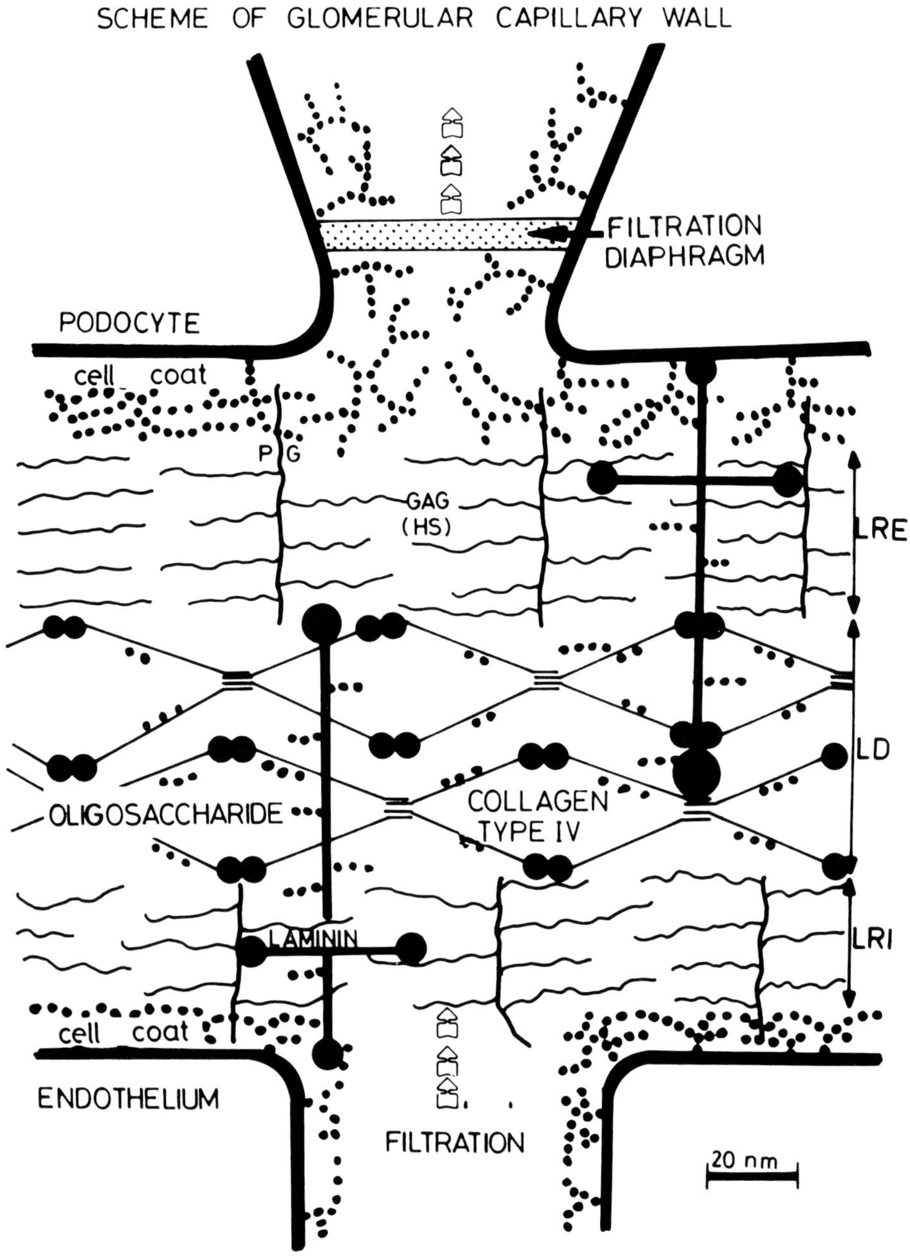

FIGURE 10. A speculative scheme of the glomerular capillary wall. Our polarization microscopic findings suggest that 1. collagen type IV micellae are tangentially arranged in the lamina densa (LD); 2. heparan sulfate (HS) side chains of the PGs located in lamina rara externa (LRE) and interna (LRI) are oriented more or less parallel to the tangent of the cell membranes; 3. different oligosaccharides of different glycoproteins (dots) also assume a spatial orientation. The molecular dimensions are not proportionally depicted. Location of laminin is speculative.

of the HS-PGs are supposed to be perpendicularly arranged to the cell membranes. Supposedly, this oriented PG structure is a crucial morphological basis of the filtration barrier. The collagen type IV molecules and their aggregates are arranged so that the long arm of the molecules or at least their great majority, form an angle less than 45° with the tangent of the basement membrane, otherwise we could not obtain a positive sign of birefringence

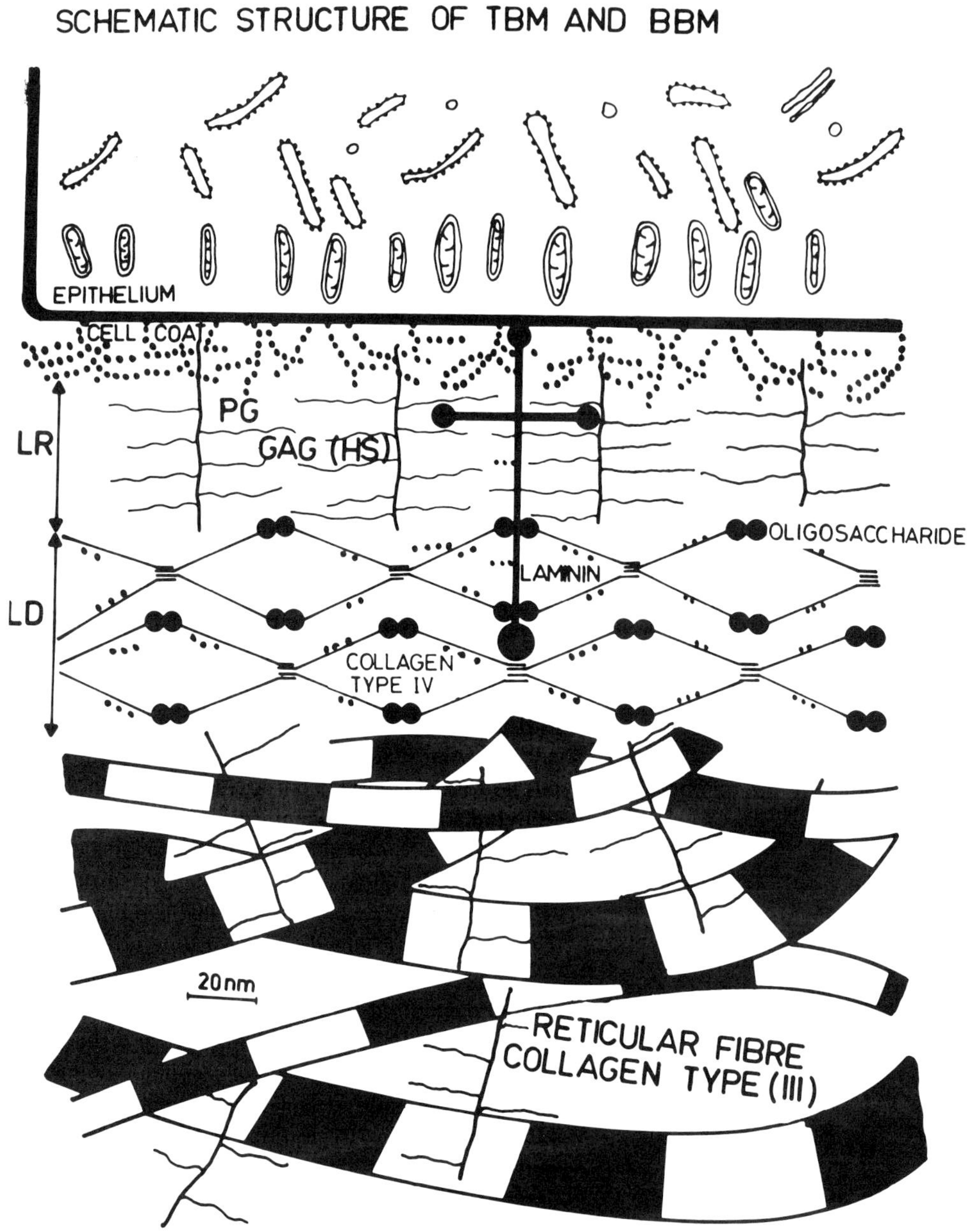

FIGURE 11. A speculative model of TBM and BBM structure. Reticular fibers with associated PGs are attached to the outer surface of the basement lamina. This explains the higher retardation values of TBMs and BBMs measured after different topo-optical reactions compared to those obtained in GBMs.

induced by picrosirius red staining. It is tempting to believe that laminin (and associated glycoproteins like nidogen) stabilize this oriented micellar system of the collagen. It is surmised that oligosaccharides of different cell coat and structural glycoproteins are also oriented.

The structure of TBM and BBM may be similar except that ''reticular'' fibers with associated PGs are attached to the outer surface of these basement laminae (Figure 11). The interpretation of optical findings obtained in TBM requires further consideration (Figure 12). If collagen and GAG would assume a preferential circular or longitudinal orientation in

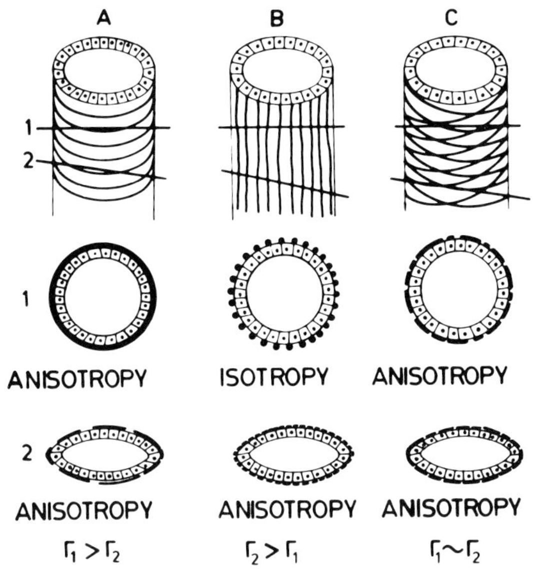

FIGURE 12. Theoretical possibilities to explain the birefringence of TBM. (A) If a structure (e.g., collagen) is circularly oriented around the tubular epithelial cell layer, in cross section a strong anisotropy can be observed (A1); in sections cut obliquely a much more moderate birefringence is seen (A2). Isotropy would appear in longitudinal section (not shown). (B) If there is a longitudinally oriented system in the tubular wall, isotropy is found in cross section (B1) and anisotropy is seen in oblique section (B2). Maximal anisotropy would appear in longitudinal section (not shown). (C) If linear macromolecules exhibit a spiral course in the wall of the tubule, anisotropy is seen in transversal (C1), oblique (C2) and longitudinal sections (not shown). This seems to be the case in BM of the renal tubules.

TBM, the birefringence obtained after relevant topo-optical reactions should change with the cutting plane. This is not the case. Peritubular basement membranes exhibit optical anisotropy in longitudinal, transversal, and also in oblique sections. It is supposed that both collagen and GAGs assume a spiral course in this basement membrane.

It is suggested that we have to reevaluate the generally believed assumption that the basement membrane is an amorphous structure. It is also clear, however, that much further work needs to be done to characterize this oriented system and its functional implications.

ACKNOWLEDGMENTS

The topo-optical investigations reported in this chapter were initiated by my friend and colleague, Miklós Szokol, M.D., Ph.D. (Associate Professor, Department of Pathology, University Medical School, Debrecen) who passed away suddenly in his most active years. The animal experiments were carried out and kidneys of rats suffering from hypertension were kindly provided by him. I am grateful to my former students, Drs. E. Tóth-Pál, Cs. Papp, and Katalin Kiss for their inestimable help in performing the polarization microscopical measurements.

REFERENCES

1. **Hay, E. D., Hasty, D. L., and Kiehnau, K. L.,** Fine structure of collagens and their relation to glycosaminoglycans (GAG), in *Collagen-Platelet Interaction,* Gastpar, H., Kuhn, K., and Marx, R., Eds., Schattauer, Stuttgart, 1978, 129.
2. **Kristic, R. V.,** *Ultrastruktur der Säugetierzelle,* Springer-Verlag, Berlin, 1976.
3. **Kefalides, N. A.,** *Biology and Chemistry of Basement Membranes.* Academic Press, New York, 1978.
4. **Bernfield, M.,** *Basement Membranes and Cell Movement,* Ciba Found. Symp. 108, Pitman, London, 1984.
5. **Wolff, J. R., Sievers, J., and Berry, M., Eds.,,** *Mesenchymal-Epithelial Interactions in Neural Development,* Springer-Verlag, Berlin, 1987.
6. **Martinez-Hernandez, A. and Amenta, P. S.,** The basement membrane in pathology, *Lab. Invest.,* 48, 656, 1983.
7. **Paulsson, M.,** Noncollagenous proteins of basement membranes, *Coll. Relat. Res.,* 7, 443, 1987.
8. **Abrahamson, D. R.,** Recent studies on the structure and pathology of basement membranes, *J. Pathol.,* 149, 257, 1986.
9. **Tryggvason, K., Höyhtyä, M., and Salo, T.,** Proteolytic degradation of extracellular matrix in tumor invasion, *Biochim. Biophys. Acta,* 907, 191, 1987.
10. **Kefalides, N. A. and Alper, R.,** Structure and organization of macromolecules in basement membranes, in *Collagen, Biochemistry and Biomechanics,* Vol. II., Nimni, M. E., Ed., CRC Press, Boca Raton, FL, 1988, 73.
11. **Timpl, R.,** Structure and biological activity of basement membrane proteins, *Eur. J. Biochem.,* 180, 487, 1989.
12. **Kleinman, H. K., McGarvey, M. L., Hassel, J. R., and Martin, G. R.,** Formation of a supramolecular complex is involved in the reconstitution of basement membrane components, *Biochemistry,* 22, 4969, 1983.
13. **Grant, D. S., Leblond, C. P., Kleinman, H. K., Inoue, S., and Hassel, J. R.,** The incubation of laminin, collagen IV, and heparan sulfate proteoglycan at 35°C yields basement membrane-like structures, *J. Cell Biol.,* 108, 1567, 1989.
14. **Risteli, J., Bachinger, H. P., Engel, J., Furthmayr, H., and Timpl, R.,** 7-S collagen: characterization of an unusual basement membrane structure, *Eur. J. Biochem.,* 108, 239, 1980.
15. **Timpl, R., Wiedemann, H., Van Delden, V., Furthmayr, H., and Kühn, K.,** A network model for the organization of type IV collagen molecules in basement membranes, *Eur. J. Biochem.,* 120, 203, 1981.
16. **Yurchenco, P. D. and Furthmayr, H.,** Self-assembly of basement membrane collagen, *Biochemistry,* 23, 1839, 1984.
17. **Timpl, R., Rohde, H., Robey, P. G., Rennard, S. P., Foidart, J. M., and Martin, G. R.,** Laminin — A glycoprotein from basement membranes, *J. Biol. Chem.,* 254, 9933, 1979.
18. **Kleinman, H. K., Cannon, F. B., Laurie, A. W., Hassel., J. R., Aumailley, M., Terranova, V. P., Martin, G. R., and Dubois-Daecq, M.,** Biological activities of laminin, *J. Cell. Biochem.,* 27, 317, 1985.
19. **Timpl, R., Rohde, H., Risteli, L., Ott, U., Gehron Robey, P., and Martin, G. R.,** Laminin, *Methods Enzymol.,* 82, 831, 1982.
20. **Engel, J., Odermatt, E., Engel, A., Madri, J. A., Furthmayr, H., Rohde, H., and Timpl, R.,** Shapes, domain organizations, and flexibility of laminin and fibronectin, two multifunctional proteins of the extracellular matrix, *J. Mol. Biol.,* 150, 97, 1981.
21. **Terranova, V. P., Rao, C. N., Kalebic, T., Margulies, I. M., and Liotta, L. A.,** Laminin receptor on human breast carcinoma cells, *Proc. Natl. Acad. Sci. U.S.A.,* 80, 2144, 1983.
22. **Edgar, D., Timpl, R., and Thoenen, H.,** The heparin-binding domain of laminin is responsible for its effects on neurite outgrowth and neuronal survival, *EMBO J.,* 3, 1463, 1984.
23. **Woodley, D. T., Rao, C. N., Hassell, S. R., Liotta, L. A., Martin, G. R., and Kleinman, H. K.,** Interactions of basement membrane components, *Biochem. Biophys. Acta,* 761, 278, 1983.
24. **Charonis, A. S., Tsilibary, E. C., Yurchenko, P. D., and Furthmayr, H.,** Binding of laminin to type IV collagen: a morphological study, *J. Cell Biol.,* 100, 1848, 1985.
25. **Hynes, R. O.,** Integrins: a family of cell surface receptors, *Cell,* 40, 549, 1987.
26. **Von der Mark, K. and Kühl, U.,** Laminin and its receptor, *Biochim. Biophys. Acta.* 823, 147, 1985.
27. **Laurie, G. W., Leblond, C. P., and Martin, G. R.,** Localization of type IV collagen, laminin, heparan sulfate proteoglycan, and fibronectin to the basal lamina of basement membranes, *J. Cell Biol.,* 95, 340, 1982.
28. **Kallioinen, M.,** Immunoelectron microscope demonstration of the basement membrane components laminin and type IV collagen in the dermal cylindroma, *J. Pathol.,* 147, 97, 1985.
29. **Abrahamson, D. R.,** Post-embedding colloidal gold immunolocalization of laminin to the lamina rara interna, lamina densa, and lamina rara externa of renal glomerular basement membranes, *J. Histochem. Cytochem.,* 34, 847, 1986.

30. **Timpl, R., Dziadek, M., Fujiwara, S., Nowack, H., and Wick, G.,** Nidogen: a new self-aggregating basement membrane protein, *Eur. J. Biochem.*, 137, 455, 1983.
31. **Paulsson, M., Deutzmann, R., Dziadek, M., Nowack, H., Timpl, R., Weber, S., and Engel, J.,** Purification and structural characterization of intact and fragmented nidogen obtained from a tumor basement membrane, *Eur. J. Biochem.*, 156, 467, 1986.
32. **Carlin, B., Jaffe, R., Bender, B., and Chung, A. E.,** Entactin, a novel basal lamina-associated sulfated glycoprotein, *J. Biol. Chem.*, 256, 5209, 1981.
33. **Paulsson, M., Dziadek, M. Suchanek, C., Huttner, W. B., and Timpl, R.,** Nature of sulphated macromolecules in mouse Reichert's membrane, *Biochem. J.*, 231, 571, 1985.
34. **Dziadek, M. and Timpl, R.,** Expression of nidogen and laminin in basement membranes during mouse embryogenesis and in teratocarcinoma cells, *Dev. Biol.*, 111, 372, 1985.
35. **Yurchenco, P. D., Tsilibary, E. E., Charonis, A. S., and Furthmayr, H.,** Laminin polymerization *in vitro*. Evidence for a two-step assembly domain specificity, *J. Biol. Chem.*, 260, 7636, 1985.
36. **Lander, A. D., Fujii, D. K., and Reichardt, L. F.,** Purification of a factor that promotes neurite outgrowth: isolation of laminin and associated molecules, *J. Cell Biol.*, 101, 898, 1985.
37. **Davis, G. E., Manthorpe, M., Engvall, E., and Varon, S.,** Isolation and characterization of rat schwannoma neurite-promoting factor: evidence that the factor contains laminin, *J. Neurosci.*, 5, 2662, 1985.
38. **Chiu, A. Y., Matthew, W. D., and Patterson, P. H.,** A monoclonal antibody that blocks the activity of a neurite regeneration-promoting factor: studies on the binding site and its localization *in vivo, J. Cell Biol.*, 103, 1383, 1986.
39. **Fujiwara, S., Wiedemann, H., Timpl, R., Lustig, A., and Engel, J.,** Structure and interactions of heparan sulfate proteoglycans from a mouse tumor basement membrane, *Eur. J. Biochem.*, 143, 145, 1984.
40. **Hassel, J. R., Leyshon, W. C., Ledbetter, S. R., Tyree, B., Suzuki, S., Kato, M., Kimata, K., and Kleinman, H. K.,** Isolation of two forms of basement membrane proteoglycans, *J. Biol. Chem.*, 260, 8098, 1985.
41. **Kanwar, Y. S., Veis, A., Kimura, J. H., and Jakubowski, M. L.,** Characterization of heparan sulfate-proteoglycan of glomerular basement membranes, *Proc. Natl. Acad. Sci. U.S.A.*, 81, 762, 1984.
42. **Paulsson, M., Yurchenco, P. D., Ruben, G. C., Engel, J., and Timpl, R.,** Structure of low density heparan sulfate proteoglycan isolated from a mouse tumor basement membrane, *J. Mol. Biol.*, 197, 297, 1987.
43. **Noonan, D. M., Horigan, E. A., Ledbetter, S. R., Vogeli, G., Sasaki, M., Yamada, Y., and Hassel, J. R.,** Identification of cDNA clones encoding different domains of the basement membrane heparan sulfate proteoglycan, *J. Biol. Chem.*, 263, 16379, 1988.
44. **Klein, D. J., Brown, D. M., Oegema, T. R., Brenchley, P. E., Anderson, J. C., Dickinson, M. A. J., Horigan, E. A., and Hassel, J. R.,** Glomerular basement membrane proteoglycans are derived from a large precursor, *J. Cell Biol.*, 106, 963, 1988.
45. **Mason, I. J., Taylor, A., Williams, J. G., Sage, H., and Hogan, B. L. M.,** Evidence from molecular cloning that SPARC, a major product of mouse embryo parietal endoderm, is related to endothelial cell "culture shock" glycoprotein of M_r = 43,000, *EMBO J.*, 5, 1465, 1986.
46. **Holland, P. W. H., Harper, S. J., McVey, J. H., and Hogan, B. L. M.,** *In vivo* expression of mRNA for the Ca^{++}-binding protein SPARC (osteonectin) revealed by *in situ* hybridization, *J. Cell Biol.*, 105, 473, 1987.
47. **Young, M. F., Bolander, M. E., Day, A. A., Ramis, C. I., Robey, P. G., Yamada, Y. and Termine, J. D.,** Osteonectin mRNA: distributon in normal and transformed cells, *Nucleic Acids Res.*, 14, 4483, 1986.
48. **Termine, J. D., Kleinman, H. K., Whitson, S. W., Conn, K. M., McGarvey, M. L., and Martin, G. R.,** Osteonectin, a bone-specific protein linking mineral to collagen, *Cell*, 26, 99, 1981.
49. **Brenner, B. M., Hostetter, T. H., and Humes, H. D.,** Molecular basis of proteinuria of glomerular origin, *N. Engl. J. Med.*, 298, 826, 1978.
50. **Rennke, H. G. and Venkatachalam, M. A.,** Structural determinants of glomerular permselectivity, *Fed. Proc.*, 36, 2619, 1977.
51. **Latta, H., Johnston, W. H., Stanley, T. M., and Sykes, E.,** Sialoglycoproteins and filtration barriers in the glomerular capillary wall, *J. Ultrastruct. Res.*, 51, 354, 1975.
52. **Kanwar, Y. S. and Farquhar, M. G.,** Anionic sites in the glomerular basement membrane. *In vivo* and *in vitro* localization to the laminae rarae by cationic probes, *J. Cell Biol.*, 81, 137, 1979.
53. **Kanwar, Y. S. and Farquhar, M. G.,** Presence of heparan sulfate in the glomerular basement membrane, *Proc. Natl. Acad. Sci. U.S.A.*, 76, 1303, 1979.
54. **Kanwar, Y. S., Hascall, V. C., and Farquhar, M. G.,** Partial characterization of newly synthesized proteoglycans isolated from the glomerular basement membrane, *J. Cell Biol.*, 90, 527, 1981.
55. **Farquhar, M. G.,** The glomerular basement membrane. A selective macromolecular filter, in *Cell Biology of Extracellular Matrix*, Hay, E. D., Ed., Plenum Press, New York, 1981, 335.

56. **Stow, J. L., Sawada, H., and Farquhar, M. G.,** Basement membrane heparan sulfate proteoglycans are concentrated in the laminae rarae in podocytes of the rat renal glomerulus, *Proc. Natl. Acad. Sci. U.S.A.*, 82, 3296, 1985.
57. **Reale, E., Luciano, L., and Kühn, K.,** Cationic dyes reveal proteoglycans on the surface of epithelial and endothelial kidney cells, *Histochemistry,* 82, 513, 1985.
58. **Kanwar, Y. S., Linker, A., and Farquhar, M. G.,** Increased permeability of the glomerular basement membrane to ferritin after removal of glycosaminoglycans (heparan sulfate) by enzyme digestion, *J. Cell Biol.*, 86, 688, 1980.
59. **Kelley, V. E. and Cavallo, T.,** Glomerular permeability. Focal loss of anionic sites in glomeruli of proteinuric mice with lupus nephritis, *Lab. Invest.*, 42, 59, 1980.
60. **Laurie, G. W., Leblond, C. P., Inoue, S., Martin, G. R., and Chung, A.,** Fine structure of the glomerular basement membrane and immunolocalization of five basement membrane components to the lamina densa (basal lamina) and its extensions in both glomeruli and tubules of the rat kidney. *Am. J. Anat.*, 169, 463, 1984.
61. **Roll, F. J., Madri, J. A., Albert, J., and Furthmayr, H.,** Codistribution of collagen types IV and AB_2 in basement membranes and mesangium of the kidney, *J. Cell Biol.*, 85, 597, 1980.
62. **Martinez-Hernandez, A., Gay, S., and Miller, E. J.,** Ultrastructural localization of type V collagen in rat kidney, *J. Cell Biol.*, 92, 343, 1982.
63. **Martinez-Hernandez, A. and Chung, A. E.,** The ultrastructural localization of two basement membrane components: entactin and laminin in rat tissues, *J. Histochem. Cytochem.*, 32, 289, 1984.
64. **Timpl, R.,** Recent advances in the biochemistry of glomerular basement membrane, *Kidney Int.*, 30, 293, 1986.
65. **Courtoy, P. J., Kanwar, Y. S., Hynes, R. O., and Farquhar, M. G.,** Fibronectin localization in the rat glomerulus, *J. Cell Biol.*, 87, 691, 1980.
66. **Drenckhahn, D. and Franke, R.-P.,** Ultrastructural organization of contractile and cytoskeletal proteins in glomerular podocytes of chicken, rat, and man, *Lab. Invest.*, 59, 673, 1988.
67. **Junqueira, L. C. U., Montes, G. S., Toledo, O. M. S., Bexiga, S. R. R., Gordilho, M. A., and Brentani, R. R.,** Evidence for collagen molecular orientation in basement membranes, *Histochem. J.*, 15, 785, 1983.
68. **Puchtler, H., Waldrop, F. S., and Valentine, L. S.,** Polarization microscopic studies of connective tissues stained with picrosirius red F3BA, *Beitr. Pathol.*, 150, 174, 1973.
69. **Deák, G.,** *Submicroscopic Structure of Basement Membranes as Revealed by Topo-Optical Reactions,* Ph.D. Thesis, University of Pécs, Hungary, 1979.
70. **Byrom, F. B.,** *The Hypertensive Vascular Crisis. An Experimental Study,* William Heineman Medical Books, London, 1969.
71. **Foidart, J. M., Nochy, D., Nusgens, B., Foidart, J. B., Mahieu, P. R., Lapiere, C. M., Lambotte, R., and Bariety, J.,** Accumulation of several basement membrane proteins in glomeruli of patients with preclampsia and other hypertensive syndromes of pregnancy. Possible role of renal prostaglandins and fibronectin, *Lab. Invest.*, 49, 250, 1983.
72. **Romhányi, G., Deák, G., and Fischer, J.,** Aldehyde-bisulphite-toluidine blue (ABT) staining as a topo-optical reaction for demonstration of linear order of vicinal –OH groups in biological structures, *Histochemistry,* 43, 333, 1975.
73. **Deák, G.,** Nodular thickening of peritubular basement membranes in diabetic kidneys, *Acta Morphol. Acad. Sci. Hung.*, 24, 191, 1976.
74. **Fischer, J.,** Demonstration of microorganisms in tissues by the ABT and KOH-ABT topo-optical reactions, *Acta Morphol. Acad. Sci. Hung.*, 24, 203, 1976.

Chapter 8

EXPERIMENTAL RECONSTRUCTION OF THE ORIENTED GLYCOSAMINOGLYCAN STRUCTURE OF THE EXTRACELLULAR MATRIX

TABLE OF CONTENTS

I. INTRODUCTION

Previous polarization-microscopic investigations revealed a predominantly axioparallel orientation of GAGs and collagen in ECM of different connective tissues. These data suggest that interactions between GAGs and collagen fibrils may be responsible for the oriented deposition of GAGs in the extra celluar space.

It has been known for a long time that sulfated GAGs and collagen electrostatically interact with each other (for review see Mathews,[1] Módis,[2,3] Scott[4,5]). *In vitro* experiments suggest that GAGs precipitate soluble collagen resulting in the formation of fibrils.[6-9] Proteoglycans behave similarly.[10,11] Different GAG molecules exhibit different capacities to interact with collagen. Hyaluronic acid and keratan sulfate fail to interact at all while iduronic acid containing GAGs (dermatan sulfate, heparan sulfate) are more interactive than chondroitin sulfates.[5,12] Charge density, degree of sulfation, and conformation of GAGs seem to be important factors for the interaction of these molecules with collagen.[13]

Based on the electrostatic interactions of GAGs with collagen, Mathews[14-16] proposed a scheme showing the spatial relationship of PGs to collagen fibrils in the ECM of connective tissue. According to this, the core protein of PGs and collagen fibrils run parallel to each other, while the GAG side chains of PGs leaving the core protein bend approximately 90° and are arranged along the surface of the collagen fibril assuming an axiparallel orientation to the fibrils.

Electron-microscopic investigations, however, revealed a predominantly axiperpendicular orientation of the PG molecules to collagen fibrils in ECM of soft connective tissues such as tendons,[17,18] skin,[19] sclera,[20] and cornea.[21] A similar orientation pattern of one population of cartilage PGs has also been demonstrated electronmicroscopically[22,23] (also see Chapter 11). Using copper containing cationic phthalocyanin dyes such as alcian blue, cuprolinic blue, cupromeronic blue for electron-microscopic visualization of PG molecules (see Chapter 11), these authors found a regular array of PG particles along the collagen fibrils, and they demonstrated an attachment of small dermatan sulfate PG molecules to the d band, or less frequently to the e band of the gap zone of the collagen type I, fibrils. Gap-zone-associated PGs were also seen in hyaline cartilage.[24] Keratan sulfate PGs were found to be located at the a and c bands.[5,21] Besides electrostatic interactions between sulfated GAGs and collagen fibrils, a protein-protein interaction is also supposed between PGs and collagen fibrils.[4,10]

Our supposition that the axiparallel arrangement of sulfated GAGs to collagen fibrils is a corollary of the interaction of these constituents in the ECM of several connective tissues required support by further experimental data. *In vitro* and *in vivo* experiments were designed and carried out to check this hypothesis. The results summarized below provide evidence that electrostatic binding of sulfated GAGs to cationic charges of the collagen fibrils is at least one of the factors which regulate the parallel orientation of GAGs to collagen fibril surface. We believe, however, that our data and conclusions explain the mechanism of oriented deposition of GAGs only in cases when the interacting GAG molecules belong to PGs associated with the collagen fibrils.

II. *IN VITRO* RECONSTRUCTION OF ORIENTED GAG-STRUCTURE IN THE HYALINE CARTILAGE MATRIX

Human fetal femoral condylar cartilages were used for our experiments (5-month-old fetus, abortus material, provided by the Department of Obstetrics and Gynecology of this university). Fetal cartilage is chosen because it contains predominantly chondroitin sulfates which can be readily extracted by enzymatic digestion. Unfixed, cryostate sections were prepared. Other parts of samples were fixed in a 4:1 (v/v) mixture of absolute ethanol and

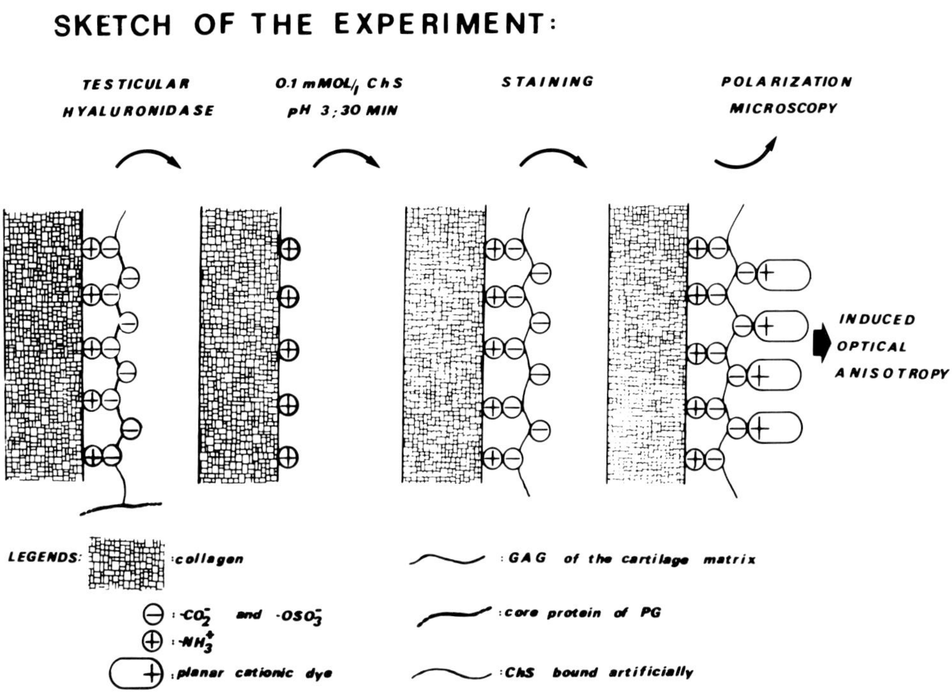

FIGURE 1. Sketch of our *in vitro* experiment carried out on fetal cartilage sections to reconstruct the oriented GAG structure. Columns indicate parts of collagen fibrils, and wavy lines are portions of GAG chains.

neutralized formaldehyde, embedded in wax and cut. Sections with 6 μm thickness both of unfixed, frozen, and fixed, embedded materials were subjected to one of the following treatments. (1) Staining with dimethylmethylene blue or toluidine blue at pH 2.5 for polarization microscopic analysis of spatial orientation of sulfated GAGs. (2) Staining with picrosirius red F3B for collagen. (3) Digestion with testicular hyaluronidase (Reanal, Budapest) (4 mg enzyme/1 ml to 0.1 mol/1 phosphate buffer, pH 6, 37°C, 18 to 24h) to remove chondroitin sulfates and hyaluronic acid from the sections. After this treatment, stainings with dimethylmethylene blue and picrosirius red F3B were performed. (4) Digested sections were incubated in solutions of different polyanions at different concentration, pH, ionic strength, and temperature. After washing, these sections were stained with dimethylmethylene blue and were evaluated under the polarization microscope (Figure 1). For quantitative analysis of the degree of orientation, anomalous dispersion curves were taken and compared with those recorded from sections stained without any pretreatments.

The following polyanions were used in our experiments: Chondroitin sulfate pract. (Fluka, Basel, $M_r \approx 7,800$), Chondroitin sulfate, Na salt, from ligamentum nuchae (SERVA, Heidelberg, $M_r \approx 17,000$), Chondroitin sulfate "C", Na salt, from shark cartilage, B grade (Calbiochem, Los Angeles, $M_r \approx 15,000$), Chondroitin sulfate B, Na salt, super special grade (Seikagaku Kogyo, Tokyo, $M_r \approx 30,000$), Heparin — Na, pharm., USP XIX (SERVA, M_r = 8,000 to 12,000), Dextran sulfate, Na salt (SERVA, $M_r \approx 250,000$) Carrageenan (SERVA, M_r = 500,000), Rumalon® (Robapharm, Basel), Arteparon® (Luitpold, Munich), proteoglycan, D_1 fraction from bovine nasal cartilage and human articular cartilage (kindly provided by Dr. T.T. Glant, Department of Anatomy, Debrecen), Carrageenan is a polysulfated polygalactose isolated from seaweed *Chondrus crispus*. It has a helical conformation, and its λ-form induces granulomas when injected into connective tissues.[25-30] Rumalon® and Arteparon® injections are used in therapy for osteoarthrosis.[31-35] Rumalon® ampullae contain extracts of cartilage and red bone marrow, while Arteparon® solution contains glycosaminoglycanpolysulfate.

Except Rumalon®, Arteparon®, and PGs, the concentration of polyanions varied between 10^{-4} and 10^{-6} mol/1. Rumalon® and Arteparon® solutions were used in undiluted form as well as in 1:1, 1:5, and 1:10 dilutions. Concentration of PGs varied between 40 μg/ml and 1 mg/ml. As solvents and diluents, phosphate buffered saline (PBS, pH 7.2) or 0.1 mol/1 Na-acetate-HCl or Na-acetate-acetic acid buffers (between pH 2.5 and 6 with 0.5 increments) were used. The ionic strength of polyanion solutions was changed by adding NaCl or $MgCl_2$, their final concentrations varied between 0.05 and 1.5 mol/l. Sections were incubated at 22°C and 37°C. The following incubation times were used: 5, 10, 20, 30 min; 1, 2, 4, 8, 24, 48 h. After incubation, sections were carefully washed in a solvent of polyanions for 3 × 5 min (when the incubation period was 30 min or less) and 3 × 10 min (when longer incubation time was used).

The following observations were made. The results obtained with frozen and paraffin sections were essentially the same. The untreated cartilage sections stained with dimethylmethylene blue or toluidine blue exhibited strong metachromasia and optical anisotropy (Figure 2). The sign of birefringence induced by both of these dyes was negative related to the long axis of the matrix trabeculae indicating that sulfated GAGs were predominantly oriented longitudinally along this axis. GAGs of the pericellular capsule were circularly arranged. Picrosirius red F3B staining revealed a similar orientation pattern of collagen fibrils.

Cartilage matrix failed to stain with metachromatic dyes after digestion with testicular hyaluronidase (Figure 3). Picrosirius red topo-optical reaction, however, became stronger as compared to undigested control samples, but it revealed the same orientation pattern of collagen when this staining was carried out on undigested sections (Figures 4 and 5). This enhancement in intensity of staining and birefringence is due to enzymatic removal of the chondroitin sulfate side chains of PGs which occupy and mask the cationic charges on the surface of collagen fibrils.[36] Furthermore, removal of chondroitin sulfates significantly decreases the number of fixed negative charges in cartilage matrix hereby decreasing the repulsion forces of the different matrix compartments against the polyanionic sirius red F3B molecule. Consequently, both the dye penetration of the matrix and dye-binding to the collagen fibrils will be increased by the digestion with testicular hyaluronidase (or other enzymes which depolymerize GAGs or PGs) when using anionic dyes.

The binding of different polyanions to the tissue could be detected by metachromatic dyes. The staining intensity, however, varied as a function of concentration and M_r of polyanions, pH and ionic strength of polyanion solutions, temperature, and time of incubation. A weak metachromasia was observed in the cartilage matrix after incubation in solutions of chondroitin sulfates, dermatan sulfate, and heparin at 10^{-5} and 10^{-6} mol/l concentrations. Binding of dextran sulfate and carrageenan could only be detected when these compounds were used in 10^{-4} mol/l concentration. Rumalon® and Arteparon® solutions failed to induce metachromasia at 1:10 dilution. PG monomers were effective at 0.5 mg/ml and 1 mg/ml concentrations. Binding of polyanionic compounds to the tissue was considerably higher below pH 4 than at higher pH values. This can be explained by the higher number of ionized $-NH_2$ (cationic) groups of the tissues at low pH values. Binding of dextran sulfate and carrageenan was more moderate as compared to other polyanions even at low pH values. Increasing the ionic strength of the incubation solution gradually decreased the binding of polyanions, and binding was abolished if the ionic strength was higher than 0.3 to 0.4 It follows that polyanions are bound to the tissues by ionic interactions. Binding of different polyanions to the tissue samples was higher at 37°C than at room temperature. Chondroitin sulfates and dermatan sulfate are bound after 5 min, and strong binding is detected at 30 min (Figure 6). Binding of heparin, Rumalon® and Arteparon® was observed after incubation for 30 min, and 1 or 2 h incubation time was enough to obtain a strong metachromasia. In case of other polyanions, these were the high molecular weight substances,

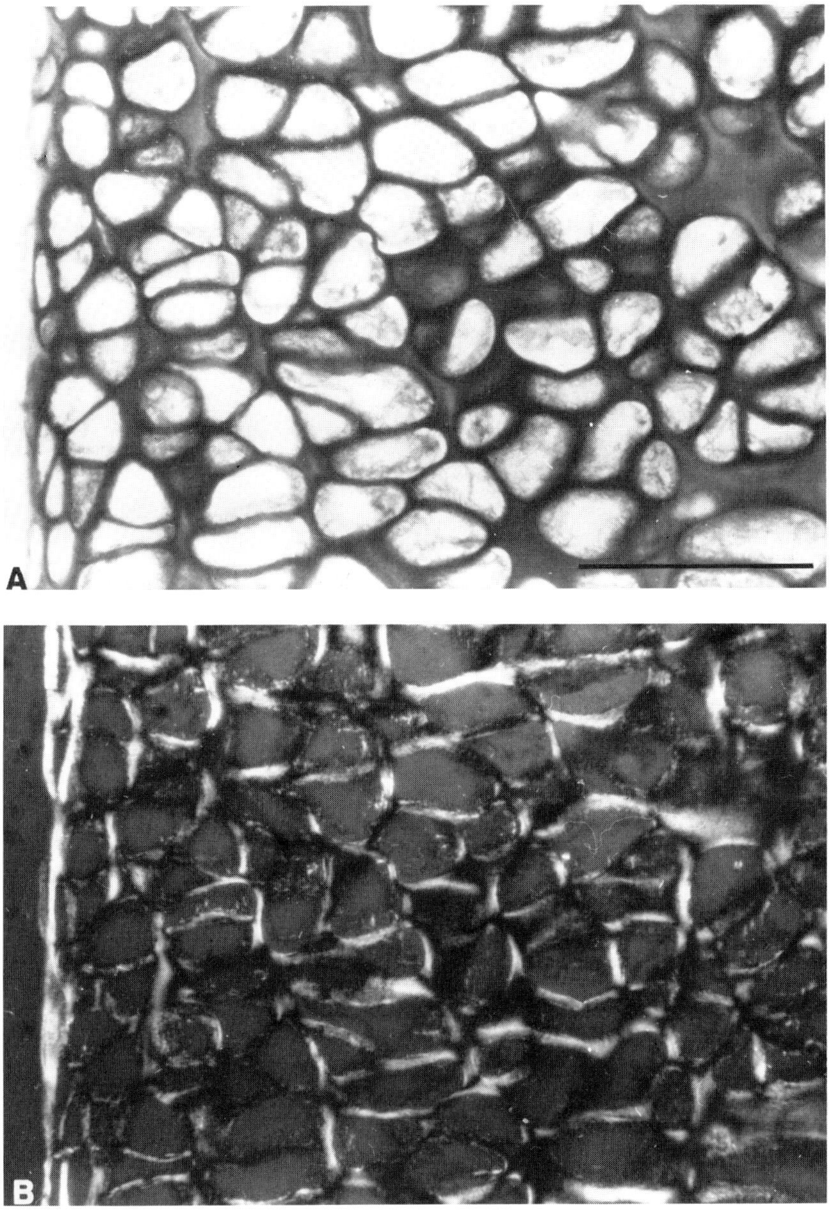

FIGURE 2. Untreated control section of human fetal cartilage (5-month-old fetus, autopsy material) after staining with dimethylmethylene blue at pH 2.5 taken in normal (A) and polarized light (B). Note the birefringent cartilage matrix in (B). Scale: 100 μm.

at least 1 h incubation time was required to detect binding, but strong binding was observed after 4 to 8 h.

To learn more about the nature of tissue chemical groups involved in interactions with polyanions, some sections digested with testicular hyaluronidase were subjected to one of two different deamination procedures (nitrosation and acetylation).[37] After these treatments, section were incubated only in chondroitin sulfate solution at pH 2.5 and stained with dimethylmethylene blue at pH 2.5. Both of these treatments completely prevented the binding of chondroitin sulfate suggesting that $-NH_2$ groups of tissue constituents are involved in the reaction.

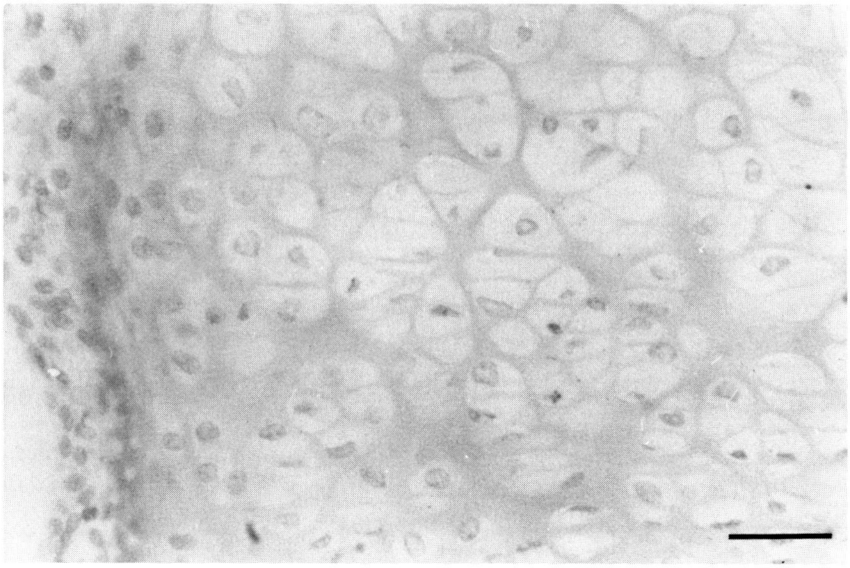

FIGURE 3. Human fetal cartilage digested with testicular hyaluronidase, stained with dimethylmethylene blue at pH 2.5 and taken in normal light. The cartilage matrix is not stained indicating the removal of chondroitin sulfate molecules by the digestion. Scale: 100 μm.

The binding of polyanions did not show any tissue specifity. Besides cartilage matrix, the cytoplasm of different cells and other structures including skeletal muscle, perichondrium, and vessel walls, became metachromatic. This is due to the presence of cationic charges in these structures. Polarization microscopic analysis of the digested, incubated and stained sections, however, reveals optical anisotropy in collagen-containing structures. The cartilage matrix was also birefringent after incubation with all polyanions (Figure 6). It follows that polyanions are bound to the cartilage matrix in an oriented manner. The sign of birefringence is negative with respect to the long axis of matrix trabeculae indicating that polyanions artifically bound to the matrix assumed the same orientation pattern as described for sulfated GAGs of untreated specimens. Furthermore, the spatial orientation of exogenous polyanions is identical to that of the collagen. These data strongly suggest that collagen fibrils of the cartilage matrix (and other collagen-containing structures) can bind polyanions, and the binding results in a predominantly axiparallel orientation of polyanionic macromolecules to the collagen fibrils. It is reasonable to suppose, that in the case of PG preparations the GAG side chains of exogenous PGs are arranged parallel to the collagen fibrils. The same holds true for Rumalon® which also contains PGs.

It is worth mentioning that structures other than collagen-containing connective tissue matrices which became metachromatic after incubation with polyanions never showed birefringence (e.g., cytoplasm of different cells). In contrast to the collagen, these structures are unable to orient the bound polyanionic molecules because their cationic constituents do not possess any spatial orientation pattern which could be revealed by polarization microscopy.

The oriented binding of different polyanions to cartilage matrix collagen was also evidenced by anomalous dispersion curves of birefringence taken (Figures 7 to 10). As shown by these curves, binding polyanions at physiological pH resulted in their spatial orientation similar to the original GAG-structure, but this similarity was only qualitative and never quantitative. If the incubation is carried out at low pH, more polyanions are bound and the density of artifically produced polyanionic structure becomes almost identical to that found in untreated specimens. Supposedly all or almost all of the cationic charges are saturated by polyanions on the collagen fibrils under this experimental condition.

We conclude from these *in vitro* experimental results that the electrostatic interactions between sulfated GAGs (chondroitin sulfates, dermatan sulfate) and collagen is one of the factors which are responsible for the formation of the spatially oriented microstructure of GAGs in the tissues. We suppose that the linearly distributed cationic charges of the collagen fibrils may explain the predominantly axiparallel arrangement of GAGs ionically bound to collagen.

III. *IN VIVO* RECONSTRUCTION EXPERIMENTS IN CONNECTIVE TISSUE OF RAT SKIN

We continued our previously described *in vitro* experiments with *in vivo* models. The aim of these investigations remained the same: we wanted to study the binding of different polyanions to collagen fibers. CFY rats weighing 300 g were used for the experiments. The local concentration of PGs is low in the connective tissue of the skin of such aging rats, and the corollary of this is that metachromatic dyes fail to stain GAGs in the skin of our untreated experimental animals (Figure 11). By contrast, the PG content is high enough in the connective tissue of the skin of young rats to obtain basophilic staining. As we wanted to visualize the presence of exogenous polyanions injected into the skin by cationic dyes, the use of young rats was avoided in these experiments. According to our polarization-microscopic observations, sulfated GAGs are arranged parallel to the long axis of the collagen fibers in the connective tissue of the skin of young rats. Our experimental data show that it is possible to temporarily reconstruct a similarly oriented polyanionic structure along the collagen fibers by single intracutaneous injections of different polyanions in the skin of aging rats.

In addition to the polyanions used in our *in vitro* experiments, the following substances were used: polygalacturonic acid $M_r \approx 25,000$ to 50,000 (SERVA, Heidelberg), polymannuronic acid (Alginic acid), $M_r \approx 100,000$ (SERVA), SP_{54}® (Na-pentosanpolysulfate), $M_r \approx 2,000$ (Benechemie, Munich), Ateroid® (SCM), (CRINOS, S.p.A., Como). Polygalacturonic acid and polymannuronic acid are polycarboxyls, while SP_{54}® and SMA ampullae contain polysulfates. SP_{54}® and Ateroid® are used in the therapy of thrombosis.[38-41]

Rumalon®, Arteparon®, Sp_{54}, and Ateroid® solutions were used without any further dilutions. Other polyanions were administered at 0.6 mmol/l concentration and were dissolved in PBS. Except original ampullae, solutions were sterilized by filtration.

Under narcosis (Nembutal, SERVA), 0.5-ml solutions were intracutaneously injected to a chemically depilated area of skin on the back. The same animal received three injections of the same polyanion to three different sites on the left side, while three sites on right side of the back were injected by 0.5 ml PBS. Sites of injections were labeled by waterproof dyes.

Injected areas were removed after 2, 4, 12, 24, 48, and 72 h. Samples were fixed in absolute ethanol-neutralized commercial formaldehyde (4:1), embedded in wax and cut perpendicularly to the surface. Sections were stained with dimethylmethylene blue or toluidine blue, both at pH 2.5 and 4.5, and evaluated under the polarization microscope.

The extracellular matrix of skin injected by PBS did not exhibit metachromatic staining. In samples injected with polycarboxylated compounds, the collagen fibers remained unstained by metachromatic dye except amorphous or granular material observed between the fiber bundles. This was the residue of the injected material which gradually disappeared from the skin during the experimental period.

The dermal connective tissue was strongly and metachromatically stained in samples removed after 2 h following the injections of different polysulfates. Collagen fibers were always metachromatic, but in some cases (e.g., after injection of PGs) metachromatic material also appeared between the fiber bundles injection of chondroitin sulfates, dermatan

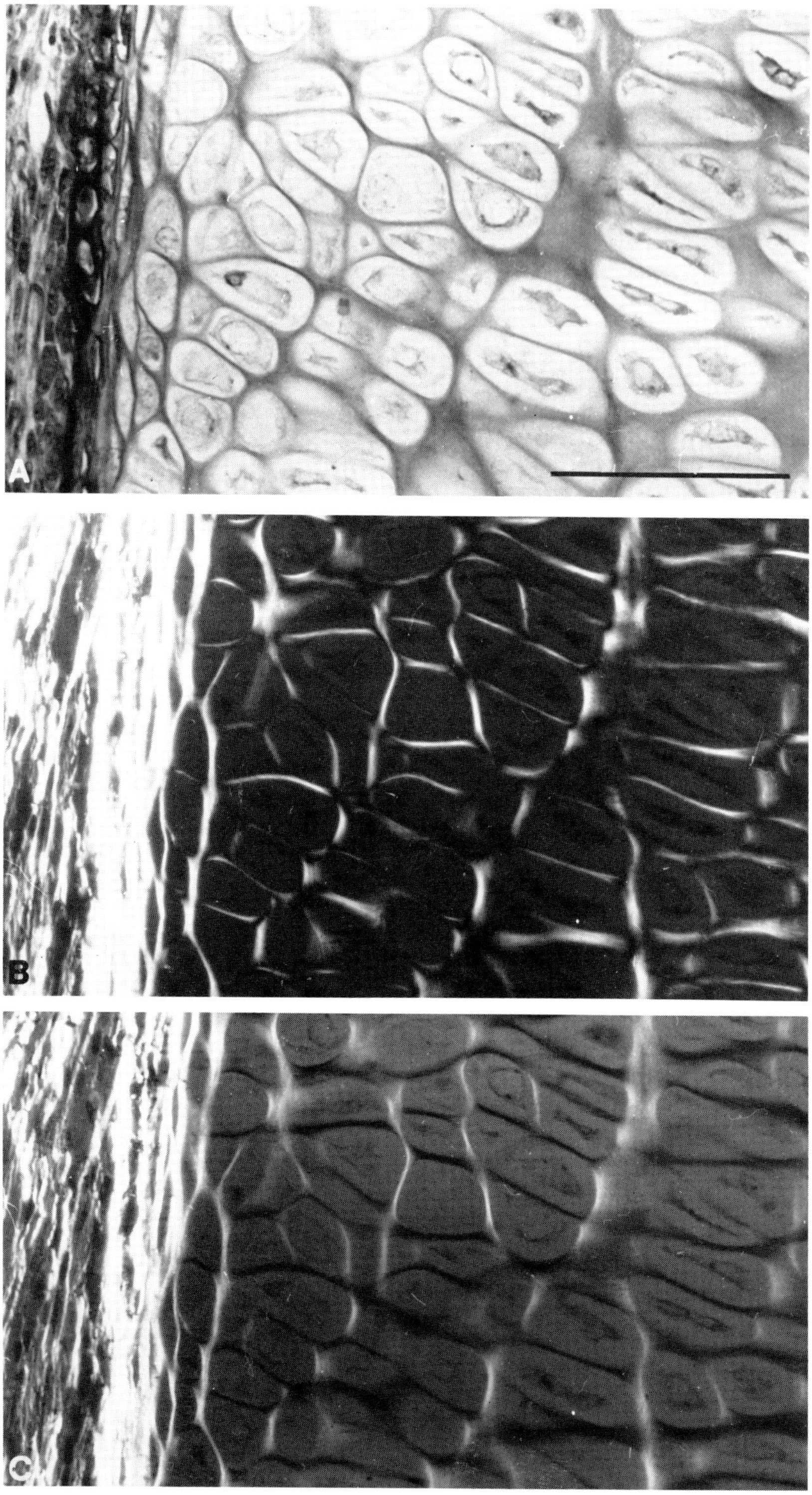

FIGURE 4. Human fetal cartilage after digestion with testicular hyaluronidase and staining with picrosirius red F3B taken in normal (A) and polarized light without (B) and after compensation (C). Collagen of perichondrium and cartilage matrix is stained and birefringent. After compensation, the more or less perpendicularly located cartilage matrix trabeculae are bright in one direction, and are dark in perpendicular direction where their birefringence is extinguished. This is due to the parallel arrangement of collagen fibrils along the matrix trabeculae. Scale: 100 μm.

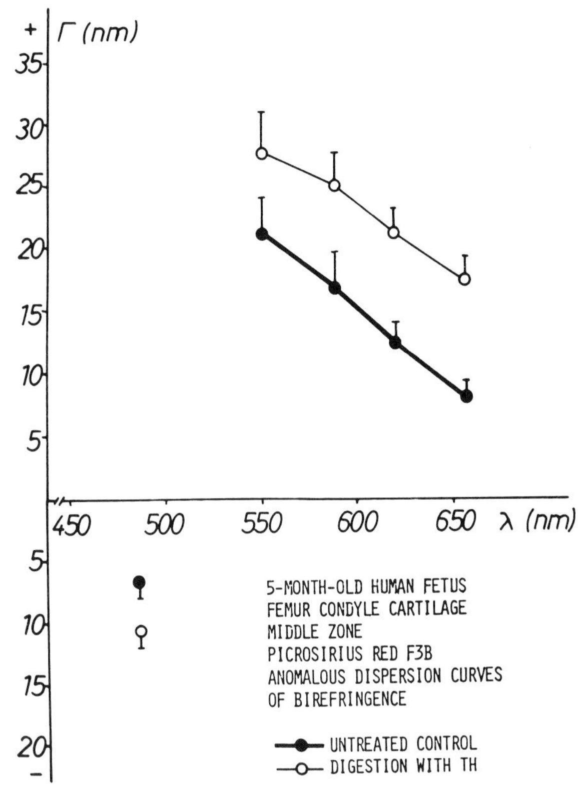

FIGURE 5. Anomalous dispersion curves of birefringence recorded from human fetal cartilage matrix after staining with picrosirius red. Note the increase of birefringence after digestion with testicular hyaluronidase (TH).

sulfate and Arteparon® resulted in the appearance of the strongest metachromasia of collagen fibers. These data indicate that polysulfates are bound to the collagen fibers but polycarboxyls failed to do so. In addition, a general finding was that the dermal connective tissue became swollen after injection of different polyanions. This is probably due to the high water-binding capacity of these materials.

The metachromatic staining of connective tissue was still intense after 4 h, but it was strongly diminished after 12 h and practically disappeared after 24 h. Some remnants of polycarboxyls were still visible after 48 or 72 h. After 12 to 24 h, macrophages appeared with metachromatic contents.

Samples taken after 2 and 4 h were analyzed under the polarization microscope. Collagen fibers stained metachromatically after injections of polysulfates, exhibited strong birefringence appearing in green or yellowish polarization colors. The sign of birefringence was negative with respect to the long axes of the fibers (Figure 12). Keeping in mind that both toluidine blue and dimethylmethylene blue molecules are perpendicularly oriented to the chromotrope, it follows that the exogenous polysulfates are in an axiparallel orientation with the collagen fibers. Interfiberly deposited exogenous PGs failed to exhibit birefringence.

We can thus conclude that polysulfates are bound to collagen in an oriented manner also under *in vivo* conditions. The parallel arrangement of different polysulfates and collagen fibrils is similar to the spatial orientation pattern of sulfated GAGs observed in different connective tissues including the connective tissue of skin of young rats. It seems that "reconstruction" of the oriented polyanionic structure along the collagen fibers is possible (Figure 13). It must be underlined, however, that reconstruction is only a temporarly event.

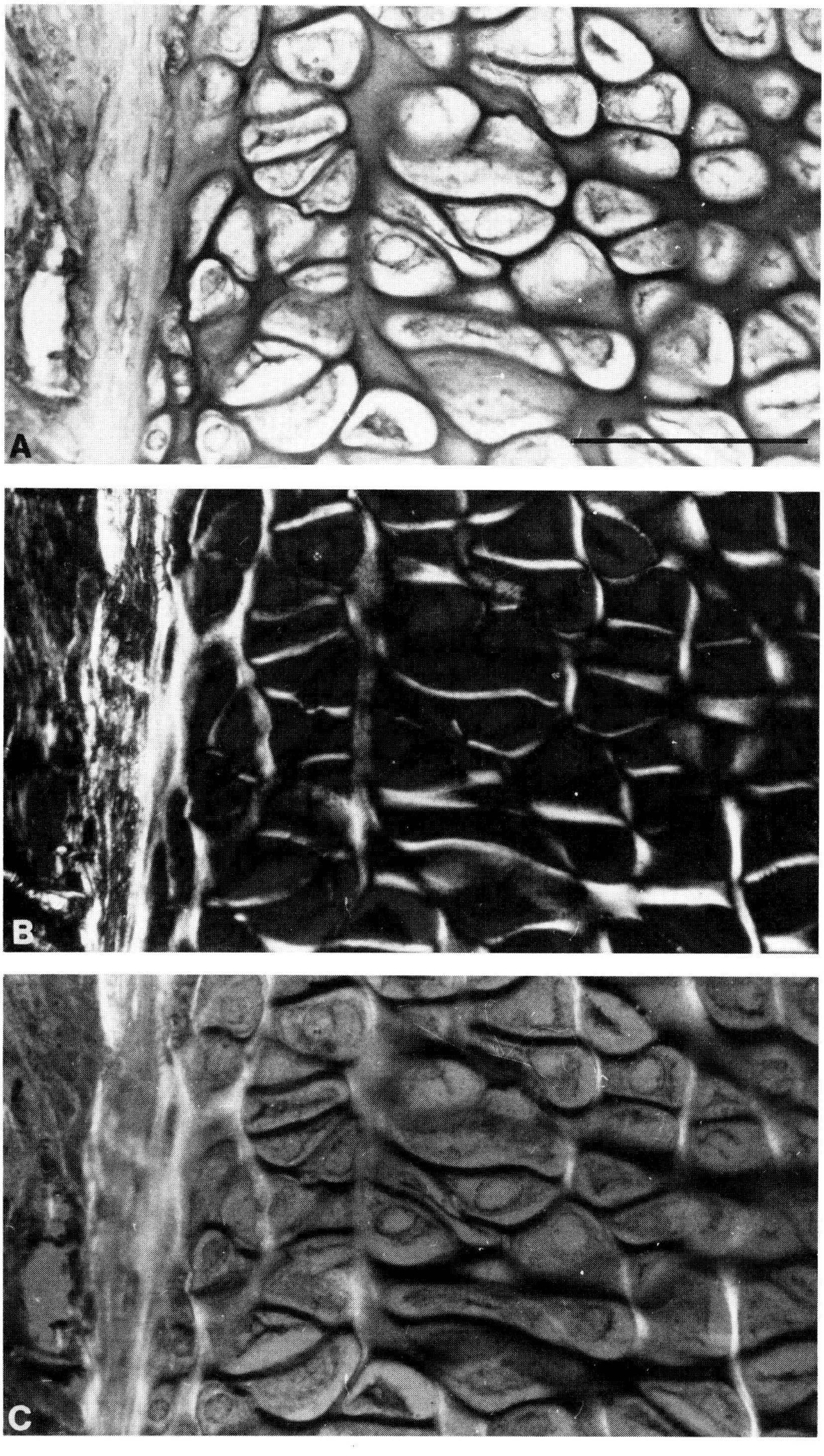

FIGURE 6. Human fetal cartilage after digestion with testicular hyaluronidase, incubation in chondroitin sulfate at pH 3 and staining with dimethylmethylene blue at pH 2.5. The visual field was taken in normal (A) and polarized light without (B) and after compensation (C). Note that the cartilage matrix is stained and birefringent (compare with Figure 2) indicating the oriented binding of exogenous chondroitin sulfate molecules to the collagen fibrils. The perpendicular appearance of dark and bright cartilage matrix trabeculae after compensation suggest that chondroitin sulfates artificially bound to collagen fibril run in the same direction as the fibrils (compare with Figure 4c). Scale: 100 μm.

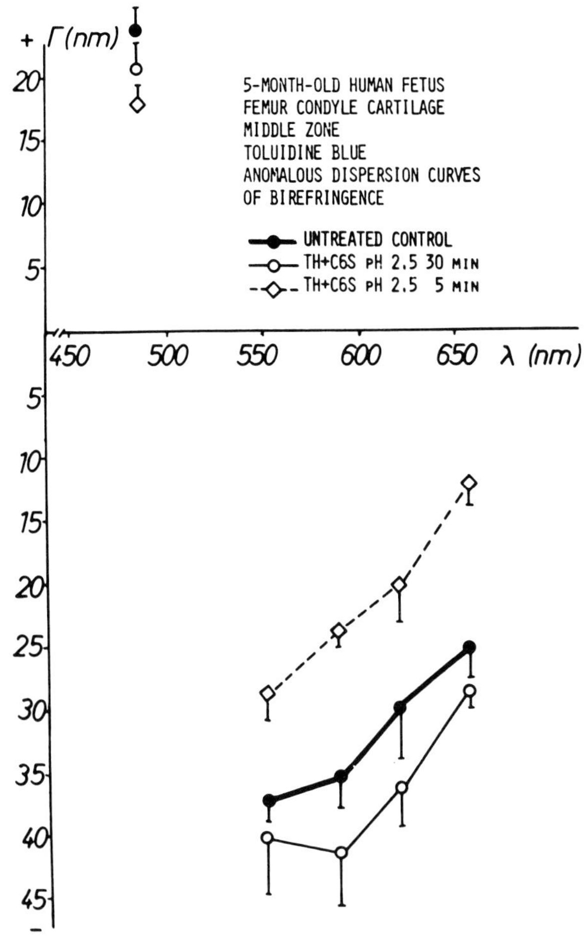

FIGURE 7. Anomalous dispersion curves of birefringence taken from human fetal cartilage matrix after staining with toluidine blue at pH 2.5. Curves obtained after digestion with testicular hyaluronidase (TH) and incubation in chondroitin 6-sulfate (C6S) at pH 2.5 for 5 and 30 min indicate a partial (after 5 min of incubation) and total (after 30 min of incubation) reconstruction of the oriented chondroitin sulfate structure of the cartilage matrix.

Exogenous polyanions disappear within a short period, they are drained by tissue fluid and taken away by macrophages.

These experimental data suggest again that the interactions between collagen and GAGs may result in an oriented deposition of GAGs in the ECM.

IV. CONCLUDING REMARKS

Both our *in vitro* and *in vivo* experiments show that polysulfates can be artificially bound to collagen structure of different tissues, and the corollary of this ionic interaction is the predominant axiparallel orientation of the exogenous polyanions along the collagen fibrils or fibers. This artificially produced orientation pattern is similar to that observed by polarization microscopical analysis of different ECMs (Figure 13).

It is reasonable to suppose that collagen fibril or fiber surfaces somehow guide the

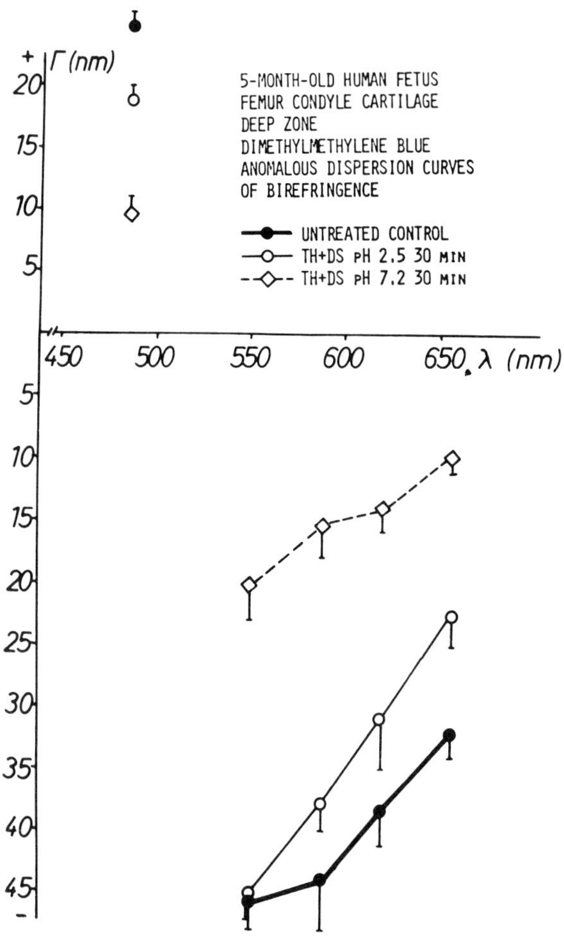

FIGURE 8. Anomalous dispersion curves of birefringence taken from human fetal cartilage marix after staining with dimethylmethylene blue at pH 2.5. Curves taken after digestion with testicular hyaluronidase (TH) and incubation in dermatan sulfate (DS) solution at different pH values show that a more complete binding of DS to collagen can be achieved at pH 2.5 than at pH 7.2.

oriented deposition of polyanions. It is known that collagen fibrils, at least type I fibrils, show axial periodicity, and several bands located within the major D-period are predominantly cationic due to the excess of basic amino acids in these bands[42-46] (Figure 14). These bands of cationic charges are close enough to each other to fix polyanions of linear configuration in an axioparallel position to the collagen fibrils. Theoretically, an arrangement of polyanions along the cationic bands would also be possible. In this case, however, GAGs of different connective tissues and polyanions artificially bound to collagen should be arranged parallel to the bands, i.e., orthogonally to the long axis of collagen fibrils. Polarization microscopy never revealed this type of orientation pattern. Electron microscopy showed a predominant orthogonal arrangement of PGs to collagen fibrils in soft connective tissues, although longitudinally oriented PGs were also shown.[4,5,17-19] The orthogonal orientation of PGs implies a parallel orientation of their GAG side chains to the collagen fibrils. We think that the GAGs and other polysulfates are present in solutions as highly hydrated molecules of linear configuration. These tiny and extended "rods" or helices may find a better scaffolding when aligning along the fibril axis on the fibril surface than turning around the fibrils. Other, still unknown sterical, factors may also operate physiologically in formation of this spatial orientation pattern of GAGs.

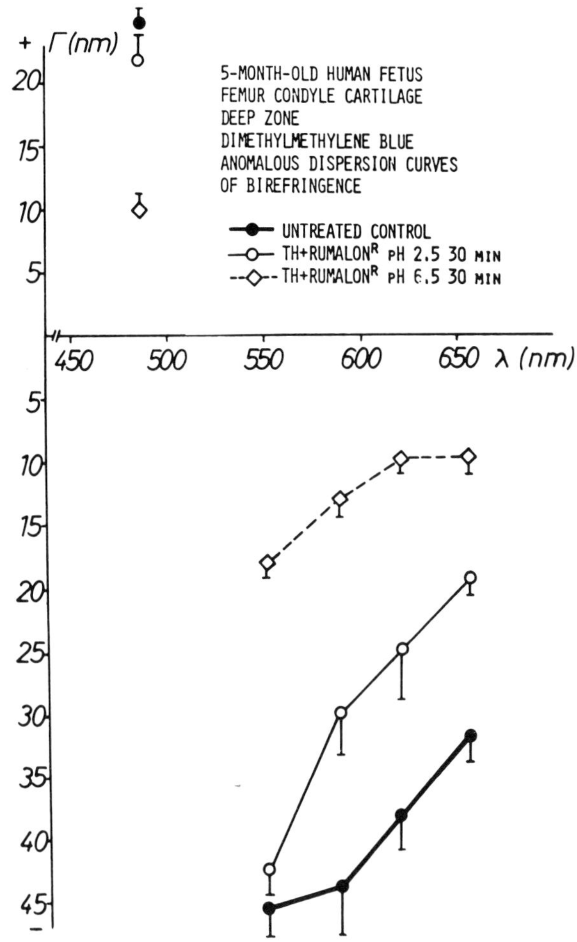

FIGURE 9. Anomalous dispersion curves of birefringence recorded from human fetal cartilage matrix after staining with dimethylmethylene blue at pH 2.5. The oriented binding of some constituent (probably proteoglycan) of Rumalon® solution to cartilage matrix digested with testicular hyaluronidase (TH) is demonstrated.

In interpreting our results obtained with PG monomers and Rumalon® (which supposedly also contains PGs), besides an electrostatic interaction between these compounds and collagen fibrils, the role of a more specific protein-protein interaction can also be supposed between PGs and collagen.[5,11]

Our experimental data would also suggest therapeutical consequences to reconstitute GAG-structure for example in GAG-depleted articular cartilage (in osteoarthritis) or in aged skin. In our opinion, such conclusions cannot be drawn from these experiments for the following reasons. Exogeneous polyanions are stored only for a short period (a few hours) on collagen fibers of the skin. The penetration of polyanions of high molecular weight into the cartilage matrix cannot be optimal. However, Enislidis[34] has demonstrated by autoradiography and scintimetry that intra-articularly injected, tritium-labeled Arteparon® penetrated the whole thickness of human articular cartilage and the compound was stored in the cartilage at decreasing concentrations up to the 12th day following injection. Single polysulfates seemingly cannot be fixed for longer periods in the cartilage matrix. It is known

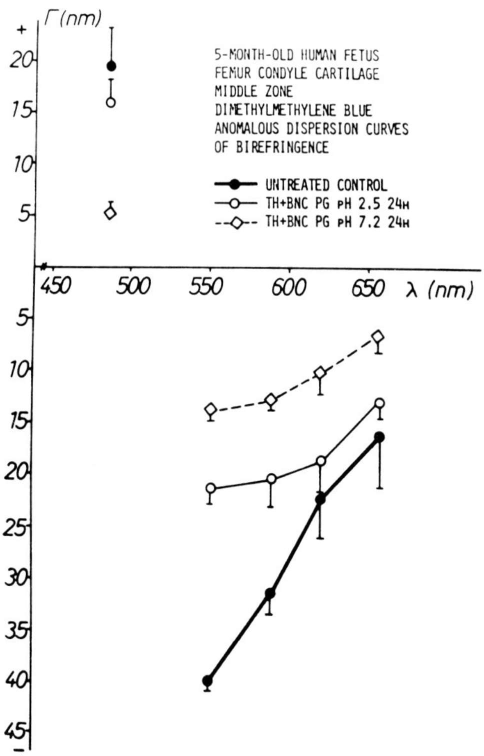

FIGURE 10. Anomalous dispersion curves of birefringence recorded from human fetal cartilage matrix after staining with dimethylmethylene blue at pH 2.5. Proteoglycan monomers isolated from bovine nasal cartilage (BNC) can also be bound to the collagen of cartilage matrix digested with testicular hyaluronidase (TH) in an oriented manner. The binding and orientation of PGs are more complete at pH 2.5 than at pH 7.2.

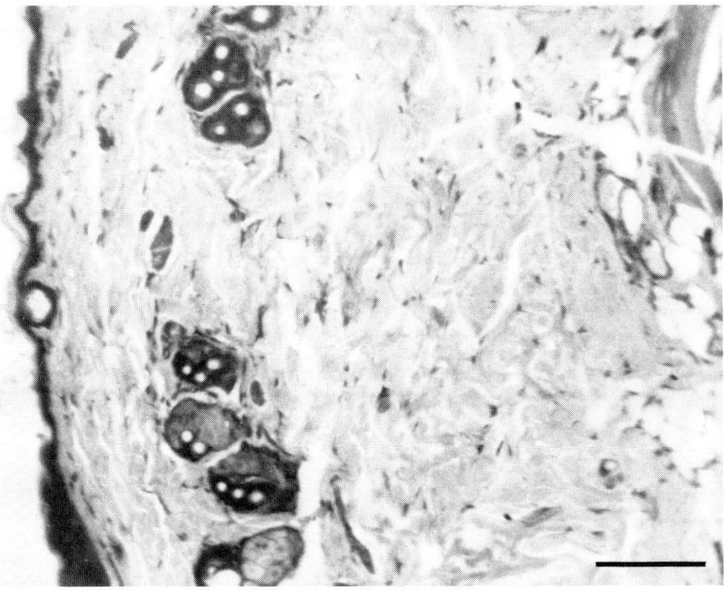

FIGURE 11. Section of skin of an aged CFY-rat after staining with dimethylmethylene blue at pH 2.5. No staining can be observed in the extracellular matrix of connective tissue. Scale: 100 μm.

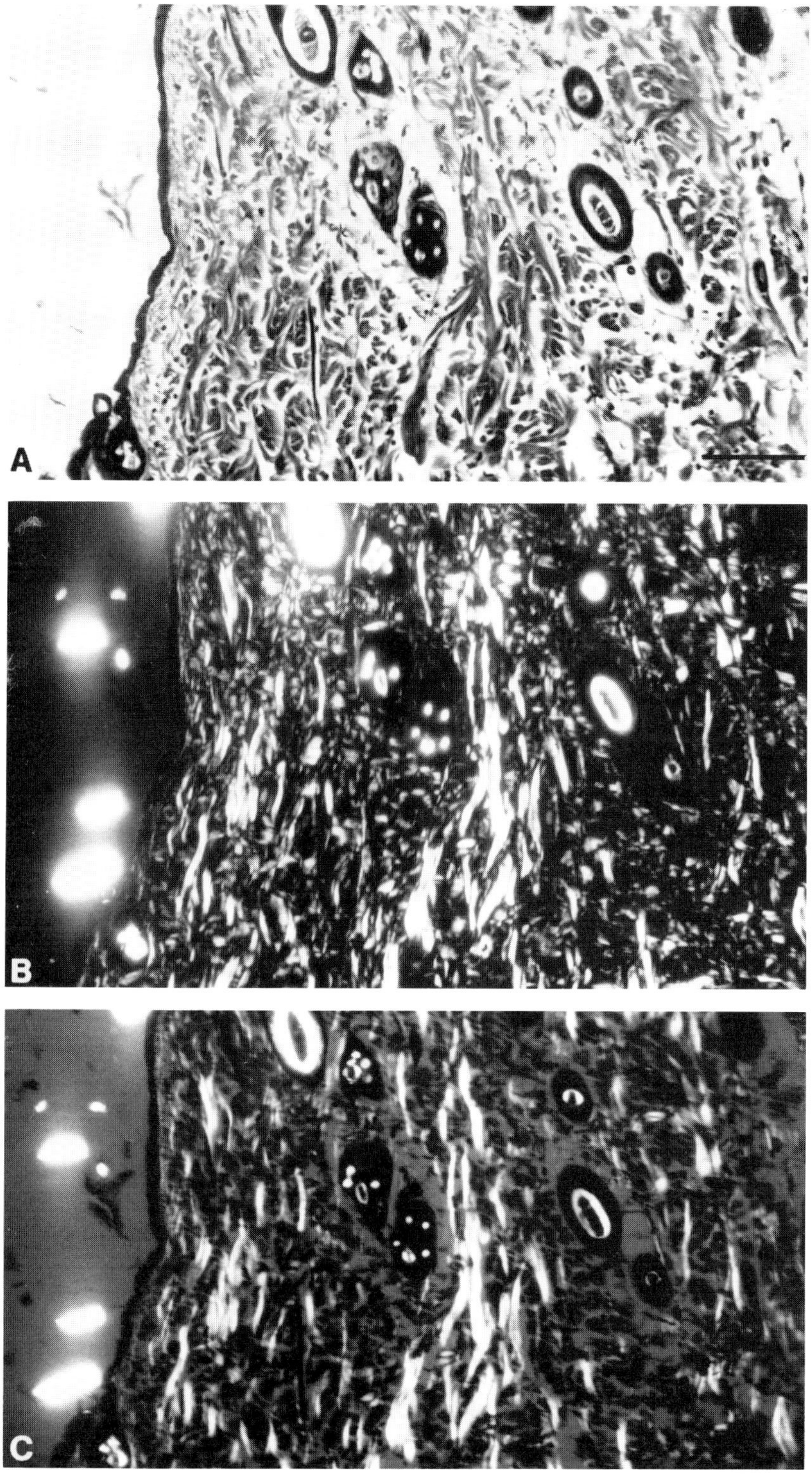

FIGURE 12. Microphotographs taken from a skin section of an aged CFY-rat 2 h after a single intradermal injection of dermatan sulfate solution and staining with dimethulmethulene blue at pH 2.5. (A) in normal light; (B) in polarized light; (C) in polarized light after compensation. Collagen fibers of the injected area of dermis are metachromatically stained and birefringent indicating the oriented binding of exogenous dermatan sulfate to the collagen fibrils. Hairs are strongly birefringent in (B) and (C). Scale: 100 μm.

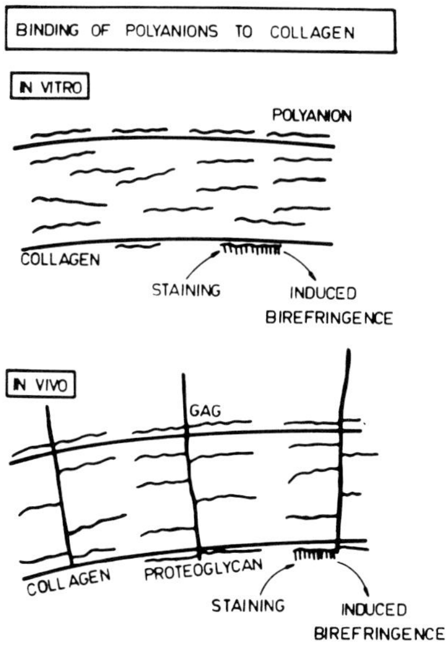

FIGURE 13. The diagram summarizes the conclusions of our reconstruction experiments. In the lower part of the figure, the orientation patterns of GAGs and PGs revealed by polarization and electron microscopy in different connective tissues are shown. Portions of collagen fibrils are depicted by parallel lines. Parallel orientation of GAGs of PGs to collagen fibrils could be inferred from polarization-microscopic examination of sections stained by cationic dyes. On the surface of one GAG molecule the perpendicularly oriented dye molecules are also shown. In the upper part of the figure, the spatial orientation of polyanions artificially bound to collagen fibrils of cartilage matrix and dermal connective tissue is schematically illustrated. As revealed by polarization microscopy after stainings with cationic dyes, the exogenous polyanions — except the polycarboxylls — assume an axioparallel orientation after binding to collagen. This parallel orientation is not necessarily an absolute parallelism: if the linear polyanions form an angle less than 45° with long axis of collagen fibrils they will show anisotropy induced by appropriate staining.

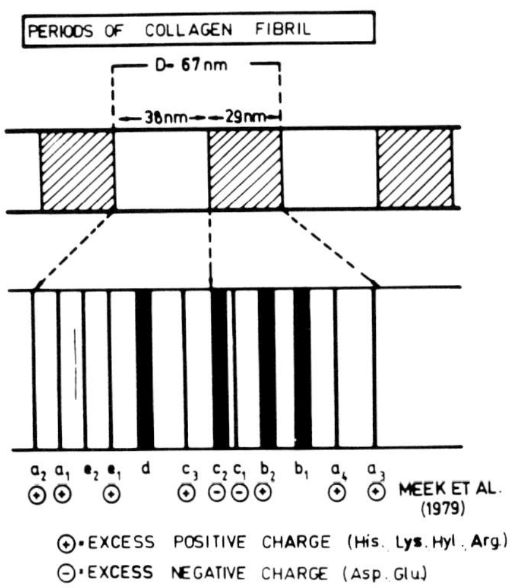

FIGURE 14. A scheme showing the transversal bands of a type I collagen fibril using the data of Meek et al.[44] Note that there are numerous bands where cationic charges predominate.

that PGs are present in PG-aggregate-mediated by hyaluronic acid and link proteins in the cartilage matrix, and their binding to hyaluronic acid prevents or slows down their dissolution from the tissue.[47] The intra-articular use of PGs in therapy, however, is not recommended. It has been shown that intra-articular or intraperitoneal injections of PGs may provoke alterations resembling arthritis.[48-51]

ACKNOWLEDGMENTS

Sincere thanks are due to my former students, Drs. I. Lakatos, Emese Szabó, Cs. Király, and Éva Bernáth for their valuable help in carrying out the experiments. I am grateful to Dr. I. Francia (Central Research Laboratory, University Medical School, Debrecen) for measuring the M_r of polyanions. Some experimental materials were kindly provided by the manufacturers of the following products (Arteparon® — Luitpold, Munich, Rumalon® — Robapharm, Basel, Ateroid® — CRINOS, Como, SP_{54}® Benechemie, Munich, proteoglycans — Dr. T. T. Glant, Debrecen). Their help is appreciated.

REFERENCES

1. **Mathews, M. B.,** *Connective Tissue. Macromolecular Structure and Evolution,* Springer-Verlag, Berlin, 1975.
2. **Módis, L.,** Topo-Optical Investigation of Mucopolysaccharides (Acid Glycosaminoglycans), *Handbuch der Histochemie,* Vol. 2, Part 4, Graumann, W. and Neumann, K., Eds., G. Fischer Verlag, Stuttgart, 1974.
3. **Módis, L.,** The molecular structure of the interfibrillar matrix in connective tissue, *Acta Biol. Acad. Sci. Hung.,* 29, 197, 1978.
4. **Scott, J. E.,** Proteoglycan-collagen interactions, *CIBA Symposia,* 124, 104, 1986.
5. **Scott, J. E.,** Proteoglycan — fibrillar collagen interactions, *Biochem. J.,* 252, 313, 1988.
6. **Wood, G. C.,** The formation of fibrils from collagen solutions. 2. A mechanism of collagen-fibril formation. *Biochem. J.,* 75, 598, 1960.
7. **Wood, G. C.,** The formation of fibrils from collagen solutions. *Biochem. J.,* 75, 605, 1960.
8. **Németh-Csóka, M.,** Untersuchungen über die Kollagenen Fasern. I. Über die submikroskopische Struktur der *in vitro* präzipitierten Kollagenen Fasern und die stabilisierende Rolle der saueren Mucopolysaccharide, *Acta Histochem. (Jena),* 8, 282, 1960.
9. **Németh-Csóka, M.,** Untersuchungen über die Kollagenfasern. II. Vergleichende chemische, polarisation- und elektronenmikroskopische Untersuchungen über die Kollagenfasern mit Chondroitinschwefelsaure als Grundsubstanz, *Acta Histochem. (Jena),* 12, 255, 1961.
10. **Öbrink, B. and Wasteson, A.,** Nature of the interaction of chondroitin 4-sulphate and chondroitin sulphate-proteoglycan with collagen, *Biochem. J.,* 121, 227, 1971.
11. **Oegema, T. R., Laislaw, J., Hascall, V. C., and Dziewiatkowski, D. D.,** The effect of proteoglycans on the formation of fibrils from collagen solutions, *Arch. Biochem, Biophys.,* 170, 698, 1975.
12. **Öbrink, B. and Sundelöf, L. O.,** Light scattering study of associating macromolecules. The binding of glycosaminoglycans to collagen, *Eur. J. Biochem.,* 37, 226, 1973.
13. **Németh-Csóka, M., Kajtár, J., and Kajtár, M.,** Biological significance of helical conformation of acid polysaccharides, *Connect. Tissue Res.,* 3, 207, 1975.
14. **Mathews, M. B.,** The interaction of collagen and acid mucopolysaccharides. A model for connective tissue, *Biochem. J.,* 96, 710, 1965.
15. **Mathews, M. B.,** The macromolecular organization of connective tissue, in *The Chemical Physiology of Mucopolysaccharides,* Quintarelli, G., Ed., Little, Brown, Boston, 1968, 189.
16. **Mathews, M. B.,** The interactions of proteoglycans and collagen-model systems, in *Chemistry and Molecular Biology of the Intercellular Matrix,* Vol. 2, Balazs, E. A., Ed., Academic Press, London, 1970, 1155.
17. **Scott, J. E.,** Collagen-proteoglycan interactions. Localization of proteoglycans in tendon by electron microscopy, *Biochem. J.,* 187, 887, 1980.
18. **Scott, J. E. and Orford, C. R.,** Dermatan sulphate-rich proteoglycan associates with rat tail-tendon collagen at the d band in the gap region, *Biochem. J.,* 197, 213, 1981.

19. **Scott, J. E. and Haigh, M.**, Proteoglycan-type I collagen fibril interactions in bone and non-calcifying connective tissues, *Biosci. Rep.*, 5, 71, 1985.
20. **Young, R. D.**, The ultrastructural organization of proteoglycans and collagen in human and rabbit scleral matrix, *J. Cell Sci.*, 74, 95, 1985.
21. **Scott, J. E. and Haigh, M.**, 'Small'-proteoglycan:collagen interactions: keratan sulphate proteoglycan associates with rabbit corneal collagen fibrils at the 'a' and 'c' bands, *Biosci. Rep.*, 5, 765, 1985.
22. **Dell'Orbo, C., Quacci, D., and Ruggeri, A.**, Nouve tecniche istochimiche per la dimostrazione dei proteoglicani estracellulari al microscopio elettronico, *Riv. Istoch. Norm. Pat.*, 21, 193, 1977.
23. **Ruggeri, A., Dell'Orbo, C., and Quacci, D.**, Electron microscopic visualization of proteoglycans with alcian blue, *Histochem. J.*, 7, 187, 1975.
24. **Orford, C. R. and Gardner, D.**, Proteoglycan association with collagen d band in hyaline cartilage articular cartilage, *Connect. Tissue Res.*, 12, 345, 1984.
25. **Robertson, V., B. and Schwartz, B.**, Ascorbic acid and the formation of collagen, *J. Biol. Chem.*, 201, 689, 1953.
26. **McCandless, E. L.**, Chemical structural requirements for stimulation of connective tissue growth by polysaccharides, *Ann. N.Y. Acad. Sci.*, 118, 860, 1965.
27. **Figueras, T. and Pardo, A.**, Collagen biosynthesis and degradation during deposit and resorptive phases of carrageenin granuloma, *Coll. Relat. Res.*, 6, 379, 1986.
28. **Anderson, N. S., Campbell, J. W., Harding, M. M., Rees, D. A., and Samuel, J. W. B.**, X-ray diffraction studies of polysaccharide sulphates: double helix models for k- and i-carrageenans, *J. Mol. Biol.*, 45, 85, 1969.
29. **Rees, D. A.**, Shapely polysaccharides. The 8th Colworth Medal lecture, *Biochem. J.*, 126, 257, 1972.
30. **Németh-Csóka, M., Kajtár, M., and Kajtár, J.**, Secondary structure of glycosaminoglycans, *Acta Morphol. Acad. Sci. Hung.*, 24, 261, 1976.
31. **Rejholec, V. and Králová, M.**, Langzeitbehandlung der Coxarthrose mit Rumalon® — ein kontrollierter Versuch über 16 Jahre, *Akt. Rheumatol.*, 9, 139, 1984.
32. **Annefeld, M. and Erne, B.**, The mode of action of a glycosaminoglycan-peptide-complex (Rumalon®) on articular cartilage of rat *in vivo, Clin. Rheumatol.* 6, 340, 1987.
33. **Adam, M., Krabcová, M., Musilová, J., Pesáková, V., Brettschneider, I., and Deyl, Z.**, Contribution to the mode of action of glycosaminoglycan-polysulphate (GAGPS) upon human osteoarthrotic cartilage. Biochemical study of the collagen and proteoglycan turnover, *Arzneimittelforschung,* 30, 1730, 1980.
34. **Enislidis, A. C.**, The distribution in tissue of tritium-labelled glycosaminoglycane polysulphate on intra-articular injection, *Med. Welt,* 23, 733, 1977.
35. **Bach, G. L., Panse, P., and Zeiller, P.**, Glykosaminoglykanpolysulfat (GAGPs, Arteparon®) zur Basistherapie der Arthrose. III. Biochemisch-diagnostische und klinische Untersuchungen zur intramuskulaeren Anwendung von GAGPs. *Z. Rheumatol.*, 36, 269, 1977.
36. **Módis, L., Ádány, R., and Lakatos, I.**, Polarisationsoptische Analyse der menschlichen embryonalen Knorpelmatrix, *Acta Histochem.* (Jena) Suppl. 26, 305, 1982.
37. **Lillie, R. D. and Fullmer, H. M.**, *Histopathologic Technic and Practical Histochemistry,* 4th. ed., McGraw-Hill, A Blakiston Publication, New York, 1976, 287.
38. **Köstering, H., König, F., Weber, S., Warmann, E., and Guerrero, M.**, Einfluss von oral und transkutan appliziertem SP_{54} auf praeformierte Thrombosen im Tierversuch, *Med. Welt,* 24, 139, 1973.
39. **Soria, C., Soria, J., Ryckewaert, J. J., Holmer, E., and Caen, J. P.**, Anticoagulant activities of a pentosane polysulphate: comparison with standard heparin and a fraction of low molecular weight heparin, *Thrombos. Res.*, 19, 455, 1980.
40. **Prino, G. and Mantovani, M.**, The *in vitro* effect of Ateroid® on the fibrinolytic activity of rat euglobulins, *Eur. J. Pharmacol.*, 6, 190, 1969.
41. **Mantovani, M.**, Platelet aggregation in rabbits. Effect of a mucopolysaccharide complex i.v. administered, *Arzneimittelforschung,* 23, 239, 1973.
42. **Woodhead-Galloway, J.**, *Collagen: the Anatomy of a Protein,* Camelot Press, Southampton, 1980.
43. **Doyle, B. B., Hukins, D. W. L., Hulmes, D. J. S., Miller, A., and Woodhead-Galloway, J.**, Collagen polymorphism: its origin in the amino acid sequence, *J. Mol. Biol.*, 91, 79, 1975.
44. **Meek, K. M., Chapman, J. A., and Hardcastle, R. A.**, The staining pattern of collagen fibrils. Improved correlation with sequence data, *J. Biol. Chem.*, 254, 10710, 1979.
45. **Tzaphlidou, M., Chapman, J. A., and Meek, K. M.**, A study of positive staining for electron microscopy using collagen as a model system-I. Staining by phosphotungstate and tungstate ions, *Micron,* 13, 119, 1982.
46. **Tzaphlidou, M., Chapman, J. A., and Al-Samman, M. H.**, A study of positive staining for electron microscopy using collagen as a model system-II. Staining by uranyl ions, *Micron,* 13, 133, 1982.
47. **Muir, H.**, Proteoglycans as organizers of the intercellular matrix, *Biochem. Soc. Trans.*, 11, 613, 1983.
48. **Boniface, R. J., Cain, P. R., and Evans, C. H.**, Articular responses to purified cartilage proteoglycans, *Arthritis Rheum.*, 31, 258, 1988.

49. **Glant, T. and Oláh, I.,** Experimental arthritis produced by proteoglycan antigens in rabbits, *Scand. J. Rheumatol.,* 9, 271, 1980.
50. **Glant, T.,** Induction of cartilage degradation in experimental arthritis produced by allogeneic and xenogenic proteoglycan antigens, *Connect. Tissue Res.,* 9, 137, 1982.
51. **Glant, T. T., Mikecz, K., Arzomanian, A., and Poole, A. R.,** Proteoglycan-induced arthritis in BALB/c mice: clinical features and histopathology, *Arthritis Rheum.,* 30, 201, 1987.

Chapter 9

FACTORS INVOLVED IN FORMATION AND MAINTENANCE OF ORIENTED MICROSTRUCTURE OF MATRIX CONSTITUENTS

TABLE OF CONTENTS

I. INTRODUCTION

Polarization and electron microscopic investigations revealed orientation patterns of some ECM constituents. The following patterns proved to be constant ultrastructural features of different embryonic and postembryonic connective tissues: (1) sulfated GAGs and collagen fibrils are oriented parallel to each other if they are in interaction with each other, (2) both sulfated GAGs and collagen are tangentially arranged in the vicinity of the cell membrane.

We supposed that the following factors might be involved in formation and maintenance of the oriented microstructural patterns in the ECM: (1) cell-matrix interactions; (2) cell shape and orientation of cells and cell groups; (3) interactions of matrix constituents with each other; (4) mechanical factors.

The role of chemical interactions between matrix constituents is not discussed in this chapter. Previously shown experimental data suggest that electrostatic interactions between sulfated GAGs and collagen is definitely involved in the formation of oriented deposition of GAGs and PGs along the collagen fibrils. It is also known that self assembly of collagen molecules is a crucial factor in the formation of the collagen fibrils which represent a highly oriented matrix compartment.[1] Due to the present limitations of polarization microscopic techniques and lack of relevant electron microscopic data, we have no information about the spatial orientation of other matrix constituents such as different glycoproteins and hyaluronic acid, although it is anticipated that some of these constituents are deposited in an oriented manner in the matrix. For example, tissue fibronectin molecules of the pericellular matrix of several connective tissues can interact with cell-membrane receptors, collagen, PGs and also with each other. An oriented fibronectic system in close vicinity of the cell membrane may be involved in formation of oriented deposition of collagen other macromolecules which interact with the collagen fibrils.

The aim of this chapter is to summarize experimental data showing the role of some cell-matrix interactions, cell orientation patterns, and mechanical factors in the formation of spatial orientation of different matrix constituents.

II. ROLE OF CELL-MATRIX INTERACTIONS

A. MODEL EXPERIMENTS IN COLLAGEN GEL

Collagen gels made from soluble collagen are widely used in tissue culture laboratories. It has been known for a long time that if cultured cells are grown on or in collagen gel they exhibit different morphological and functional features compared to cells cultured on plastic or glass.[2-14] Cells cultured in collagen gel, show parameters more similar to those observed *in vivo*. Generally, hydrated gels are used to provide a "natural" environment for the cells. Hydrated collagen gels contain fibrils 50 to 500 nm in diameter,[10] and the interfibrillar spaces are filled by water.

We prepared hydrated collagen gels according to Elsdale and Bard[9] from soluble collagen isolated from rat tail tendons. Collagen gel was prepared in small glass Petri dishes by adding 1.6 ml 0.14 mol/l NaOH and 0.8 ml 10× concentrated MEM (Minimal Essential Medium, GIBCO) to 5 ml dialyzed collagen solution. Before adding NaOH and MEM, "cell models" such as glass pearls with 5 mm diameter, or plastic rods with 4 mm diameter, or silanized controlled-pore glass (CPG-10, SERVA, 1400 Å pore sizes, 120 to 200 mesh). (SERVA), or Sephadex® beads (G-50, G-100, diameters: 40 to 300 μm, Pharmacia) were put into the solution. After the gel has been formed at 37°C, the Petri dishes were investigated under polarization microscope. The same experiments were repeated with 10% aqueous solution of gelatin: in this case, gel was formed at 4°C.

After a few hours, intensive birefringence could be observed in the vicinity of the artificial "cell-models" put both in collagen or gelatin solutions (Figures 1 to 3). Strong birefringence

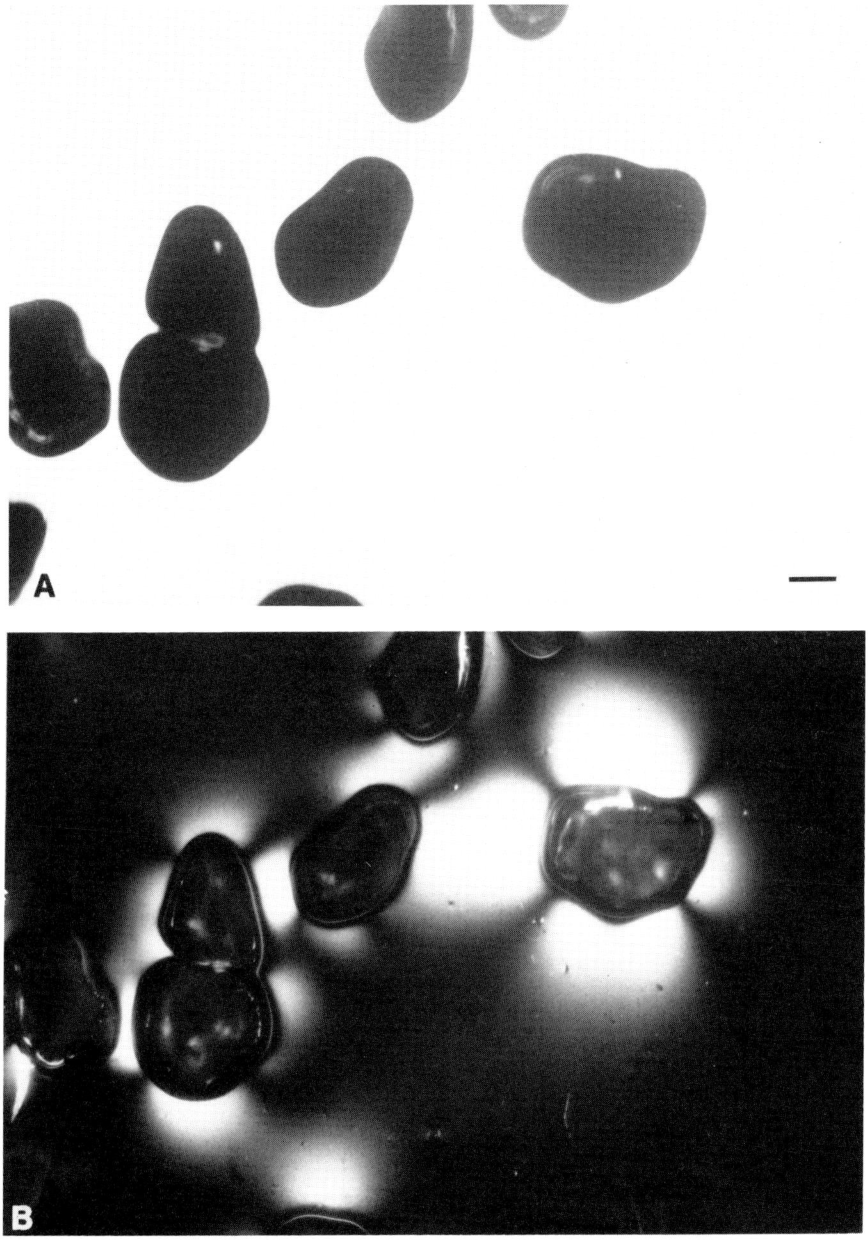

FIGURE 1. Controlled pore glass (CPG-10) in collagen gel taken in normal (A) and polarized light (B). Note the strong birefringence around the glass particles put into collagen solution before the gel is formed. The appearance of birefringence as Maltese crosses indicates the spherulite structure of collagen around the glass. Scale: 100 μm.

could also be detected near the wall of the Petri dishes. The gel in areas remote from surfaces was either moderately anisotropic or isotropic. The sign of ''pericellular''birefringence was positive relating to the tangent of the surface the spheroidal models or rods indicating a circular orientation of the collagen micellae around the bodies used as ''cell models''. Birefringence around Sephadex® beads appeared gradually; it reached the maximum intensity when the beads were maximally swollen.

FIGURE 2. Polarized light micrograph showing the birefringence of collagen gel around a glass pearl. The periphery of the glass pearl is seen in the bottom of the figure. The birefringence gradually decreases moving off the pearl surface. Scale: 100 μm.

Retardation values of birefringence were determined around the glass pearls and plastic rods using a Zetopan-pol (Reichert) polarization microscope equipped with λ/10 Brace-Köhler compensator. Measurements were carried out in three locations: in the immediate vicinity of the surfaces (zone "A"), in 200 to 300 μm distance from the surface (zone "B"), and in remote areas ("ground matrix"). As the Figure 4 and 5 show, the highest retardations values were measured in close vicinity to "cell models", and the birefringence gradually decreased towards the "cell-free" regions. Using the so-called "ambiguity"-test (see Chapter 12) to compare the histograms constructed from the data measured in three different locations, the decrease of birefringence proved to be significant.

We conclude that highly aspecific interactions which may operate between collagen and artificial surfaces are sufficient to induce an ordered deposition of this matrix constituent. The tangential arrangement of the first layer of collagen micellae on the surfaces may somehow also cause a similar arrangement of filaments which are not in contact with the surface, but are located relatively close to the surface.

We think that such simple and aspecific interactions may also contribute to the formation of oriented pericellular collagen structure under *in vivo* conditions. This oriented microstructure can be modulated by many other factors. One of these is the movement of cells. It is known that fibroblasts cultured in collagen gel can contract the gel located around the cell membrane resulting in the reorganization of the collagen structure.[15-17] Fibroblasts may exert forces larger than those actually needed for locomotion. This strong traction (= force used to propel themselves) distorts the collagen gel, creating a pattern similar to tendons and organ capsules. Beyond the cell membrane collagen is aligned parallel to the long axis of the cells.[15] When muscle cells are co-cultured with aligned tracts of collagen formed by the traction of fibroblasts, they also form aligned layers along the oriented collagen structure imitiating the development of the muscles.[18] The orientation of fibers of collagen gel by tension, developed by contracting fibroblasts, is dependent on the structure of the cytoskeleton. Both microtubules and microfilaments are necessary for the aligment of cells and the traction of collagen gel.[19]

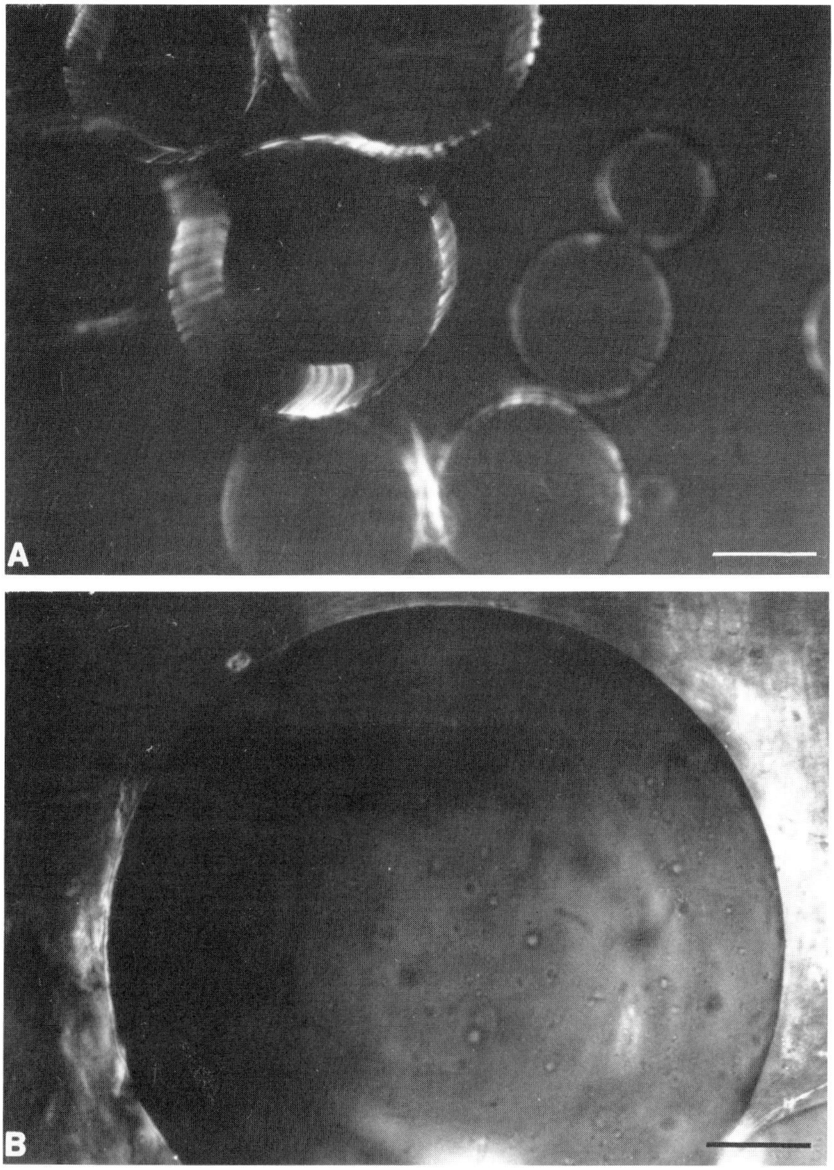

FIGURE 3. Sephadex® beads in collagen gel taken in polarized light. (A) At the beginning of swelling of the beads, birefringence can already be detected. The birefringent rings around the beads are not sharp because of the movement of the surface caused by swelling during the exposure. (B) Swollen bead with "pericellular" birefringence. Scale: 100 μm.

Our experiments carried out with Sephadex® beads clearly indicated that the swelling of "cells" makes the "pericellular" matrix structure mechanically more compact as revealed by the increase of birefringence of the collagen gel during swelling of the beads. It is known that the matrix structure is more compact around the cells being in the hypertrophic zone of the growth plate than around the proliferating cells (see Section III, of this Chapter). Hypertropic cells are swollen cells. It is reasonable to suppose that our model experiments contribute to the explanation of the difference found in matrix organization between the proliferation and hypertrophic cell zones of the epiphyseal cartilage.

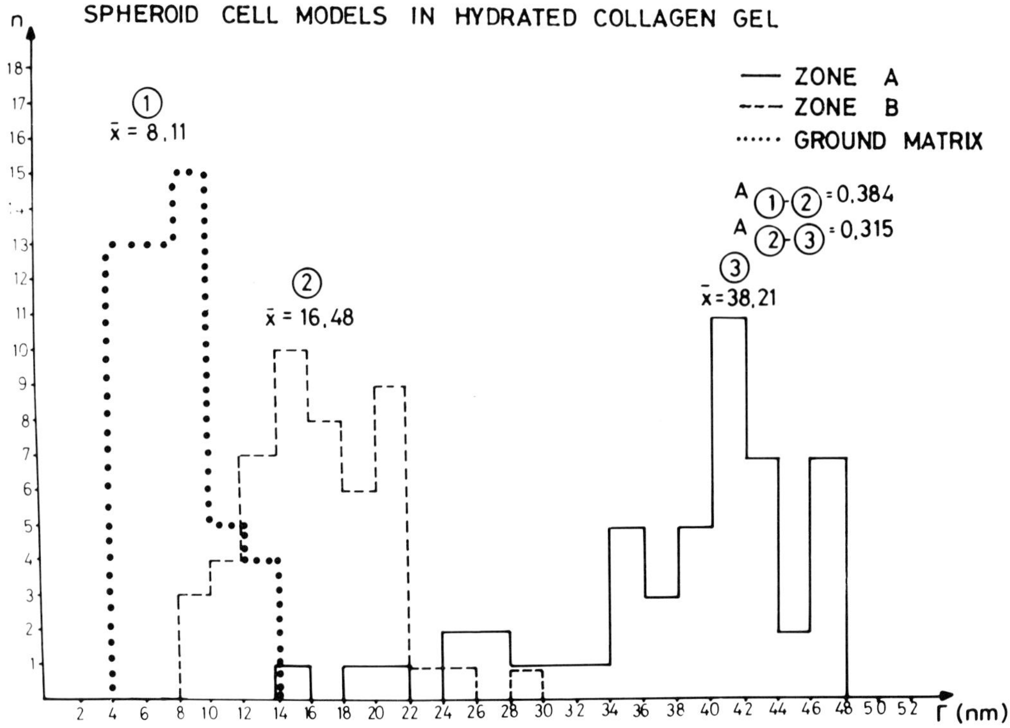

FIGURE 4. Distribution of retardation values measured in collagen gel in close vicinity of a glass pearl (Zone A), in an approximately 300 μm distance (zone B), and in an area free of pearls (ground matrix). Note the gradual decrease of birefringence moving off the glass pearl surface. As the ambiguity-values (A) indicate, significantly higher retardation values can be measured in close vicinity to the pearl surface than in other parts of the gel.

B. CYTOSKELETON-CELL MEMBRANE-MATRIX INTERACTIONS

A number of data accumulated in the last decade by the investigation of different cells grown in culture suggested a structural interconnection between different extracellular matrix constituents and cytoskeleton via the cell membrane. This discovery created a new bridge between cell biology and matrix biology which strongly stimulated scientists of both disciplines to make further efforts to understand how extracellular matrix can influence the cell machinery.

It has been described by Singer and his group[20-22] that focal contacts between cells and substratum contain a transmembrane complex of fibronectin fibers and microfilaments called fibronexus. Subsequent investigations revealed that the molecular composition of these adhesion plaques[23] is rather complex both on the extracellular and cytoplasmic face. Due to the different conditions of cell culture used in these experiments, there are some contradictory data especially concerning the extracellular components, e.g., the presence of fibronectin is questioned.[23-25] Heparan sulfate proteoglycans is supposed to stablize fibronectin-containing cell-to-substrate contacts.[22] Fibronectin is bound by the fibronectin receptor of the cell membrane. This is a 140 kDa protein termed integrin.[26-29] The presence of other receptors is also possible.[30] The next component of this highly organized system is a cytoskeletal protein, talin[31] which may be attached to both integrin and a 130 kDa submembraneously located cytoskeletal protein, vinculin. Vinculin may bind to talin and α-actinin which certainly binds to actin and cross-links it forming the microfilaments[25] (Figure 6). Other cytoskeletal proteins may also bind to the cytoplasmic part of the focal contacts such as tensin, fimbrin, 200, 175, and 82 kDa proteins.[35,36] Microfilaments appear to interact with other major constituents of the cytoskeleton, with the microtubules and intermediate

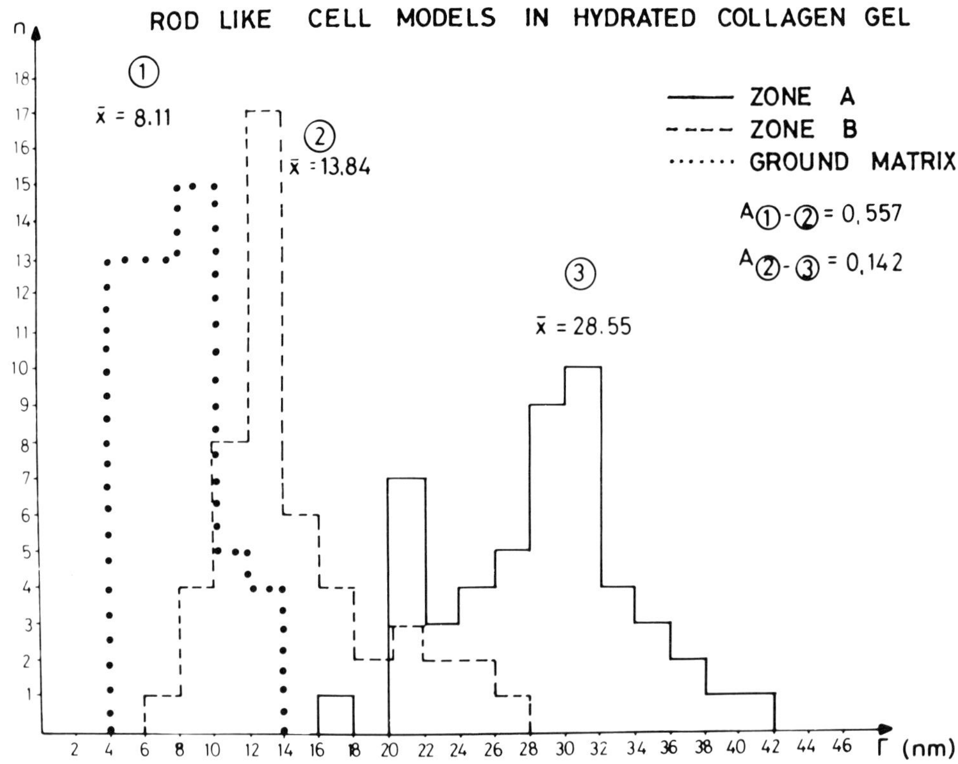

FIGURE 5. Distribution of retardation values measured in collagen gel in close vicinity of a plastic rod (zone A), in an appr. 300 μm distance from the rod surface (zone B) and in areas free of any "cell models" (ground matrix). Highest retardation values are recorded from the gel in close vicinity of the plastic surface indicating that the highest degree of collagen orientation is in this part of the collagen gel.

filaments.[37] Intermediate filaments interact with the nuclear surface, where they appear to terminate at the level of the nuclear envelope.[37] An interaction between intermediate filaments and nuclear matrix is supposed.[37,38] According to the speculations of Bissel and Barcellos-Hoff[38] and Bissell et al.[39] extracellular matrix may direct gene expression by this interconnected molecular system.

These intricate interactions may explain why the alteration of the cytoskeleton results in modification of organization of the extracellular matrix.[40,41] Matrix can also modulate cytoskeleton organization and hereby the cell shape.[7, 42] Changes in cell shape may result in changes in numerous cell functions including expression of the phenotype. For example, disruption of microfilaments enhances the phenotypic expression of cartilage cells and mesenchymal cells resulting in an increased biosynthesis and secretion of PGs.[43-47] Exogenous matrix components such as glycosaminoglycans can also influence the cartilage differentiation under *in vitro* conditions.[48]

Several cytoskeletal constituents have been immunohistochemically detected in cultured cartilage cells such as microfilaments,[45] microtubules[45] vimentin-containing intermediate filaments.[45,49] and vinculin.[50] Glial fibrillary acid protein (GFAP) containing intermediate filaments has been demonstrated in chondrocytes of elastic cartilage in the human epiglottis.[51] Abundance of vimentin-containing intermediate filaments characterizes not only cultured cells but also cells of the cartilage tissue (Figure 7). We could also show vinculin in cartilage tissue both immunohistochemically and immunochemically (Figures 8, 9). This finding suggests that microfilaments are also anchored to the inner surface of the cell membrane of chondrocytes enclosed in their lacunae. The other components of a fibronexus-like structure

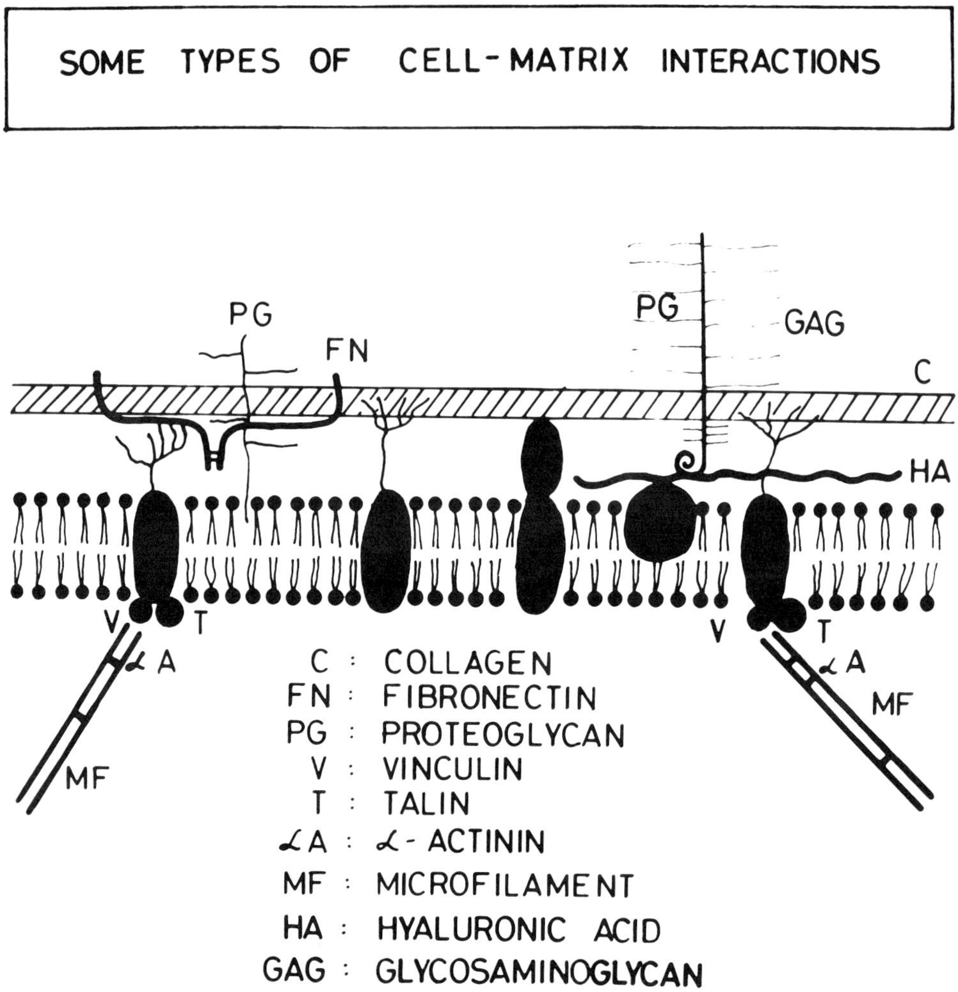

FIGURE 6. A hypothetical scheme of different cell-matrix interactions. In the left part of the figure, the fibronexus (microfilament-vinculin/talin-integrin-fibronectin) is shown. Collagen binding proteins and hyaluronic acid receptors are depicted in the middle and right parts of the diagram. Heparan sulfate PG is shown on the left side, while large, aggregating, chondroitin sulfate-PG characteristic for cartilage is seen on the right side.

in cartilage tissue are not yet known. Our preliminary data indicate that integrin may be present in chondrocyte membrane. Data showing the presence or absence of fibronectin in differentiated cartilage tissue are controversial.[52-54] Anyway, if fibronectin is present in cartilage its function remains to be elucidated. It has been shown that chondronectin binds type II collagen fibrils to the cell membrane of chondrocytes[55-57] (Figure 10). As far as we know, the relationship of chondronectin to the cytoskeleton has not been studied yet. At this moment, we can only speculate that — like other cell types — an interconnection between matrix-membrane-cytoskeleton may also exist in cartilage cells providing a structure for ''messages'' arriving from the extracellular matrix. If such structure exists, it would be the ''chondronexus''.

Cells may bind to different matrix constituents by molecules other than nectins. For example cell membranes of chondrocytes and fibroblasts have receptors for collagen called anchorin CII.[58] This 34 kDa collagen-binding protein has sequences showing considerable homology with sequences in human and bovine calpactin, lipocortin, endonexin, and protein

II, which are members of a family of Ca^{2+}- and phospholipid-binding proteins, as well as major substrates of tyrosine kinases.[59] While these proteins are located at the inner side of cell membrane and they are associated with the actin-binding protein, fodrin, anchorin CII is located at the outer cell surface.[59] Furthermore, a hyaluronate receptor is also detected in chondrocytes.[60] It binds hyaluronic acid and, indirectly, also proteoglycans linked to hyaluronic acid forming huge aggregates.[61] The hyaluronate receptor seems to be attached to actin filaments of the cytoskeleton.[62]

These data indicate several possible mechanisms by which extracellular matrix constituents and cytoskeleton can interact with each other. This molecular interaction may imply that the matrix and cytoskeleton can mutually influence each other's organization. However, further experimental data are needed for the better understanding of structural and functional relationships between extracellular matrix and cytoskeleton. This intriguing and fascinating field challenges both cell biologists and matrix experts.

III. ROLE OF ORIENTATION OF CELLS AND CELL GROUPS

A. ANALYSIS OF CHICKEN EMBRYONIC CARTILAGE

Young embryonic cartilage is already a heterogenous tissue. Cartilage of chicken embryonic limb buds is formed at about stage 25 according to Hamburger and Hamilton.[63] At stage 27 to 29, three zones can be distinguished within the cartilage anlage of long bones: (1) superficial, (2) middle and (3) hyperthropic zones (Figure 11). These zones differ from each other both in matrix composition and cell shape. Different proteoglycans and collagens are deposited in the matrix of different zones.[64-67] The entire cartilage matrix contains "heavy" proteoglycans (PG-H), while "light" PGs (PG-Lb) are localized in zone 2.[66] Collagen type II and IX are present in each zone but collagen type X is specifically located in zone 3.[67]

Small spherical and polygonal cells are found in zone 1 and large spherical cells predominate in zone 3. No preferential orientation of cells can be detected in these zones. In zone 2, however, flat, discoidal cells are seen, and they are oriented with their long diameter perpendicular to the long axis of the cartilage.

For our investigations tibia cartilages were isolated from stage 35 (9-day-old) Hybro Hunniahybrid chicken embryos, were fixed in a 4:1 mixture of absolute ethanol, and neutralized with formaldehyde, and embedded in wax. Longitudinal and serial cross sections were made and numbered. Alternating sections were stained with picrosirius red F3B combined with antecedent digestion with testicular hyaluronidase for collagen (see Chapter 12) and with *N,N'*-diethylpseudoisocyanine for sulfated GAGs. Retardation values of birefringence were determined within each zone both in longitudinal and cross sections at $\lambda = 589$ nm using a Zetopan-pol polarization microscope equipped with Brace-Köhler $\lambda/10$ compensator. Ninety measurements were made after each staining in each zone and cutting plane. Data are depicted in histograms. Histograms of data recorded from longitudinal and cross sections are compared with the so-called "ambiguity"-probe (Chapter 12). The ambiguity value (A) can be from 0 to 1. If A $\leqq 0.7$, the difference between histograms is significant.

The intensity of birefringence obtained in cross and longitudinal sections was similar both after collagen- and GAG-specific topo-optical reactions in zones 1 and 3 (Figures 12,13). Both GAGs and collagen fibrils exhibit a predominantly circular orientation around the chondrocytes within these zones. In zone 2, however, the intesity of birefringence was significantly higher than in cross sections (Figures 12, 14). These observations suggest that both collagen fibrils and sulfated GAGs associated with the collagen run more or less transversally to the long axis of the cartilage anlage in the extracellular matrix of zone 2. Seemingly, these constituents have no preferential orientation axis within the transverse plane, therefore the collagen fibrils or GAGS running in different directions extinguish the

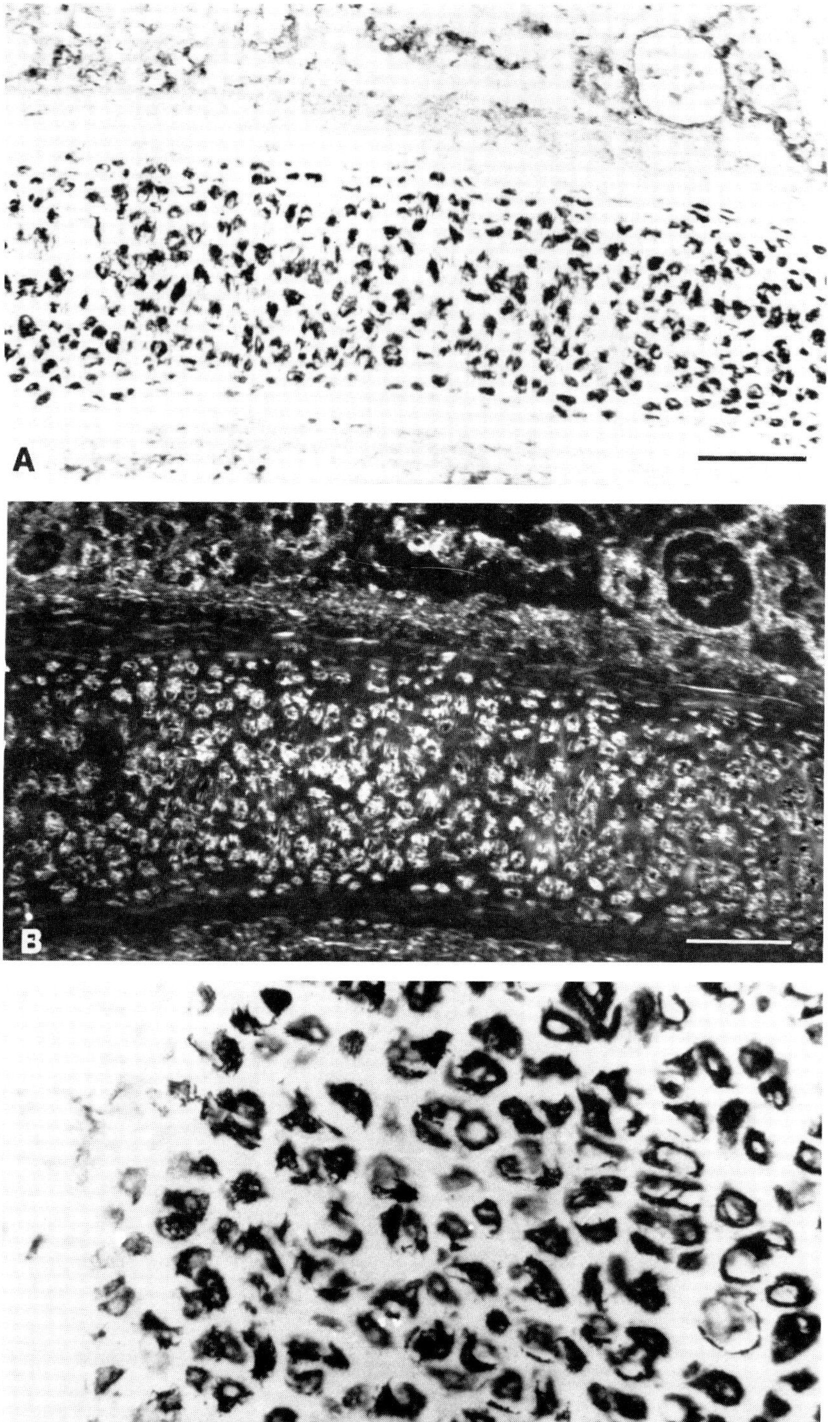

FIGURE 7. Frozen sections of trachea cartilages of a young adult CFY-rat after immunohistochemical decoration for vimentin. Vimentin containing intermediate filaments are detected by monoclonal antibody (Amersham) and AuroProbe LM® (Janssen). The reaction is intensified by silver staining (IntenSEM®, Janssen). (A) and (B) are taken from the same visual field in normal light (A) and polarized light (B), while (C) is a normal light micrograph at higher magnification. Note that after silver enhancement (B), the reaction product is birefringent in cartilage and connective tissue cells. Scales: 100 μm.

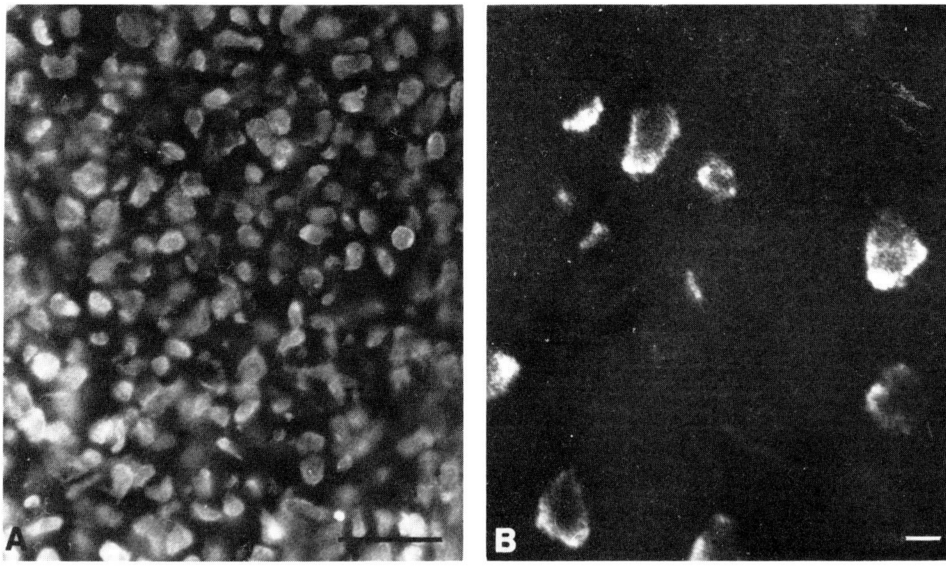

FIGURE 8. Immunofluorescent detection of vinculin in cartilage cells. Specimen: sternal cartilage of 15-day-old chicken embryo; antibody : monospecific polyclonal, kindly provided by Dr. V. E. Koteliansky (Moscow). Note the reactions of the cell membranes of the chondrocytes. Higher manification (B) reveals discrete localization of the vinculin. Scales: 100 μm (A), 10 μm (B).

optical anisotropy of each other when observed in cross section. It seems that the discoidal cells oriented perpendicularly to the long axis of cartilage analage direct the spatial orientation of these matrix constituents. This is probably due to different cell-matrix interactions.

B. ANALYSIS OF RAT GROWTH PLATE

It is well known that epiphyseal cartilage can also be divided into different zones. Except for the first zone (zone of resting cells), cells of the proliferative and hypertrophic zones form columns. The long axis of these cell columns is more or less orthogonal to the plane of the growth plate i.e., parallel to the long axis of the bone. The extracellular matrix of the growth plate is divided into pericellular and intercolumnar compartments.[68] A longitudinal orientation of collagen fibrils has been revealed by electron- and polarization microscopy in the intercolumnar matrix.[69,70] We wanted to investigate the relationship of spatial orientation of cell columns and matrix constituents (collagen, sulfated GAGs) by quantitative polarization microscopic methods.

Proximal portions of tibiae of CFY-rats weighing 150 g were isolated, fixed, and demineralized in a Susa-mixture and embedded in wax. Longitudinal and serial cross sections were cut, and they were stained with picrosirius red F3B combined with antecedent digestion with testicular hyaluronidase for collagen, and with dimethylmethylene blue for sulfated GAGs. Retardation values of birefringence were measured in the pericellular and intercolumnar matrices in proliferative and hypertrophic zones both in longitudinal and cross sections using a $\lambda/4$ compensator plate at $\lambda = 589$ nm. Sixty measurements were made in each compartment and each zone both of longitudinal and cross sections. Data were plotted in histograms and histograms obtained from pericellular and intercolumnar compartments of identical zones were compared both in cross and longitudinal sections using the "ambiguity"-test.

The retardation values measured in pericellular matrix compartments were similar to each other after both of the topo-optical reactions in longitudinal and cross sections (Figures 15 to 19). It follows that collagen and sulfated GAGs are circularly oriented in the pericellular

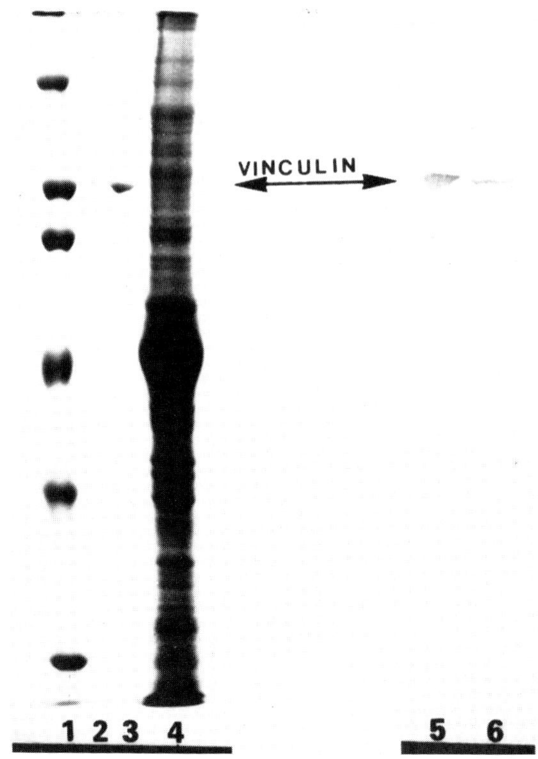

FIGURE 9. Electrophoretic patterns of denatured extract of cartilage cells isolated from sternal cartilages of 15-day-old chicken embryos (track 4). Chicken gizzard vinculin (tracks 2,3) and molecular weight standards (track 1) (SDS-6H, Sigma) were also submitted to SDS polyacrylamide electrophoresis. Two parallel runs were performed on the same gel, which were separated following electrophoresis. One part was stained by coomassie blue and destained. The other part was electroblotted to nitrocellulose paper and developed for vinculin using monospecific polyclonal antibody raised against chicken gizzard vinculin (produced and kindly provided by Dr. V. E. Koteliansky, Moscow), biotinylated anti-rabbit IgG, and avidin conjugated with peroxidase (Vector Labs.) Peroxidase activity was developed by diaminobenzidine (Sigma). (Track 5: vinculin standard, track 6: extract of chondrocytes). It is clearly seen that vinculin is present in cartilage cells.

matrix. The intensity of birefringence was higher in the pericellular matrix of the hypertropic zone than that of the proliferative zone (Figures 20 to 21). This is probably due to the swelling of hyperthrophic cells which may contribute to the formation of a more compact structure. A significant difference of retardation values was found between longitudinally and transversally cut intercolumnar matrix compartments both after collagen-and GAG-specific topo-optical reactions within each zone (Figures 16 to 19). Higher retardation values were recorded from the longitudinal sections than from cross sections. These data clearly indicate that both collagen fibrils and sulfated GAG molecules run more or less parallel to the long axis of cell columns in the intercolumnar matrix. The intensity of birefringence was significantly higher in intercolumnar matrix of the hypertrophic zone than in that of the proliferative zone (Figures 20 to 21). It seems that the gradual expansion of cells within the hypertrophic zone makes the packing of collagen and GAGs more dense also in the intercolumnar matrix.

Thus we can conclude that the orientation of cell groups (cell columns) may also regulate the spatial orientation pattern of the matrix constituents deposited in the intercolumnar space. Collagen and GAGs assume a similar orientation pattern in different matrix compartments. This is probably due to their chemical interaction which results in their axiparallel arrangement.

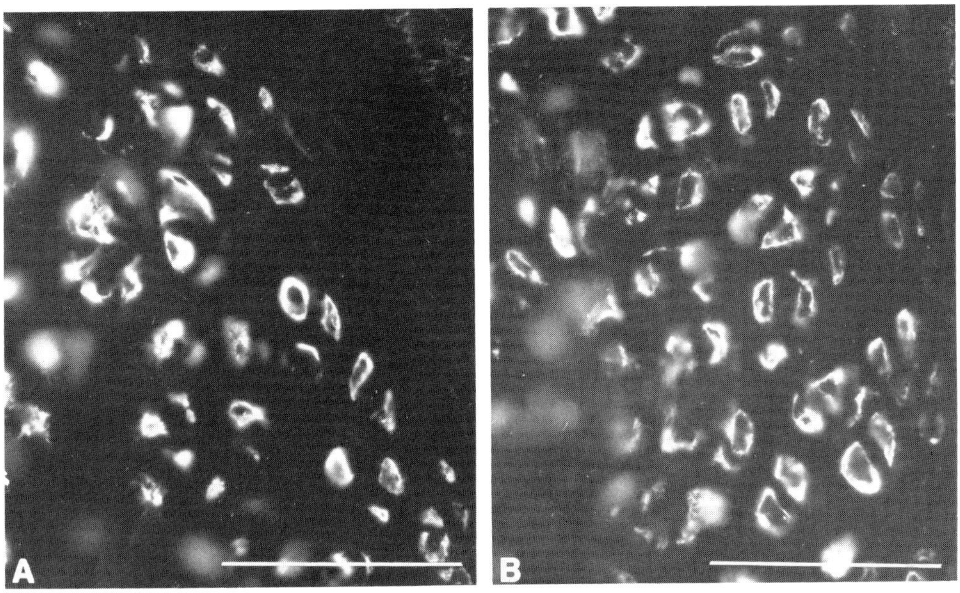

FIGURE 10. Superficial (A) and deep (B) zones of sternal cartilage of a 17-day-old chicken embryo. Indirect immunofluorescent reaction of chondronectin carried out on frozen sections. Monoclonal antibody against chondronectin was kindly provided by Dr. H. K. Kleinman (NIH, Bethesda:). Observe the pericellular localization of the chondronectin. Scales: 100 μm.

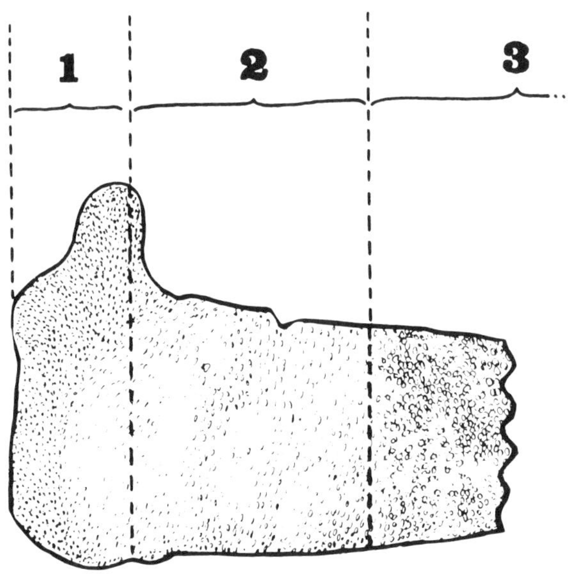

FIGURE 11. Zones of tibia cartilage of a 9-day-old chicken embryo. Camera lucida drawing. In zone 2, discoidal cells are oriented perpendicularly to the long axis of cartilage primordium.

The aforementioned data indicate that orientation of cells may influence orientation of some matrix constituents. A similar conclusion has been drawn by Trelstad.[71] He studied chondrogenesis of vertebral and limb anlages, and he found that spatial orientation of mesenchymal cells prior to collagen fiber formation was reflected in the organization of the

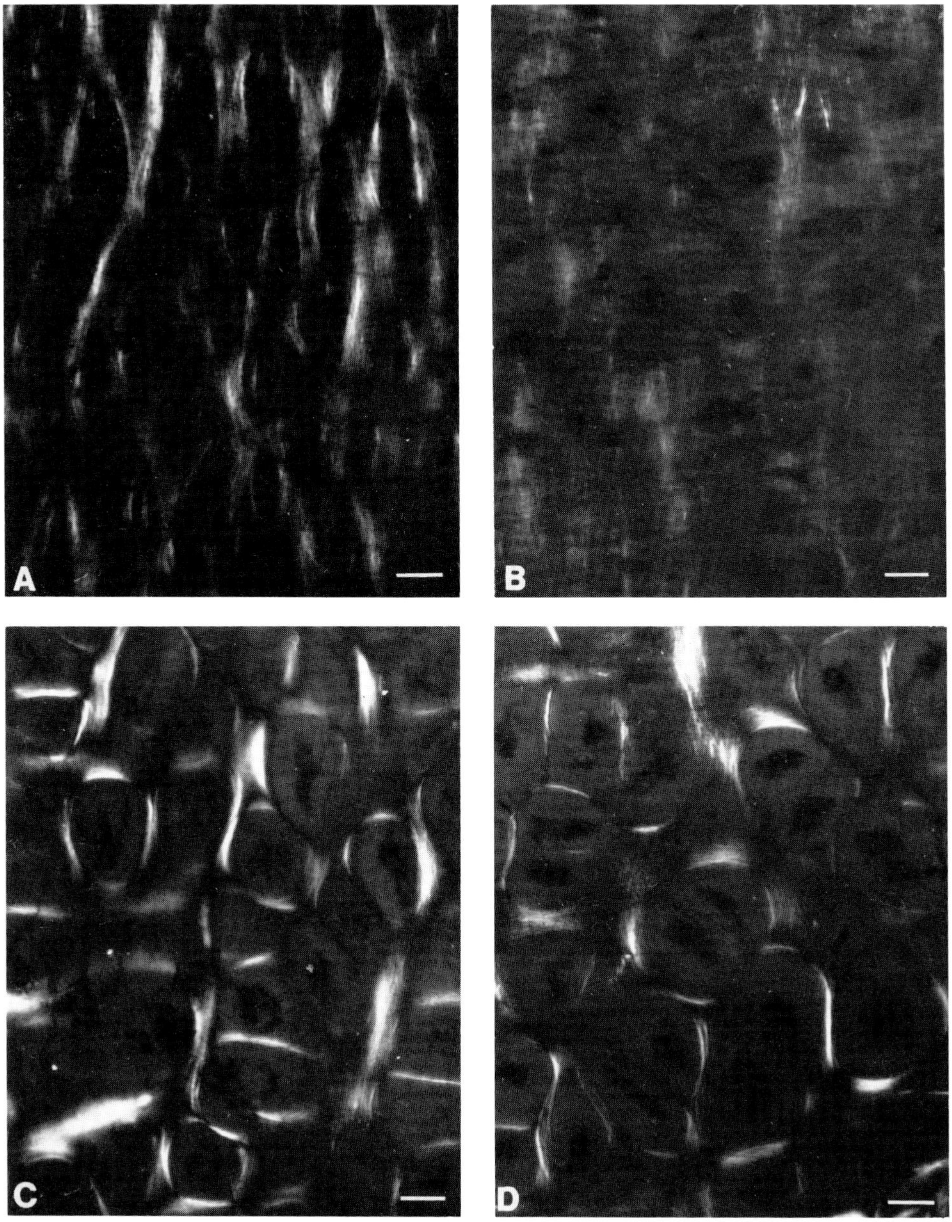

FIGURE 12. Polarized light micrographs taken from tibia cartilage of a 9-day-old chicken embryo after picrosirius red F3B staining. (A, B) zone 2, (C, D) zone 3;(A, C) longitudinal sections; (B, D) cross sections. Note the moderate birefringence of cartilage matrix in cross section of zone 2 compared to the longitudinal section of the same zone. The pericellular birefringence of zone 3 is similar in longitudinal and cross sections. Scales: 10 μm.

fully formed cartilage tissue. He argued that the orientation of collagen in the extracellular matrix is influenced by cells. Shah et al.[72] studied the development of crimps in chick tendons. Crimps are undulations of dense collagenous tissues such as tendons and their existence explains the relatively high extensibility of tendons at low tension. According to the findings of Shah et al.[72] at the onset of crimp formation the bending of collagen fibers occurs simultaneously with changes in cell morphology. At 14 d of development of the chick embryos, collagen fibrils without undulations and elongated cells are oriented parallel

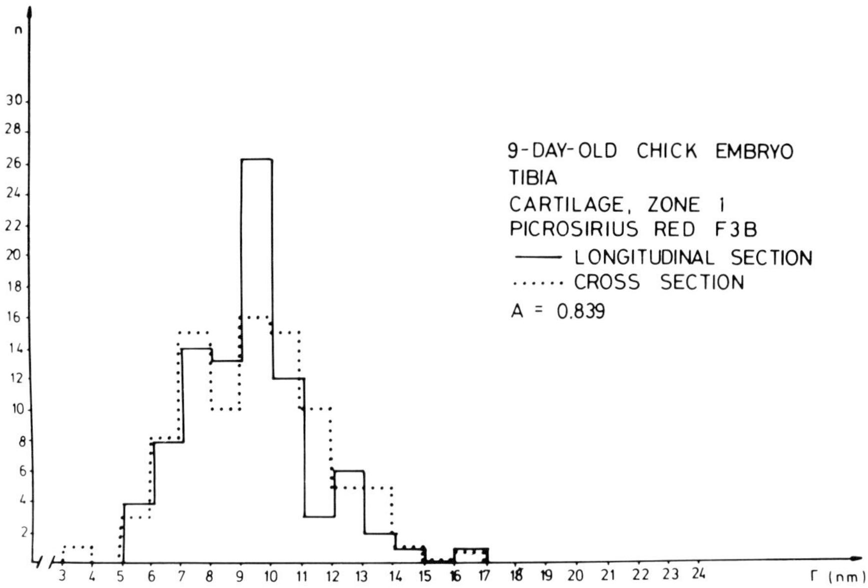

A

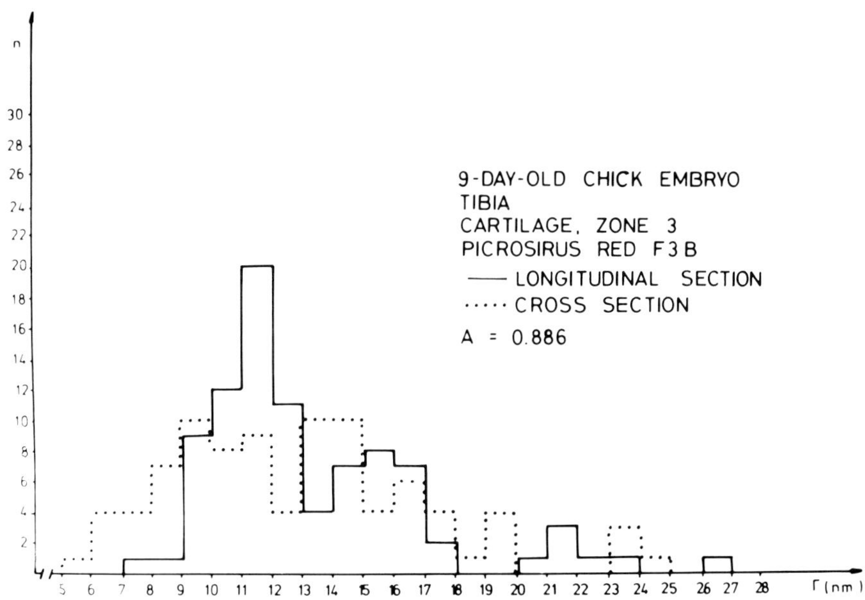

B

FIGURE 13. Retardation values measured in matrix of tibia cartilage zones 1(A) and 3(B) of a 9-day-old chicken embryo after picrosirius red F3B staining. As the ambiguity values (A) show, there are no significant differences between retardation values measured in longitudinal and cross sections. This is due to the tangential arrangement of collagen around the cells.

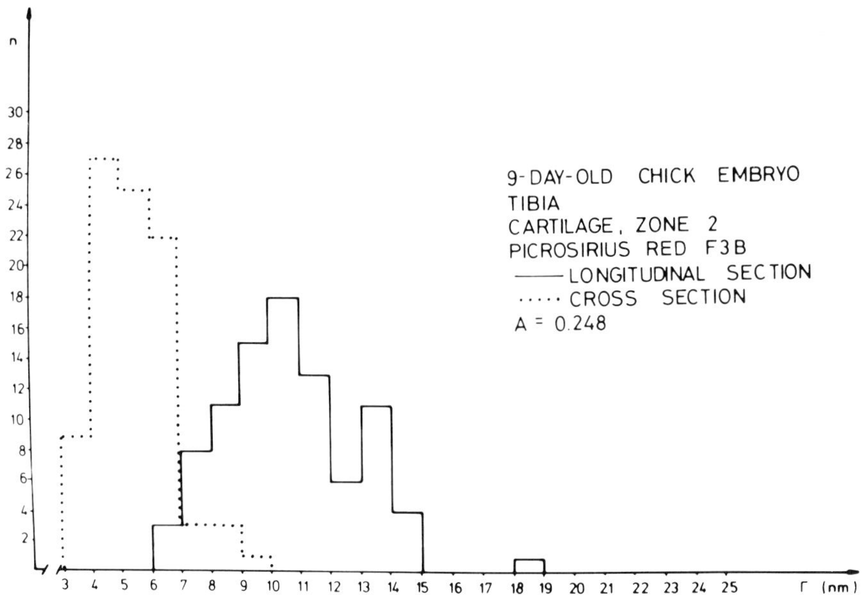

FIGURE 14. Retardation values recorded from the matrix of tibia cartilage zone 2 of a 9-day-old chicken embryo after picrosirius red F3B staining. As the ambiguity value (A) shows, significantly higher retardation values are measured in longitudinal than in cross section. This finding indicates a preferentially transversal orientation of collagen to the long axis of the cartilage anlage in zone 2. The directions of collagen and cell orientation are similar within this zone (compare with Figure 11).

to each other and to the tendon axis. At 15 d of development cells show bending along with the bending of fibrils. At 16 d , with further changes in cell shape, fibrils undulate to generate a band pattern characteristic of crimps as revealed by polarization microscopy. Shah et al.[72] conclude from these experimental data that fibroblasts which synthesize collagen are also "programmed" in some way to interact actively in the genesis of crimps.

Other experimental data, however, indicate that oriented collagen fibrils of the ECM are the major determinants of cell orientation. Fibroblasts and myoblasts plated *in vitro* on to straight laid collagen fibers have been found to elongate and align themselves along the fibers.[73,74] Similar observations were made studying the development of cornea[75,76] and cartilage.[77]

We think that both the orientation of cells and matrix constituents can mutually influence each other under *in vivo* conditions. Surfaces of cells and cell groups may influence the oriented deposition of some matrix components. If an oriented matrix structure is laid down, cells can be attached to it and they can align themselves along the preferential orientation axis of a matrix constituent, e.g., collagen fibril. The biosynthetic products of orderly arranged cells are secreted into the extracellular space, and the newly formed matrix components will assume an oriented arrangement along the cells. Elongated fibroblasts, for example, are regularly arranged in longitudinal columns and they are parallel to the collagen fibers in the tendons.[78] Longitudinally arranged membrane bound recesses are found at the ends of longitudinally oriented fibroblast processes. It is supposed that these orderly arrayed membrane bound recesses are the sites of formation and elongation of longitudinally oriented fibrils, and, furthermore, such an arrangement would provide means for the cell control of the length and orientation of the tendon bundles as they form.[1,79]

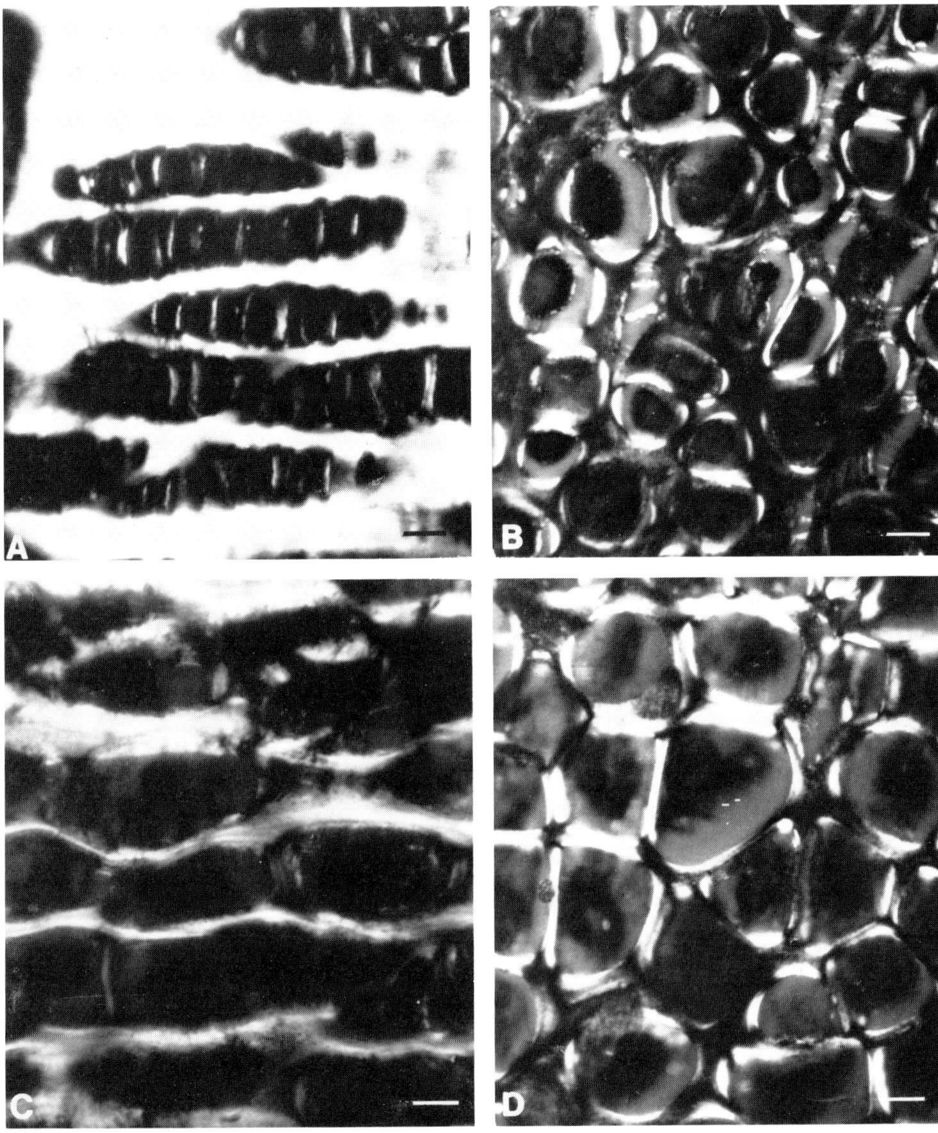

FIGURE 15. Proliferative (A,B) and hypertrophic (C,D) zones of rat epiphyseal cartilage in longitudinal (A,C) and cross sections (B,D), Polarized light micrographs after dimethylmethylen blue staining. Note that while the intercolumnar matrix is strongly anisotropic in longitudinal sections, it is hardly seen in cross sections. In cross sections, the birefringence of pericellular matrix prevails. Scales: 10 μm.

IV. ROLE OF MECHANICAL FACTORS

One of the most important functions of different connective tissues is the mechanical function. With some simplification, hyaline cartilage tissue, for example, is adapted to load bearing, while dense connective tissues, such as tendons, are specialized to withstand to tension. It is known that proper mechanical forces may stimulate the differentation of different connective tissues.[80,81] We wanted to investigate the possible role of mechanical factors in the formation of spatially oriented cartilage structure of the ECM.

For our experiments, a model has been chosen where cartilage is differentiated in a relatively indifferent mechanical environment, i.e., in the absence of loading.

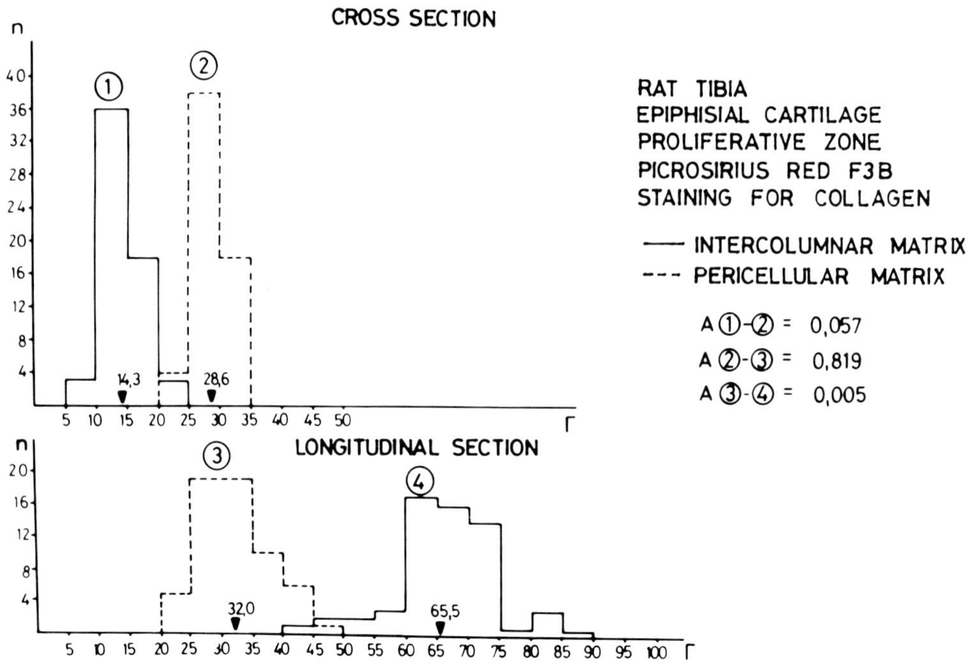

FIGURE 16. Retardation values measured in the proliferative zone of rat epiphyseal cartilage after staining with picrosirius red F3B. Significantly higher retardation values of birefringence are recorded from the longitudinally cut intercolumnar matrix as compared to the cross section of the same matrix compartment. As the ambiguity values (A) show, the pericellular birefringence does not differ significantly in cross and longitudinal sections.

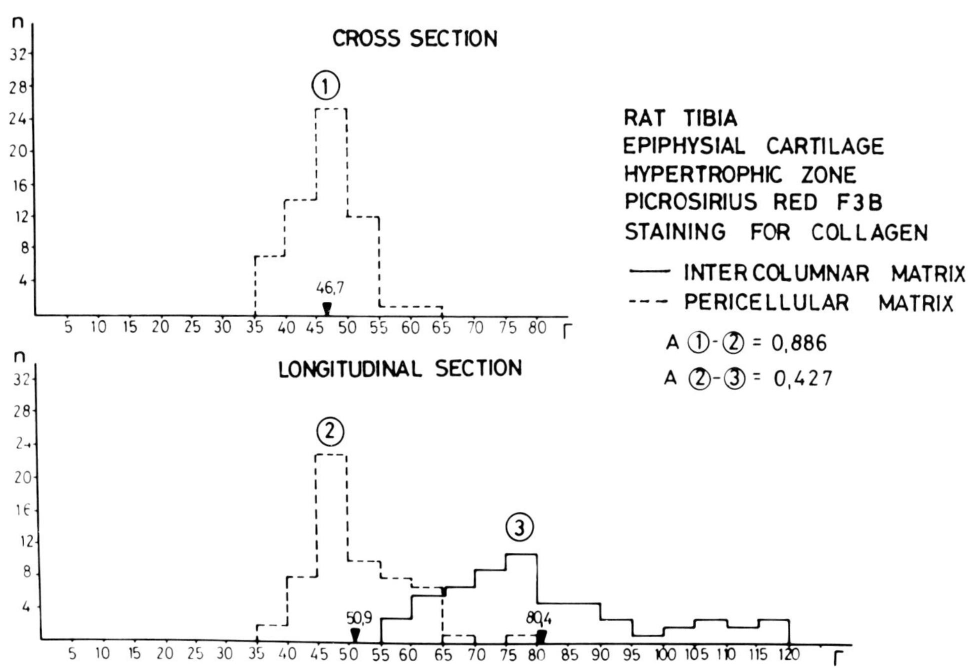

FIGURE 17. Retardation values measured in the hypertrophic zone of rat epiphyseal cartilage after picrosirius red F3B staining. The birefringence of the intercolumnar matrix cannot be measured in cross section because of its low intensity. Note the significantly higher values of the pericellular matrix in longitudinal sections.

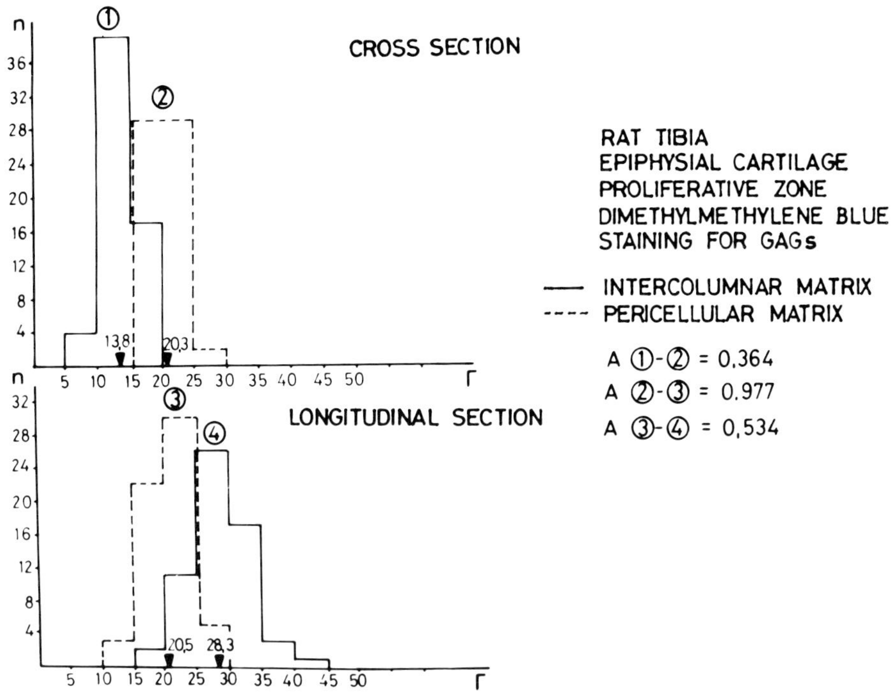

FIGURE 18. Retardation values of birefringence measured in the proliferative zone of rat epiphyseal cartilage after staining for GAGs. GAGs exhibit higher retardation values in the longitudinally cut intercolumnar matrix trabeculae than in cross sections, because they are oriented preferentially along the long axis of the trabeculae. The pericellular matrix, where GAGs are circularly arranged, show similar retardation values both in cross and longitudinal sections.

Subcutaneous or intramuscular implantation of demineralized bone or dentin has been shown to evoke the induction of responding mesenchymal cells into chondroblasts and bone cells.[82-87] The demineralized bone matrix induces the attraction, transformation and proliferation of mesenchymal cells. It has been propsed that this cascade of events occurs through the action of bone morphogenetic proteins present in demineralized bone matrix.[88,89]

CFY-rats weighing 200 g were used in our experiments. Demineralized bone pieces were prepared from diaphyseal segments of femora and tibiae. Samples were collected from freshly sacrified animals, cleaned, washed in phosphate buffered saline, and demineralized in 0.6 mol/l HCl (Figure 22). After washing several times in sterile distilled water, demineralized bone pieces were liophylized and stored at −20°C.

Twenty 200 g CFY-rats were operated on under anesthesia of Nembutal. 4 to 5 mm long demineralized bone matrix pieces, after cutting fine grooves in their edges, were implanted into intramuscular pouches made in the rectus abdominis muscle. Rats were sacrified after 2 and 3 weeks, and implants with small surrounding tissue were removed, fixed in a 4:1 mixture of absolute ethanol and neutralized formaldehyde, and embedded in wax. Sections were stained with picrosirius red F3B for collagen and with dimethylmethylene blue for sulfated GAGs, and then analyzed under the polarization microscope.

After 2 and 3 weeks of implantation, differently sized groups of cartilage cells or single cartilage cells were seen in the grooves or Haversian canals or lacunae of the implants. Cartilage cells were surrounded by a matrix which stained metachromatically indicating the presence of sulfated GAGs in the newly synthesized and deposited matrix. Metachromatic matrix areas were birefringent. It follows that GAGs are spatially oriented around the cartilage cells (Figure 23).

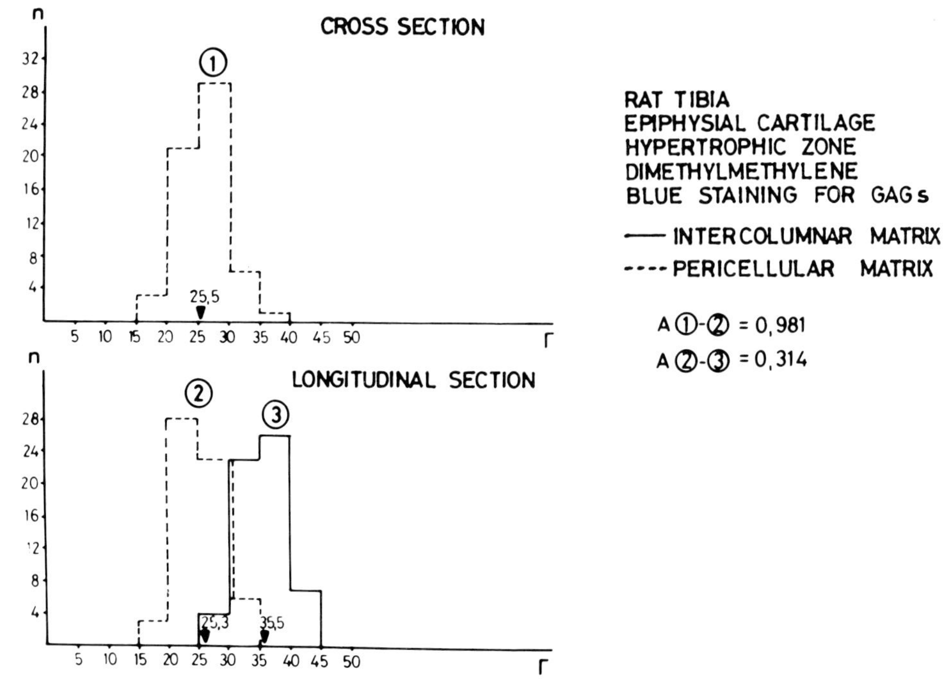

FIGURE 19. Retardation values measured in the hypertrophic zone of rat epiphyseal cartilage after staining with dimethylmethylene blue. Due to the low intensity of birefringence, retardation values cannot be measured in the intercolumnar marix when cut transversely. The birefringence of intercolumnar matrix is more intense in longitudinally cut intercolumnar matrix than in pericellular matrix.

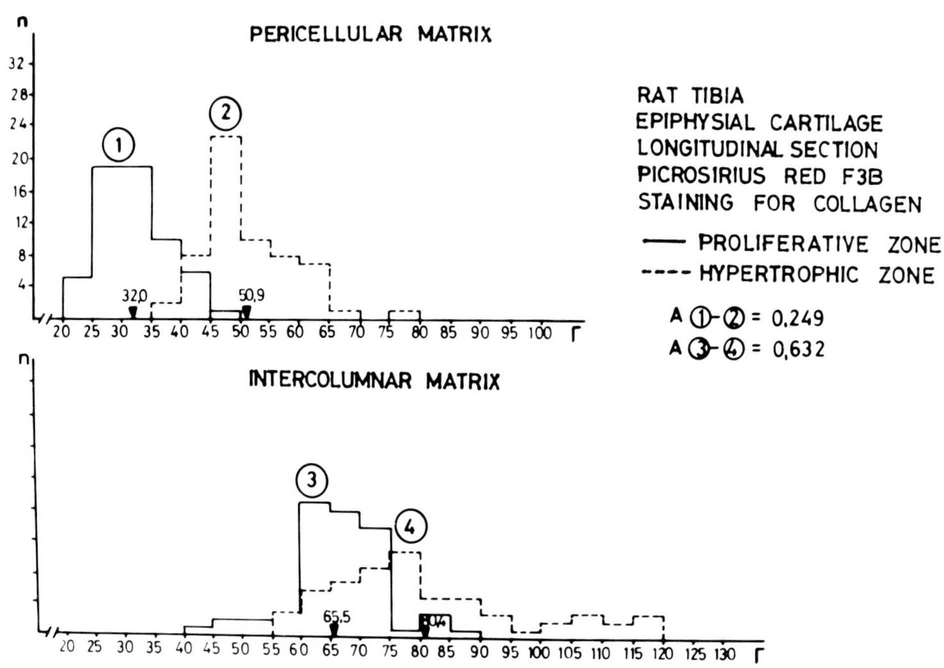

FIGURE 20. Distribution of retardation values in different zones and matrix compartments of rat epiphyseal cartilage after staining with picrosirius red F3B. Note the increase of birefringence towards the hypertropic zone.

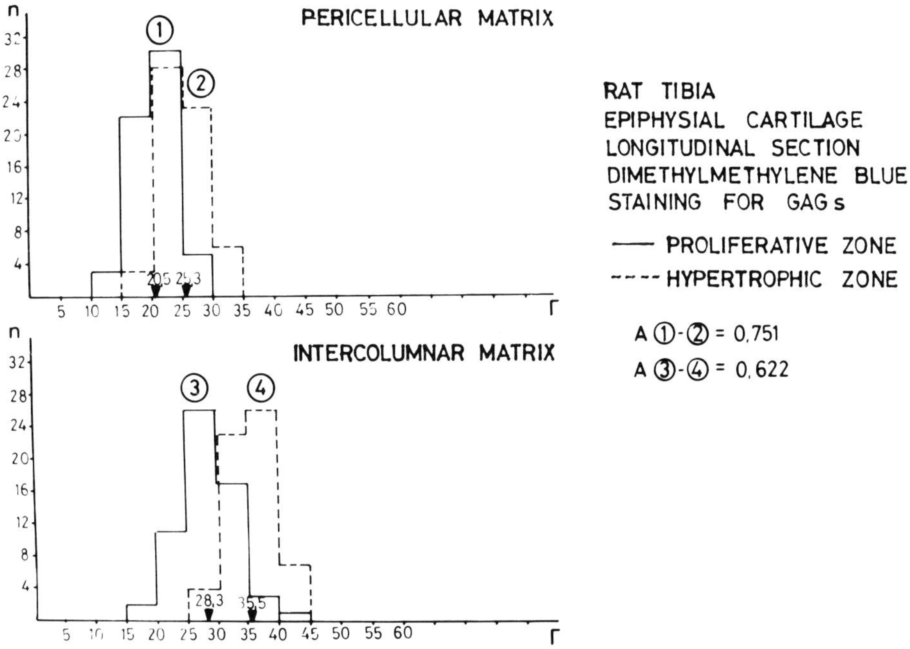

FIGURE 21. Distribution of retardation values in different zones and matrix compartments of rat epiphyseal cartilage after dimethylmethylene blue staining. GAGs of intercolumnar matrix are significantly more densely packed in the hypertrophic zone than in the proliferative zone as shown by the retardation values and ambiguity (A)-test.

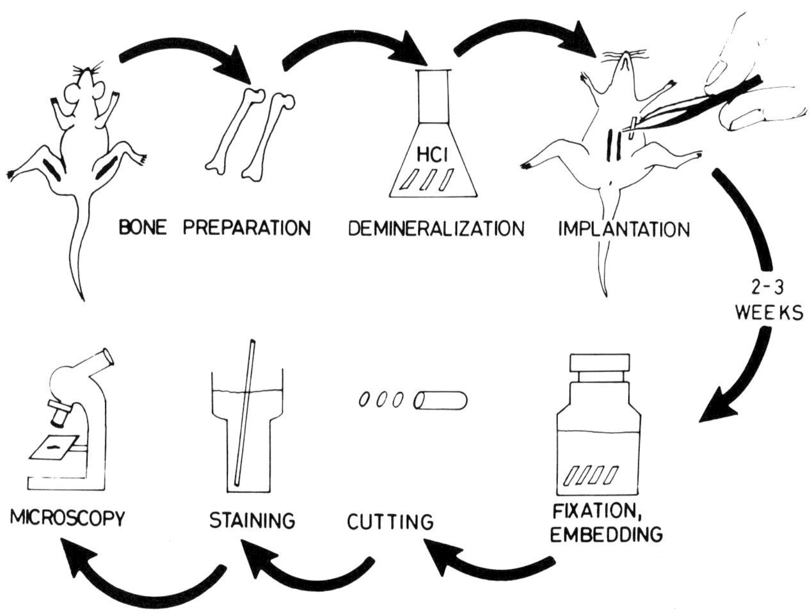

FIGURE 22. Sketch of our experiment carried out to induce cartilage by demineralized bone matrix implanted into the rectus abdominis muscle of rats.

Metachromasia and birefringence of the cartilage matrix could be completely abolished by antecedent digestion with testicular hyaluronidase (Figure 23). This suggests that predominantly chondroitin sulfate PGs are deposited. Picrosirius red F3B staining revealed the

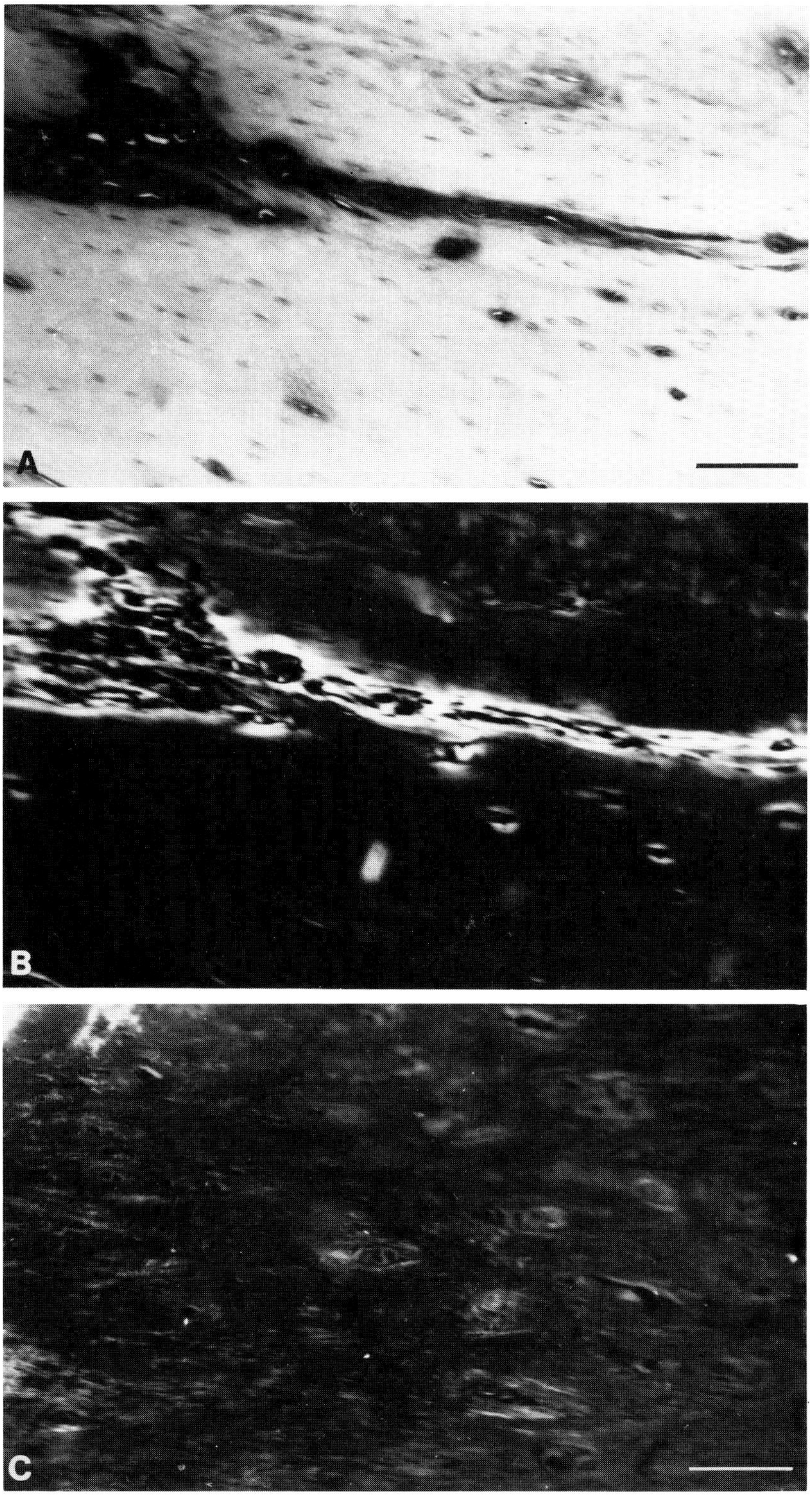

FIGURE 23. Three-week-old bone matrix implant after staining with dimethylmethylene blue for sulfated GAGs (A) Normal light micrograph; (B) the same visual field as seen in (A) taken in polarized light. Note the basophilic and birefringent matrix of cartilage formed in the implant. (C) Polarized light micrograph taken after the same staining carried out now after a digestion with testicular hyaluronidase. The birefringence of the pericellular cartilage matrix is abolished. Scales: 100 μm.

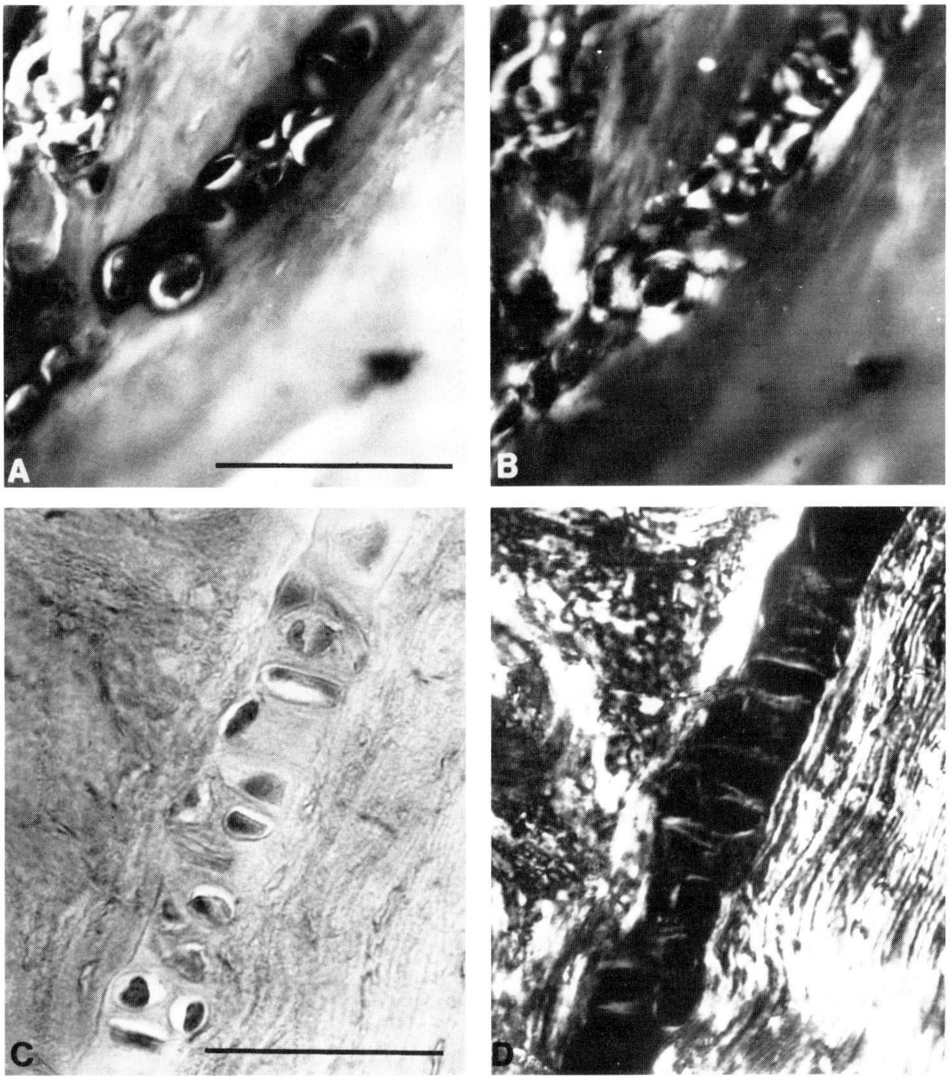

FIGURE 24. Cartilage islets in demineralized bone matrix developed after 3 weeks of implantation. (A, B) Dimethylmethylene blue; (C, D) Picrosirius red F3B staining. (A, C) normal light micrographs; (B, D) the same visual fields in polarized light. Note the pericellular birefringence in cartilage marix which is stronger after dimethylmethylene blue reaction than after staining with picrosirius red. Collagen lamellae of the bone matrix are also birefringent in (D). Scales: 100 μm.

presence of an oriented collagenous structure in the cartilage matrix (Figure 24). As can be inferred from the sign of birefringence, both chondroitin sulfate molecules and collagen fibrils are arranged circularly around the cartilage cells. The retardation value of birefringence measured in the pericellular matrix was higher after dimethylmethylene blue reaction ($\Gamma = 30.76 \pm 4.33$ nm) than after picrosirius red staining ($\Gamma = 22.01 \pm 4.95$). These data suggest that chondroitin sulfate molecules form a more compact structure in the pericellular matrix than collagen fibrils do.

A similar orientation pattern both of sulfated GAG molecules and collagen fibrils has been found in embryonic cartilage tissues (see Chapter 4). Thus, we can conclude that the elementary spatial orientation pattern, i.e., the circular arrangement of GAGs and collagen in the pericellular matrix can be differentiated without any specific mechanical stimuli in

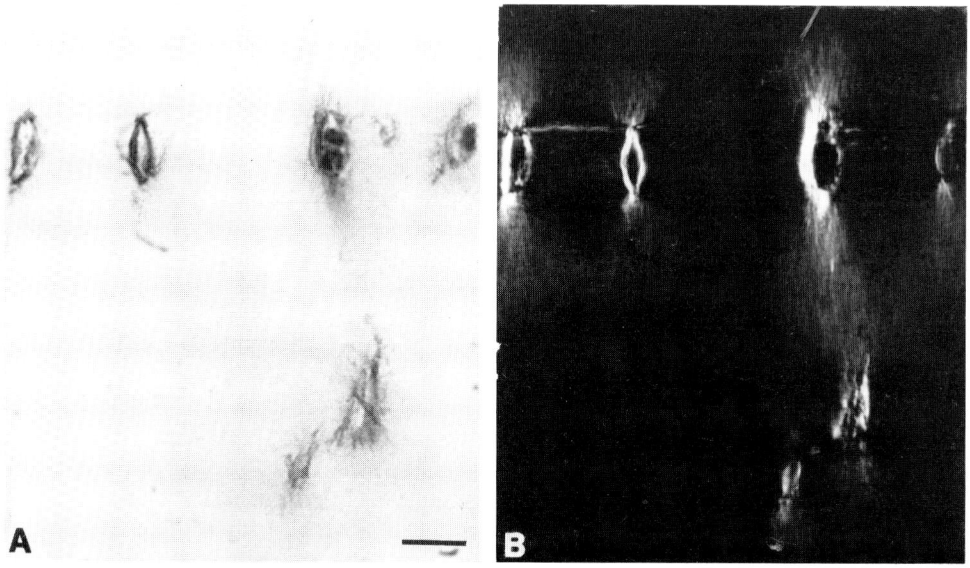

FIGURE 25. Culture of adult bovine chondrocytes maintained in agarose gel for 63d (kindly provided by Dr. M. B. Aydelotte, Chicago). Transversal section after staining with picrosirius red F3B is taken in normal (A) and polarized light (B). Chondrocytes scattered in the gel are surrounded by circularly oriented collagen as indicated by the birefringence seen in (B). Scale: 10 μm.

cartilage tissue. With high probability, the formation of this oriented microstructure is regulated by both cell-matrix and GAG-collagen interactions.

Our conclusions are further supported by the findings obtained in cell culture studies. Polarization microscopy reveals a circular orientation of chondroitin sulfate molecules and collagen in pericellular matrix formed by cells in chondrifying high density cultures (see Chapter 4) and by chondrocytes cultured in agarose gel (Figure 25). Except for atmospheric pressure, no specific mechanical stimuli effect cartilage cells in these models. Furthermore, polarization microscopic analysis of cartilage islets formed in different tumors also reveals the spatially oriented structure of GAGs and collagen around the cartilage cells (Figures 26, 27).

Finally, we would like to underline that our suggestion that elementary orientation patterns can be formed in the pericellular matrix of cartilage differentiated in an environment free of specific mechanical effect does not mean the denial of the role of mechanical factors in differentiation and modification of the structure of cartilage matrix. A number of *in vivo* and *in vitro* experimental evidences have demonstrated that both loading and immobilization may profoundly alter the composition and structure of the hyaline cartilage matrix.[90-101] In our opinion, however, specific exogeneous mechanical stimuli are not a prerequisite of the differentiation of a spatially oriented pericellular cartilage matrix structure, but they are (especially a proper loading) *sine qua non* for the formation of a cartilage matrix with a high load-bearing capacity.

ACKNOWLEDGMENTS

I wish to thank my former students, Drs. Z. Szekanecz, Andrea Kulcsár, and Erika Tóth for their invaluable help in carrying out the experiments and polarization microscopic measurements. Monospecific polyclonal antibody against vinculin was prepared and kindly provided by Dr. V. E. Koteliansky (Laboratory of Molecular and Cellular Cardiology,

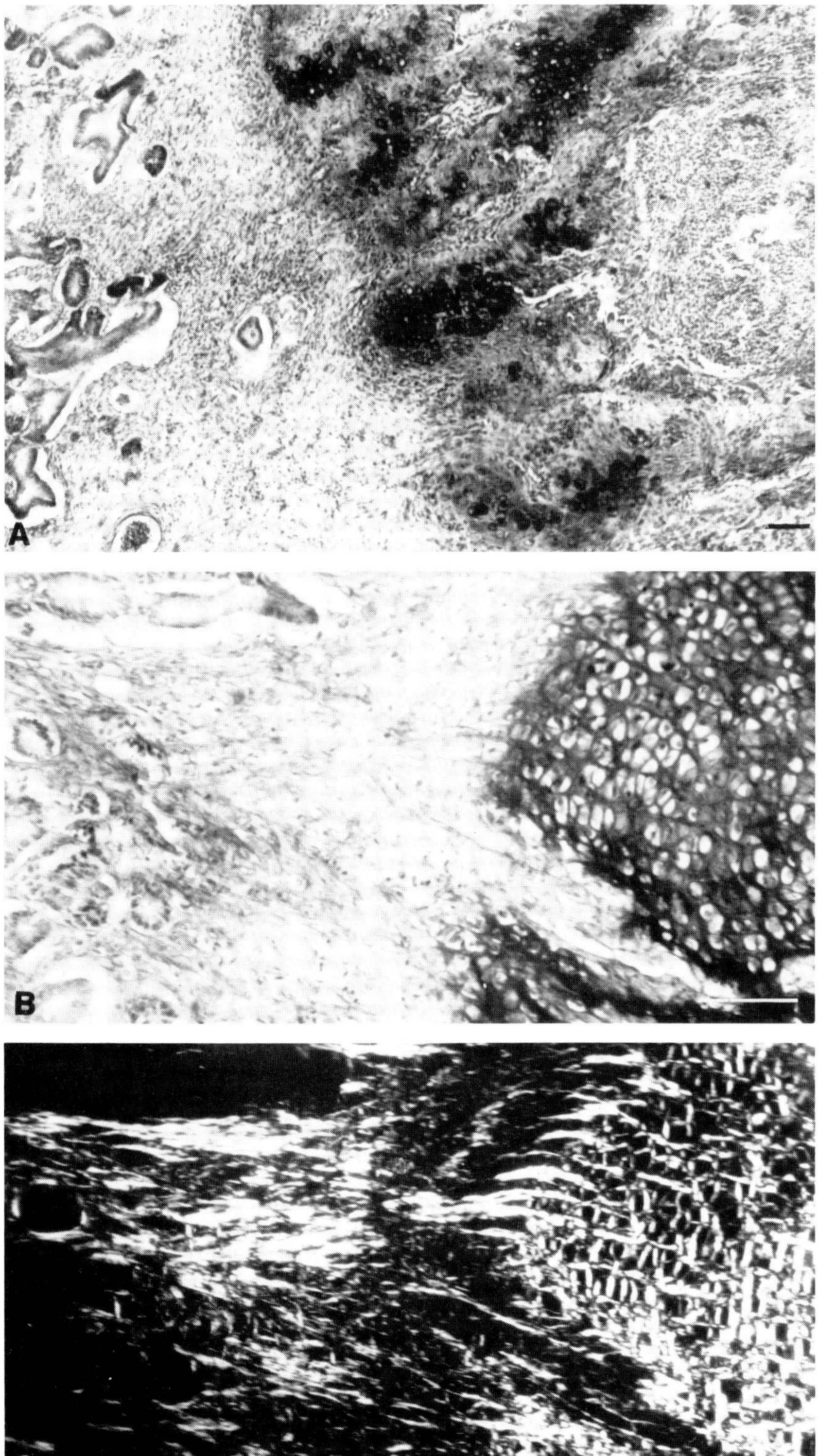

FIGURE 26. Cartilage islets in carcinoma ventriculi after staining with toluidine blue at pH 3.5 (A, B) Normal light micrographs showing the basophilic cartilage matrix. (C) The same visual field as seen in (B) taken in polarized light. Note the strong pericellular birefringence in cartilage. Connective tissue fibers are also birefringent. Scales: 100 μm.

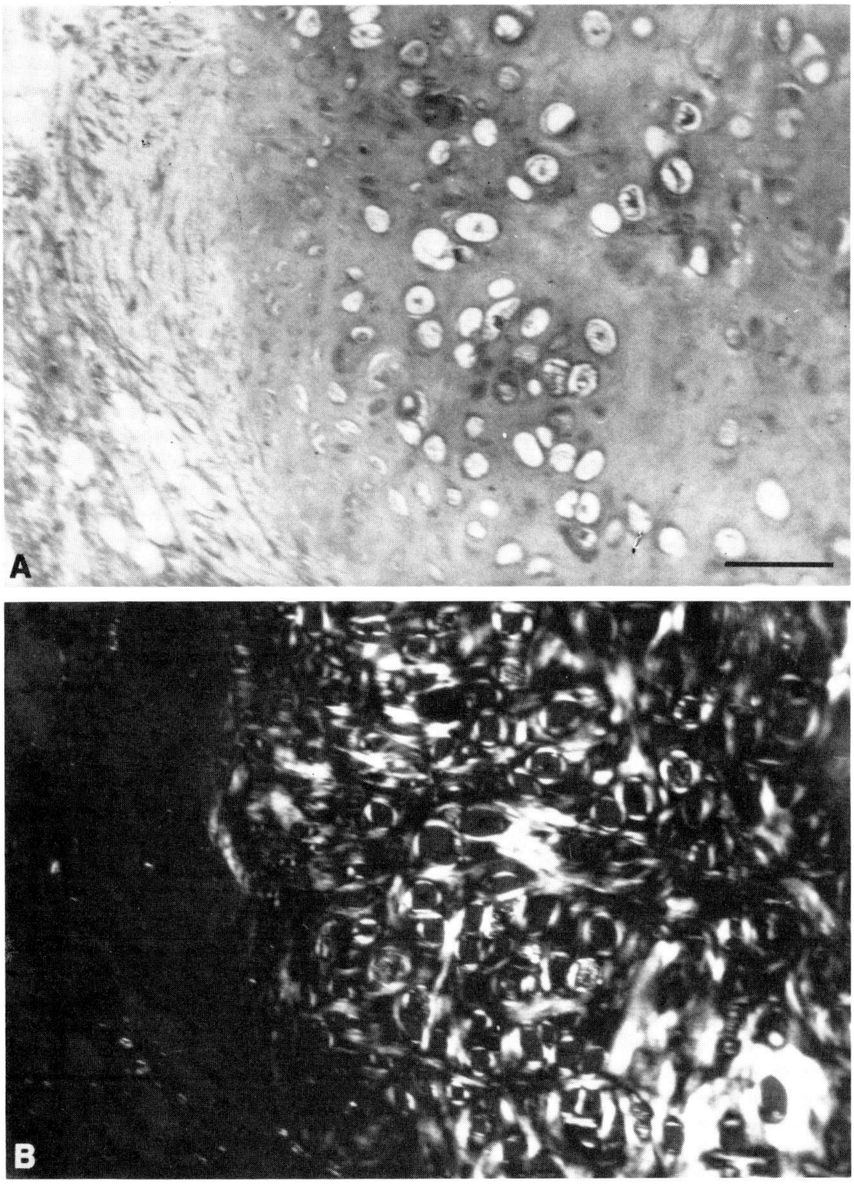

FIGURE 27. Hamartochondroma bronchii after staining with dimethylmethylene blue taken in normal (A) and polarized light (B). Observe the strong pericellular birefringence in (B) which indicates the oriented deposition of sulfated GAGs. Scale: 100 μm.

Institute of Experimental Cardiology, Cardiology Research Center, Moscow). Monoclonal antibody against chondronectin was a gift of Dr. Hynda K. Kleinman (Laboratory of Developmental Biology and Anomalies, NIDR, National Institute of Health, Bethesda). I am grateful to Professor L. Muszbek and Dr. Róza Ádány (Department of Clinical Chemistry, University Medical School, Debrecen) for their help in immunoblotting experiments. Chondrocyte cultures grown in agarose gel were prepared by Dr. Margaret B. Aydelotte (Department of Biochemistry, Rush-Presbyterian (St. Luke's Medical Center, Chicago). Blocks of different tumors containing cartilage were provided by Drs. Z. Szentirmay (National Research Institute for Oncology, Budapest), and V. Thomázy (Department of Pathology, University Medical School, Debrecen).

REFERENCES

1. **Trelstad, R. L. and Silver, F. H.**, Matrix assembly, in *Cell Biology of Extracellular Matrix*, Hay, E. D., Ed., Plenum Press, New York, 1981, chap. 7.
2. **Ehrmann, R. L. and Gey, G. O.**, The growth of cells in transparent gel or reconstituted rat tail collagen, *J. Natl. Cancer Inst.*, 16, 1375, 1956.
3. **Elsdale, T. and Bard, J.**, Collagen substrate for studies on cell behavior, *J. Cell Biol.*, 54, 626, 1972.
4. **Yang, J., Richards, J., Bowman, P., Guzman, R., Enami, J., McCormick, K., Hamamoto, S., Pitella, D., and Nandi, S.**, Sustained growth and 3D organization of primary mammary tumor epithelial cells embedded in collagen gels, *Proc. Natl. Acad. Sci. U.S.A.*, 76, 3401, 1979.
5. **Chambard, M., Gabrion, J., and Mauchamp, J.**, Influence of collagen gel on the orientation of epithelial cell polarity: follicle formation from isolated thyroid cells and from preformed monolayers, *J. Cell Biol.*, 91, 157, 1981.
6. **Yasui, N., Ohsawa, S., Ochi, T., Nakashima, H., and Ono, K.**, Primary culture of chondrocytes embedded in collagen gels, *Exp. Cell Biol.*, 50, 92, 1982.
7. **Tomasek, J. J. and Hay, E. D.**, Ultrastructure and stability of cytoskeleton of fibroblasts grown in collagen gels, *Anat. Rec.*, 202, 192A, 1982.
8. **Tomasek, J. J. and Hay, E. D.**, Analysis of the role of microfilaments and microtubules in acquisition of bipolarity and elongation of fibroblasts in hydrated collagen gels, *J. Cell Biol.*, 99, 536, 1984.
9. **Kimura, T., Yasui, N., Ohsawa, S., and Ono, K.**, Chondrocytes embedded in collagen gels maintain cartilage phenotype during long-term cultures, *Clin. Orthop.*, 186, 231, 1984.
10. **Allen, T. D., Schor, S. L., and Schor, A. M.**, An ultrastructural review of collagen gels, a model system for cell-matrix, cell-basement membrane and cell-cell interactions, *Scan Electron Microsc.*, 1, 375, 1984.
11. **Chen, J., Stuckey, E. C., and Berry, C. L.**, Three-dimensional culture of rat exocrine pancreatic cells using collagen gels, *Br. J. Exp. Pathol.*, 66, 551, 1985.
12. **Hodges, G. M., Rowlatt, C., Howlett, A. R., and Trejdosiewicz, L. K.**, An ultrastructural analysis of the physical organization of collagenous (type I) matrices: one determinant of urothelium maintenance *in vitro, Scan. Electron Microsc.*, 1, 215, 1985.
13. **Parry, G., Lee, E. Y., Farson, D., Koval, M., and Bissel, M. J.**, Collagenous substrata regulate the nature and distribution of glycosaminoglycans produced by differentiated cultures of mouse mammary epithelial cells, *Exp. Cell Res.*, 156, 487, 1985.
14. **Coates, P. W.**, Quantitation and morphological characterization of rapid axon and dendritic growth from single cerebral hemispheric neurons in hydrated collagen lattice culture, *Brain Res.*, 25, 11, 1986.
15. **Harris, A. K., Stopak, D., and Wild, P.**, Fibroblast traction as a mechanism for collagen morphogenesis, *Nature*, 290, 249, 1981.
16. **Dodd, N. J. F., Schor, S. L., and Rushton, G.**, The effects of a collagenous extracellular matrix on fibroblast membrane organization, *Exp. Cell Res.*, 141, 421, 1982.
17. **Allen, T. D. and Schor, S. L.**, The contraction of collagen matrices by dermal fibroblasts, *J. Ultrastruct. Res.*, 83, 205, 1983.
18. **Stopak, D. and Harris, A. K.**, Connective tissue morphogenesis by fibroblast traction. I. Tissue culture observations, *Dev. Biol.*, 90, 383, 1982.
19. **Bellows, C. G., Melcher, A. H., and Aubin, J. E.**, Association between tension and orientation of periodontal ligament fibroblasts and exogenous collagen fibres in collagen gels *in vitro, J. Cell Sci.*, 58, 125, 1982.
20. **Singer, I. I.**, The fibronexus: a transmembrane association of fibronectin-containing fibers and bundles of 5 nm microfilaments in hamster and human fibroblasts, *Cell*, 16, 675, 1979.
21. **Singer, I. I. and Paradiso, P. R.**, A transmembrane relationship between fibronectin and vinculin (130 kDa protein): serum modulation in normal and transformed hamster fibroblasts, *Cell*, 24, 481, 1981.
22. **Singer, I. I., Scott, S., Kawka, D. W., and Hassell, J. R.**, Extracellular matrix fibers containing fibronectin and basement membrane heparan sulfate proteoglycan coalign with focal contacts and microfilament bundles in stationary fibroblasts, *Exp. Cell Res.*, 173, 558, 1987.
23. **Abercombie, M., Heaysman, E. M., and Pegrum, S. M.**, The locomotion of fibroblasts in culture. IV. Electron microscopy of the leading lamellae, *Exp. Cell Res.*, 67, 359, 1971.
24. **Badley, R. A., Woods, A., Smith, C. G., and Rees, D. A.**, Actomyosin relationships with surface features in fibroblast adhesion, *Exp. Cell Res.*, 126, 263, 1980.
25. **Geiger, B., Volk, T., Volberg, T., and Bendori, R.**, Molecular interactions in adherens-type contacts, *J. Cell Sci.* Suppl., 8, 251, 1987.
26. **Buck, C. A. and Horwitz, A. F.**, Integrin, a transmembrane glycoprotein complex mediating cell-substratum adhesion, *J. Cell Sci.* Suppl., 8, 231, 1987.
27. **Ginsberg, M. H., Loftus, J. C., and Plow, E. F.**, Cytoadhesins, integrins, and platelets, *Thromb. Haemost.*, 59, 1, 1988.

28. **Fath, K. R., Edgell, C.-J. S., and Burridge, K.,** The distribution of distinct integrins in focal contacts is determined by the substratum composition, *J. Cell Sci.*, 92, 67, 1989.
29. **De Stroper, B., van der Schueren, B., Jaspers, M., Saison, M., Spaepen, M., van Leuven, F., van den Berghe, H., and Cassiman, J.-J.,** Distribution of the β_1 subgroup of the integrins in human cells and tissues, *J. Histochem. Cytochem.*, 37, 299, 1989.
30. **Burridge, K., Molony, L., and Kelly, T.,** Adhesion plaques : sites of transmembrane interaction between the extracellular matrix and the actin cytoskeleton, *J. Cell. Sci.* Suppl. 8, 211, 1987.
31. **Horwitz, A., Duggan, K., Buck, C., Beckerle, M. C., and Burridge, K.,** Interaction of plasma membrane fibronectin receptor with talin - a transmembrane linkage, *Nature,* 320, 531, 1986.
32. **Geiger, B.,** A 130-K protein from chicken gizzard : its localization at the termini of microfilament bundles in cultured chicken cells, *Cell,* 18, 193, 1979.
33. **Geiger, B.,** Involvement of vinculin in contact-induced cytoskeletal interaction, *Cold Spring Harbor Symp. Quant. Biol.*, 46, 671, 1982.
34. **Geiger, B.,** Membrane-cytoskeleton interaction, *Biochim, Biophys. Acta.* 737, 305, 1983.
35. **Burridge, K. and Fath, K.,** Focal contacts : transmembrane links between the extracellular matrix and the cytoskeleton, *Bioassays,* 10, 104, 1989.
36. **Koteliansky, V. E., Gneushev, G. N., and Belkin, A. M.,** Purification of a 175-kDa membrane protein, its localization in smooth and cardiac muscles. Interaction with cytoskeletal protein-vinculin, *FEBS Lett.*, 182, 67, 1985.
37. **Goldman, R. D., Goldman, A. E., Green, K. J., Jones, J. C. R., Jones, S. M., and Yang, H.-Y.,** Intermediate filament networks : organization and possible functions of a diverse group of cytoskeletal elements, *J. Cell Sci.* Suppl. 8, 327, 1987.
38. **Bissell, M. J. and Barcellos-Hoff, M. H.,** The influence of extracellular matrix on gene expression : is structure the message ?, *J. Cell Sci.* Suppl. 8, 327, 1987.
39. **Bissell, M. J., Hall, H. G., and Parry, G.,** How does the extracellular matrix direct gene expression, *J. Theor. Biol.*, 99, 31, 1982.
40. **Hynes, R. O.,** Fibronectin and its relation to cellular structure and behavior, in *Cell Biology of Extracellular Matrix,* Hay, E. D., Ed., Plenum Press, New York, 1981, chap. 10.
41. **Bornstein, P., Duksin, D., Balian, G., Davidson, J. M., and Grouch, E.,** Organization of extracellular proteins on the connective tissue cell surface : relevance to cell-matrix interactions *in vitro* and *in vivo, Ann. N.Y. Acad. Sci.*, 312, 93, 1978.
42. **Hay, E. D.,** Cell and extracellular matrix : their organization and mutual dependence, *Mod. Cell Biol.*, 2, 509, 1983.
43. **Zanetti, N. C. and Solursh, M.,** Induction of chondrogenesis in limb mesenchymal cultures by disruption of the actin cytoskeleton, *J. Cell Biol.* 99, 115, 1984.
44. **Rosen, D. M., Stempien, S. A., Thompson, A. Y., Brennan, J. E., Ellingsworth, L. R., and Seyedin, S. M.,** Differentiation of rat mesenchymal cells by cartilage-inducing factor. Enhanced phenotypic expression by dihydrocytochalasin B, *Exp. Cell Res.*, 165, 127, 1986.
45. **Brown, P. D. and Benya, P. D.,** Alterations in chondrocyte cytoskeletal architecture during phenotypic modulation by retinoic acid and dihydrocytochalasin B-induced reexpression, *J. Cell Biol.*, 106, 171, 1988.
46. **Benya, P. D., Brown, P. D., and Padilla, S. R.,** Microfilament modification by dihydrocytochalasin B causes retinoic acid-modulated chondrocytes to reexpress the differentiated collagen phenotype without a change in shape, *J. Cell Biol.*, 106, 161, 1988.
47. **Newman, P. and Watt, F. M.,** Influence of cytochalasin D-induced changes in cell shape on proteoglycan synthesis by cultured articular chondrocytes, *Exp. Cell Res.*, 178, 199, 1988.
48. **Hadházy, Cs., Szöllösi, J., and László, M. B.,** In vitro cartilage differentiation : modification of division cycle and effect of some polyanions, *Acta Histochem. (Jena)* Suppl. 27, 285, 1983.
49. **Linss, W., Neupert, G., and Langbein, L.,** Intermediärfilamente in den Zellen des hyalinen Knorpels, *Acta Histochem. (Jena),* 80, 29, 1986.
50. **Marchisio, P. C., Capasso, O., Nitsch, L., Cancedda, R., and Gionti, E.,** Cytoskeleton and adhesion patterns of cultured chick embryo chondrocytes during cell spreading and Rous sarcoma virus transformation, *Exp. Cell Res.*, 151, 332, 1984.
51. **Kepes, J. J and Perentes, E.,** Glial fibrillary acidic protein in chondrocytes of elastic cartilage in the human epiglottis : an immunohistochemical study with polyvalent and monoclonal antibodies, *Anat. Rec.*, 220, 296, 1988.
52. **Dessau, W., von der Mark, H., von der Mark, K., and Fischer, S.,** Changes in the patterns of collagens and fibronectin during limb-bud chondrogenesis, *J. Embryol. Exp. Morphol.*, 57, 51, 1980.
53. **Clemmensen, I., Holund, B., Johansen, N., and Andersen, R. B.,** Demonstration of fibronectin in human articular cartilage by an indirect immunoperoxidase technique, *Histochemistry,* 76, 51, 1982.
54. **Glant, T. T., Hadházy, Cs. Mikecz, K., and Sipos, A.,** Appearance and persistence of fibronectin in cartilage. Specific interaction of fibronectin with collagen type II, *Histochemistry,* 82, 149, 1985.

55. **Hewitt, A. T., Kleinman, H. K., Pennypacker, J. P., and Martin, G. R.,** Identification of an attachment factor for chondrocytes *Proc. Natl. Acad. Sci. U.S.A.,* 77, 385, 1980.
56. **Hewitt, A. T., Varner, H. H., Silver, M. H., Dessau, W., Wilkes, C. M., and Martin, G. R.,** The isolation and partial characterization of chondronectin, an attachment factor for chondrocytes, *J. Biol. Chem.,* 257, 2330, 1982.
57. **Hewitt, A. T., Varner, H. H., Silver, M. H., and Martin, G. R.,** The role of chondronectin and cartilage proteoglycan in the attachment of chondrocytes to collagen, in *Limb Development and Regeneration,* Part B, Kelley, R. O., Goettinck, P. F., and MacGabe, J. A., Eds., Alan R. Liss, New York, 1983, 25.
58. **Moellenhauer, J. and von der Mark, K.,** Isolation and characterization of a collagen-binding glycoprotein from chondrocyte membranes, *EMBO J.,* 2, 45, 1983.
59. **Pfäffle, M., Ruggiero, F., Hofmann, H., Fernández, M. P., Selmin, O., Yamada, Y., Garrone, R., and von der Mark, K.,** Biosynthesis, secretion and extracellular localization of anchorin CII, a collagen-binding protein of the calpactin family, *EMBO J.,* 7, 2335, 1988.
60. **Roughley, P. J., White, R. J., and Poole, A. R.,** Identification of a hyaluronic acid-binding protein that interferes with the preparation of high-buoyant-density proteoglycan aggregates from adult human articular cartilage, *Biochem. J.,* 231, 129, 1985.
61. **Heinegard, D. and Sommarin, Y.,** Macromolecules and interactions at the chondrocyte cell surface, *J. Clin. Chem. Clin. Biochem.,* 24, 937, 1986.
62. **Lacy, B. E. and Underhill, C. B.,** The hyaluronate receptor is associated with actin filaments, *J. Cell. Biol.,* 105, 1395, 1987.
63. **Hamburger, V. and Hamilton, L.,** A series of normal stages in the development of the chick embryo, *J. Morphol.,* 88, 49, 1951.
64. **Oohira, A., Kimata, K., Suzuki, S., Takata, K., Suzuki, I., and Hoshino, M.,** A correlation between synthetic activities for matrix macromolecules and specific stages of cytodifferentiation in developing cartilage, *J. Biol. Chem.,* 249, 1637, 1974.
65. **Stocum, D. L., Davis, R. M., Leger, M., and Conrad, H. E.,** Development of the tibiotarsus in the chick embryo : biosynthetic activites of histologically distinct regions, *J. Embryol. Exp. Morphol.,* 54, 155, 1979.
66. **Shinomura, T., Kimata, K., Oike, Y., Maeda, N., Yano, S., and Suzuki, S.,** Appearance of distinct types of proteoglycan in a well-defined temporal and spatial pattern during early cartilage formation in the chick limb, *Dev. Biol.,* 103, 211, 1984.
67. **Schmid, T. M. and Linsenmayer, T. F.,** Developmental acquisition of type X collagen in the embryonic chick tibiotarsus, *Dev. Biol.,* 107, 373, 1985.
68. **Eggli, P., Herrmann, W., Hunziker, E. B., and Schenk, R. K.,** Matrix compartments in the growth plate of the proximal tibia of the rats, *Anat. Rec.,* 211, 246, 1985.
69. **Speer, D. P.,** Collagenous architecture of the growth plate and perichondrial ossification groove, *J. Bone Joint Surg. (Am.)* 64, 399, 1982.
70. **Dallek, M., Jungbluth, K. H., and Holstein, A. F.,** Studies on the arrangement of the collagenous fibers in the infant epiphyseal plates using polarized light and the scanning electron microscope, *Arch. Orthop. Trauma. Surg.,* 101, 239, 1983.
71. **Trelstad, R. L.,** Mesenchymal cell polarity and morphogenesis of chick cartilage, *Dev. Biol.,* 59, 153, 1977.
72. **Shah, J. S., Palacios, E., and Palacios, L.,** Development of crimp morphology and cellular changes in chick tendons, *Dev. Biol.,* 94, 499, 1982.
73. **Elsdale, E. and Bard, J. B. L.,** Cellular interactions in mass culture of human diploid fibroblasts, *Nature,* 236, 152, 1972.
74. **Yoshizato, K., Obinata, T., Huang, H., Matsuda, R., Shioya, N., and Miyata, T.,** *In vitro* orientation of fibroblasts and myoblasts on aligned collagen film, *Dev. Growth Differ,* 23, 175, 1981.
75. **Coulombre, A. J. and Coulombre, J. L.,** The development of the structural and optical properties of the cornea, in *The Structure of the Eye,* Smelser, G. K., Ed., Academic Press, New York, 1961, 405.
76. **Bard, J. and Higginson, K.,** Fibroblast collagen interaction in the formation of the secondary stroma of the chick cornea, *J. Cell Biol.,* 74, 816, 1977.
77. **Gould, R. P., Selwood, L., Day, A., and Wolpert, L.,** The mechanism of cellular orientation during early cartilage formation in the chick limb and regenerating amphibian limb, *Exp. Cell Res.,* 83, 287, 1974.
78. **Squier, C. A. and Bausch, W. H.,** Three-dimensional organization of fibroblasts and collagen fibrils in rat tail tendon, *Cell Tissue Res.,* 238, 319, 1984.
79. **Trelstad, F. L. and Hayashi, K.,** Tendon collagen fibrillogenesis : intracellular subassemblies and cell surface changes associated with fibril growth, *Dev. Biol.,* 71, 228, 1979.
80. **Krompecher, St.,** Die Beinflussbarkeit der Gewebsdifferenzierung der Granulierenden Knochenober-flächen, insbesondere die der Kallusbildung, *Arch. Klin. Chir.,* 281, 472, 1956.
81. **Krompecher, St.,** *Form und Funktion in der Biologie,* Akademische Verlagsgesellschaft, Geest und Portig K.-G., Leipzig, 1966.

82. **Urist, M. R.**, Bone formation by autoinduction, *Science,* 150, 893, 1965.
83. **Reddi, A. H. and Anderson, W. A.**, Collagenous bone matrix-induced endochondral ossification and hemopoiesis, *J. Cell Biol.*, 69, 557, 1976.
84. **Seyedin, S. M., Thompson, A. Y., Rosen, D. M., and Piez, K. A.**, *In vitro* induction of cartilage-specific macromolecules by a bone extract, *J. Cell Biol.*, 97, 1950, 1983.
85. **Inoue, T., Deporter, D. A., and Melcher, A. H.**, Induction of cartilage and bone by dentin demineralized in citric acid, *J. Periodont. Res.*, 21, 243, 1986.
86. **Inoue, T., Deporter, D. A., and Melcher, A. H.**, Induction of chondrogenesis in muscle, skin, bone marrow, and periodontal ligament by demineralized dentin and bone matrix *in vivo* and *in vitro, J. Dent. Res.*, 65, 12, 1986.
87. **Guterman, I. A., Boman, T. E., Wang, G.-J., and Balian, G.**, Bone induction in intramuscular implants by demineralized bone matrix : sequential changes of collagen synthesis *Coll. Relat. Res.*, 8, 419, 1988.
88. **Nakagawa, M. and Urist, M. R.**, Chondrogenesis in tissue cultures of muscle under the influence of a diffusible component of bone matrix, *Proc. Soc. Exp. Biol. Med.* 154, 568, 1977.
89. **Seyedin, S. M., Thomas, T. C., Thompson, A. Y., Rosen, D. M., and Piez, K. A.**, Purification and characterization of two cartilage-inducing factors from bovine demineralized bone, *Proc. Natl. Acad. Sci. USA,* 82, 2267, 1985.
90. **Sääf, J.**, Effect of exercise on adult articular cartilage. An experimental study on guinea-pigs with relevance to the continous regeneration of adult cartilage, *Acta Orthop. Scand.* Suppl. 7, 1, 1950.
91. **Kostenszky, K. S. and Oláh, E. H.**, Effect of increased functional demand on the glycosaminoglycan (mucopolysaccharide) content of the articular cartilage, *Acta Biol. Acad. Sci. Hung.*, 23, 75, 1972.
92. **Oláh, E. H. and Kostenszky, K. S.**, Effect on altered functional demand on the articular cartilage of dogs, *Acta Biol. Acad. Sci. Hung.*, 23, 195, 1972.
93. **Caterson, B. and Lowther, D. A.**, Changes in the metabolism of the proteoglycans from sheep articular cartilage in response to mechanical stress, *Biochim. Biophys. Acta,* 540, 412, 1978.
94. **Palmoski, M. J., Colyer, R. A., and Brandt, K. D.**, Joint motion in the absence of normal loading does not maintain normal articular cartilage, *Arthritis Rheum.*, 23, 325, 1980.
95. **Palmoski, M. J. and Brandt, K. D.**, Running inhibits the reversal of atrophic changes in canine knee cartilage after removal of a leg cast, *Arthritis Rheum.*, 24, 1329, 1981.
96. **Tammi, M., Säämänen, A.-M., Jauhiainen, A., Malminen, C., Kiviranta, J., and Helminen, H. J.**, Proteoglycan alterations in rabbit knee articular cartilage following physical exercise and immobilization, *Connect. Tissue Res.*, 11, 45, 1983.
97. **Paukkonen, K., Selkäinaho, K., Jurvelin, J., and Helminen, H. J.**, Morphometry of articular cartilage: a stereological method using light microscopy, *Anat. Rec.*, 210, 675, 1984.
98. **Jurvelin, J., Kiviranta, I., Tammi, M., and Helminen, H. J.**, Effect of physical exercise on indentation stiffness of articular cartilage in the canine knee, *Int. J. Sports Med.*, 7, 106, 1986.
99. **Säämänen, A-M., Tammi, M., Kiviranta, I., Jurvelin, J., and Helminen, H. J.**, Maturation of proteoglycan matrix in articular cartilage under increased and deccreased joint loading. A study in young rabbits, *Connect. Tissue Res.*, 16, 163, 1987.
100. **Klein-Nulend, J., Veldhuijzen, J. P., van der Stadt, R. J., van Kampen, G. P. J., Kuijer, R., and Burger, E. H.**, Influence of intermittent compressive force on proteoglycan content in calcifying growth plate cartilage *in vitro, J. Biol. Chem.*, 262, 15490, 1987.
101. **Veldhuijzen, J. P., Huisman, A. H., Vermeiden, J. P. W., and Prahl-Andersen, B.**, The growth of cartilage cells *in vitro* and the effect of intermittent compressive force. A histological evaluation, *Connect. Tissue Res.*, 16, 187, 1987.

Chapter 10

POLARIZATION MICROSCOPY OF AMYLOID

TABLE OF CONTENTS

I. INTRODUCTION

Amyloid is a proteinaceous material of diverse composition and origin deposited in the extracellular space of different tissues under various pathologic conditions. Although this volume is devoted to the description of certain ultrastructural properties of the extracellular matrix in normal tissues, this chapter is also included for two reasons.

One of them is the well known fact that polarization microscopy is widely used for routine histopathologic diagnosis of amyloid, and amyloid was one of substances in which the ultrastructure had been revealed by the polarization microscopy prior to the first relevant electron microscopic observations.

The other reason for discussing polarization microscopy of amyloid here is that the analysis of a peculiar topo-optical staining reaction of amyloid allowed us the modification of the theory of metachromatic staining reactions. This reaction was the toluidine blue staining below pH 3.5 resulting in an orthrochromatic (blue) staining and strong optical anisotropy with a positive sign. Based on the experimental data described in this chapter, we suggest that the old assumption that orthochromatically stained tissue constituents are isotropic, consequently amorphous materials, is no longer valid.

II. COMPOSITION AND STRUCTURE OF AMYLOID

Amyloid deposits contain fibrils and nonfibrillar materials. Chemical analyses of isolated amyloid fibrils reveal the presence of different classes of proteins in the fibrils.

The first proteins were purified from fibrils obtained predominantly from the tissues of patients suffering from the primary form of amyloidosis. The chemical analyses showed that in this group of amyloid fibrils the major protein component, designated AL, can be characterized by several subgroups and it was composed of a homogeneous intact immunoglobulin light immunoglobulin polypeptide chain, the N-terminal fragment or fragments of a light polypeptide chain, or both an intact light polypeptide chain and its homologous N-terminal fragment or fragments.[1-5] These analytical data suggest that the cellular source of the amyloid fibrils of immunoglobulin origin is probably a plasma cell clone whose synthetic products are either locally deposited or disseminated by the bloodstream.[4,6]

Analyses of the major protein obtained from amyloid fibril preparations, derived predominantly from patients with the secondary form of amyloidosis, revealed the existence of a further class of amyloid fibril proteins designated protein AA.[7-11] Unlike the major amyloid fibril protein of immunoglobulin origin (AL), the amyloid fibril proteins of the AA type shared a common N-terminal amino acid sequence.[8-10] The AA proteins have different molecular weights ranging from 5 to 9 kDa. This suggests that, as in the case of amyloid fibril proteins of immunoglobulin origin, this class of amyloid fibril proteins may also be derived by proteolysis of a serum precursor.[5,9] The cell type which produces these proteins or their precursors is not known. It has been demonstrated that, while the AA protein has not been found in amyloid fibrils of immunoglobulin origin, immunoglobulins or their fragments invariably constitute a lesser part of the AA amyloid fibril.[9,12]

Other chemical types of amyloid fibrils have also been identified. Amyloid deposits are frequently associated with endocrine tumors, e.g., thyroid medullary carcinoma or insulinomas. In these hormone-secreting apudomas, the amyloid protein is the hormone or prohormone secreted by the tumor cells.[13-17] Other amyloid fibril constituents have also been found.[4]

Nonfibrillar components are also present in amyloid deposits. A glycoprotein with a pentagonal structure (P-component) having a 3.5 nm wide outer shell and a 2.5 nm inner core has been found to be a small (approx. 10%) but a constant constituent of all amyloid deposits.[18-20] It is not a constituent of the fibril, and it is not involved in fibril formation.[5] Presumably the P-component explains the PAS-reaction of the amyloid deposits. However, carbohydrates have also been found in the fibrils although in smaller amounts.[21] Mucopolysaccharides (glycosaminoglycans) have also been shown in amyloid in low local concentrations.[22,23]

It has been shown by electron microscopic investigations that amyloid deposits were composed of a felt-like array of 7.5 nm to 10 nm nonbranching, aggregated fibrils.[24,25] These fibrils have a hollow core profile with a 2.5 nm center. The amyloid fibrils are composed of aggregates of two to five 2.5 to 3.0 nm wide filaments arranged parallel to one another and disposed more or less longitudinally or obliquely to the long axis of the fibril. The filaments are arranged in the fibrils showing a twisted-ribbon appearance. Cross-sectioning of such a fibril would produce the 2.5 nm hollow core profile.[5,15] The fibrillar structure of amyloid explains its anisotropic properties in polarized light after different stainings.

Based on X-ray crystallography, it has been found that the polypeptide chains of amyloid fibrils are arranged perpendicularly to the axis of the fibril with the conformation of a meridional or cross-β pleated sheet.[26] This conformation characterizes a wide variety of amyloid fibril proteins.[4,5,15] Infrared spectroscopy has revealed an antiparallel β-pleated sheet structure of the amyloid fibril protein.[27] The above data have been incorporated into a model of the amyloid fibril.[28,29] According to this, the amyloid fibril is a macromolecular double helix with 100 nm periodicity, formed from two twisted β-pleated sheet micelles. Each of the paired filaments is 4 nm in diameter, and the filaments are separated by a 2 nm gap from each other. Fibrils with other parameters have also been identified.[4,5] The gap between the filaments is filled by glycoproteins (P-component) forming a tubular protein core which is the proposed binding site of the iodine particles after ''staining'' with iodine which also enhances the optical anisotropy of the amyloid.[29] The filaments are composed of the fibril proteins with β-pleated sheet conformation running perpendicular to the axis of the fibril. The cross-β conformation explains the presence of the axial grooves on the filament surfaces which are the suggested places for axiparallel binding of dye molecules after a staining of the amyloid with congo red, toluidine blue, and other dyes.[29]

Although the amyloids have different chemical compositions, they have a common structural feature in all cases, namely the twisted β-pleated sheet fibril. Because of this inherent and specific nature of amyloid, amyloid deposits, and amyloidosis are designated as β-fibrilloses.[4,30]

It is worth mentioning here that there are structural similarities between the amyloid fibrils and cytoskeletal intermediate filaments.[31] Based on these, a hypothesis has been advanced that material with features of amyloid may form as the result of an accumulation of intermediate filaments extracellularly.[31] As an example, a frequently occuring pathological alteration of the human brain called Alzheimer's disease can be mentioned. In this disease, congophil and birefringent materials are accumulated in three different locations: in perikaryons of degenerated neurons, in the neuritic plaques, and in the walls of vessels. In all of these places, paired helical β-pleated sheet filaments were identified indicating the amyloid nature of these materials.[32-35] The perykaryons of the degenerating neurons contain high amounts of paired helical filaments forming the so-called neurofibrillary tangles. Electron microscopically, the filaments bear resemblence to normal intermediate filaments (neurofilaments) except for their helical configuration. It is supposed that the first event of this disease is the formation of these abnormal neurofilaments intracellularly. The paired helical filaments accumulate progressively in neuronal cell bodies. These filaments have dimensions differing from those typically described. They are composed of two 12 nm filaments paired to give a 24 nm fibril with an 80 nm helical repeat.[5] After cell death, these filaments are deposited extracellularly forming the amyloid cores of the plaques. Using Congo red staining and polarization microscopy to identify senile plaques and neurofibrillary tangles in Alzheimer's degeneration, it has long been supposed that neurofibrillary tangles could be intracellular forms of the amyloid.[36]

Other morphologically different fibers are seen in corpore amylacea of the prostate and in other concretions.[5,37]

III. STAINING REACTIONS OF AMYLOID

A number of stainings and histochemical reactions has been proposed to demonstrate amyloid in tissue sections.[5,38-46] The vast majority of them are highly empirical techniques with unknown staining mechanisms and specificity.

The most classical stain, currently and widely used for identification of amyloid, is congo red. Divry and Florkin[36] were the first who applied polarization microscopy to demonstrate amyloid deposits stained with congo red. Numerous efforts have been made to elucidate the staining mechanism and the optical behavior of this topo-optical reaction.[5,29,38,39,47-52] Different modifications have been proposed to obtain green polarization color appearing selectively in amyloid. Generally, alkaline congo red solutions are used.[29,39,53] Due to the fact that collagen can also be reactive with congo red,[54] a further modification has been suggested, namely the use of a polar mounting medium (gum arabic) instead of the apolar Canada balsam.[55] In gum arabic, amyloid and collagen fibers exhibit birefringence with opposite signs in a polarization microscope, the sign of the amyloid being positive and that of the collagen negative with respect to the long axis of the fibers.

The anionic congo red molecules of planar configuration bind to the amyloid fibrils by non-specific and non-ionic interactions, probably by close range, hydrophobic forces, and the assymetrical and relative long dye molecules are oriented with their long axes parallel to that of the fibrils.[4,5,29] This orientation pattern of the congo red molecules is dependent upon the β-conformation of the fibril proteins, and it results in a strong birefringence with positive sign (Figure 1). Due to the structural similarity described previously between the amyloid and neurofilament tangles as well as prostatic corpora amylacea, these structures exhibit the same optical behavior (Figures 2 and 3).

Congo red staining is the most important histochemical reaction of amyloid. The appearance of a birefringence with positive sign and in green polarization color after this staining is a generally accepted morphological criterion for histopathological diagnosis of amyloid. Using this technique and polarization microscopic analysis, Romhányi[56-58] was the first who suggested the oriented fibrillar submicroscopic structure of the amyloid. His statements have been supported by subsequent electron microscopic observations. An attempt has been made to differentiate primary (immunocytic) and secondary (reactive) amyloidosis by a sequential permanganate oxidation and trypsin digestion prior to the staining of amyloid with congo red.[59] These pretreatments are said to inhibit the development of green birefringence in secondary amyloid deposits.

Other anionic dyes can also be used to stain amyloid and to increase its birefringence. Among others, the polyanionic sirius red molecules can also be bound to amyloid fibrils in an oriented manner[60,61] (Figure 4). Based on this observation and our earlier finding that a staining with picrosirius red results in an optical anisotropy of the cytoskeletal intermediate filaments (see in Chapter 3), the "collagen-specificity" of the popular picrosirius red staining must be questioned.

The iodine reaction of amyloid has only historical significance, because this method was used by Virchow[62,63] to show the presence of a "starch-like" substance in animal tissues. Iodine reaction of amyloid induces a strong birefringence with positive sign, and it is supposed that interfilamentous material, probably the P-component, is responsible for the oriented binding of iodine particles[29] (Figure 5).

Amyloid can be stained metachromatically by certain cationic dyes such as crystal violet and toluidine blue.[5,29,64] This staining pattern may be attributed to the presence of sulfated glycosaminoglycans in the amyloid deposits. Heparan sulfate[65,66] and some dermatan sulfate[23]

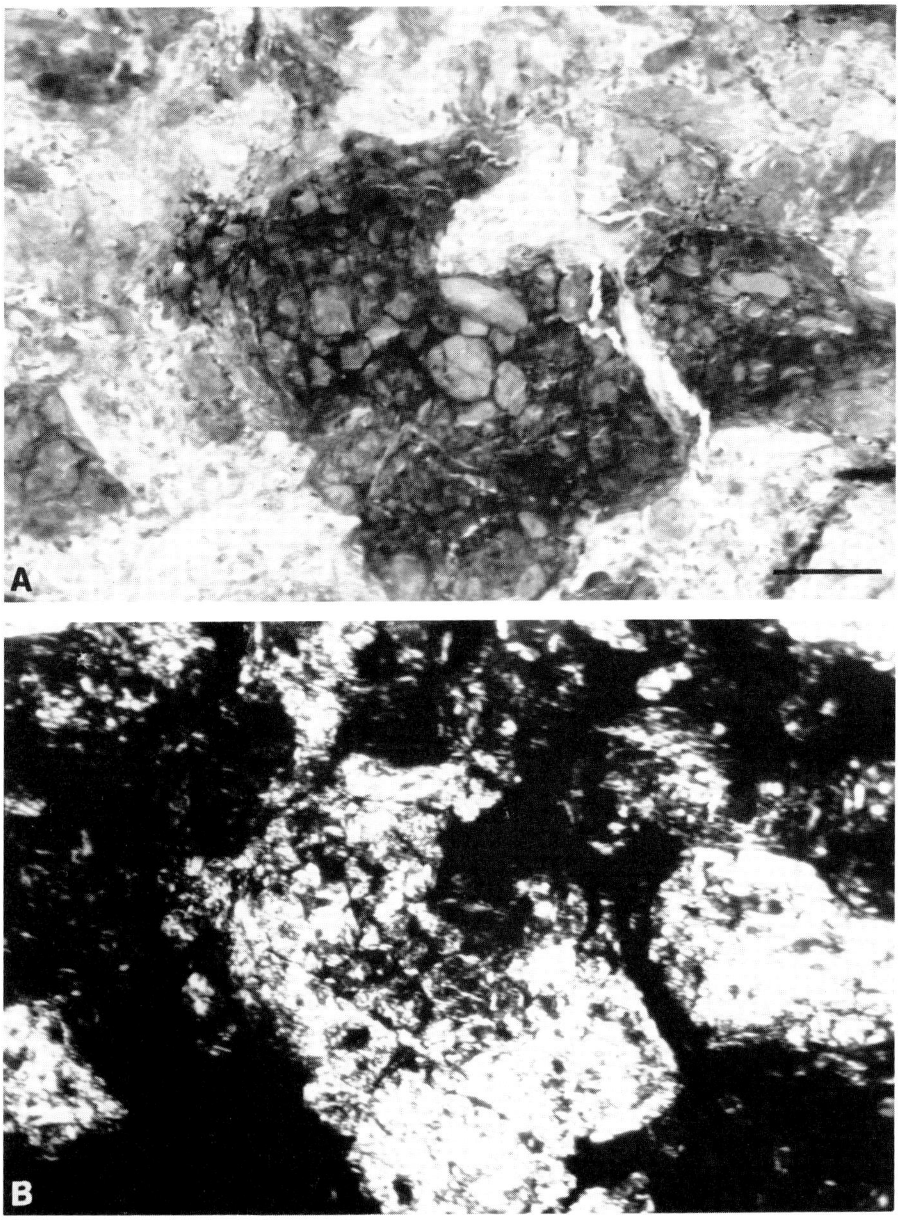

FIGURE 1. Apudamyloid deposits after staining with congo red and taken in normal (A) and polarized light (B). Scale: 100 μm.

were identified biochemically in crude amyloid deposits. In our observations, metachromasia can only be obtained with toluidine blue in amyloid deposits if the pH of the staining solution is higher than 3.5 or if an antecedent digestion with pepsin unmaskes the dye-binding groups. In these sections an induced birefringence appearing in green or yellow polarization colors can be detected. The sign of birefringence is negative with respect to the axis of amyloid fibers indicating an axiperpendicular orientation of the dye molecules to the fibers. Considering the fact that not only sulfates but also other tissue anionic groups, e.g., carboxyls of sialic acid residues and acid proteins, are ionized in this pH range, this metachromatic staining cannot be regarded as a specific reaction for sulfated GAGs with high certainty in the amyloid deposits.

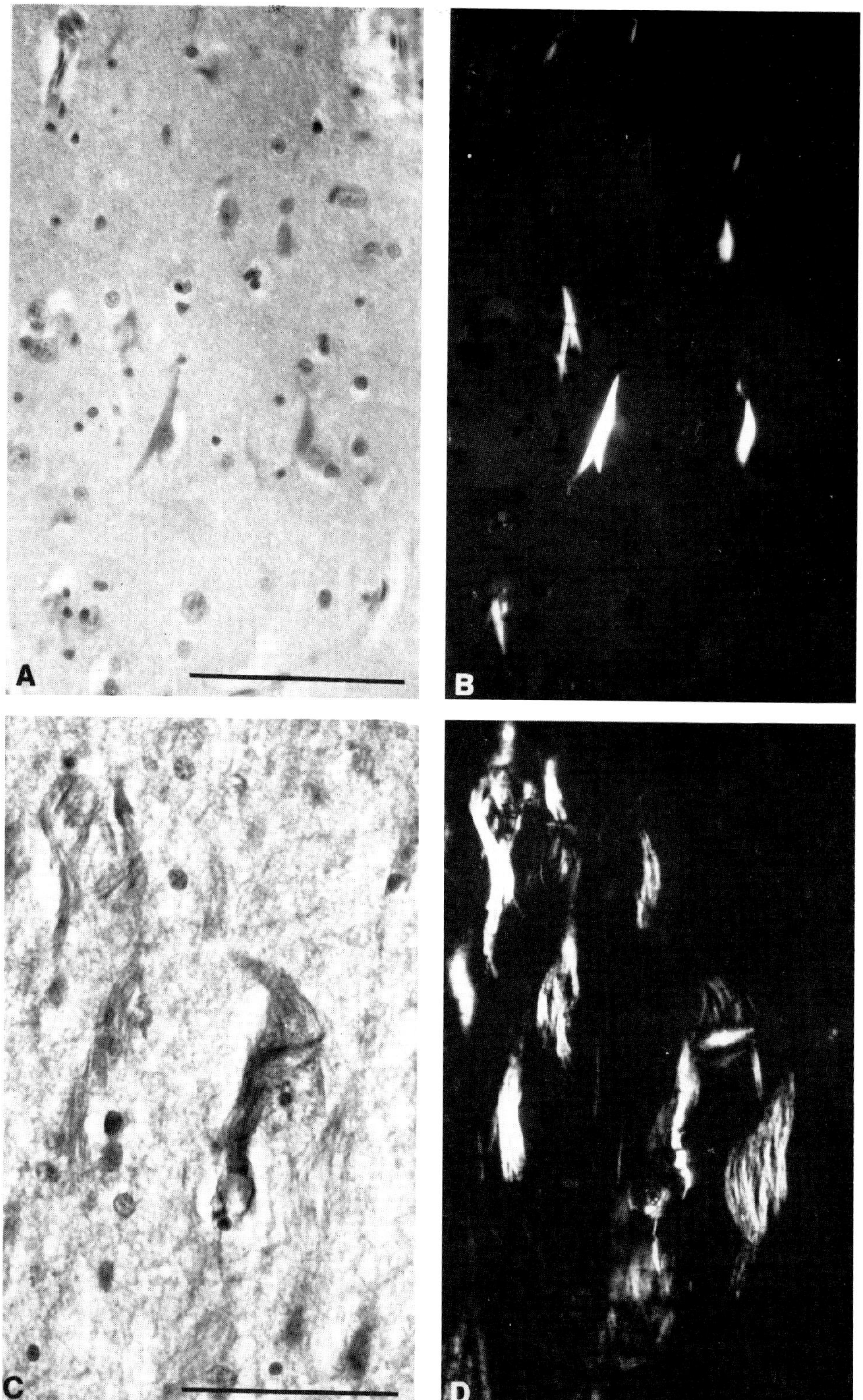

FIGURE 2. Human brain in Alzheimer's disease. Sections were stained with congo red and taken in normal (A,C) and polarized light (B,D). Observe the birefringent neurofibrillary tangles and note in D that the birefringent filaments assume a loose structure in some cells. These are probably completely degenerated cells and their paired helical filaments will subsequently form the core of the amyloid plaques. Scale: 50 μm.

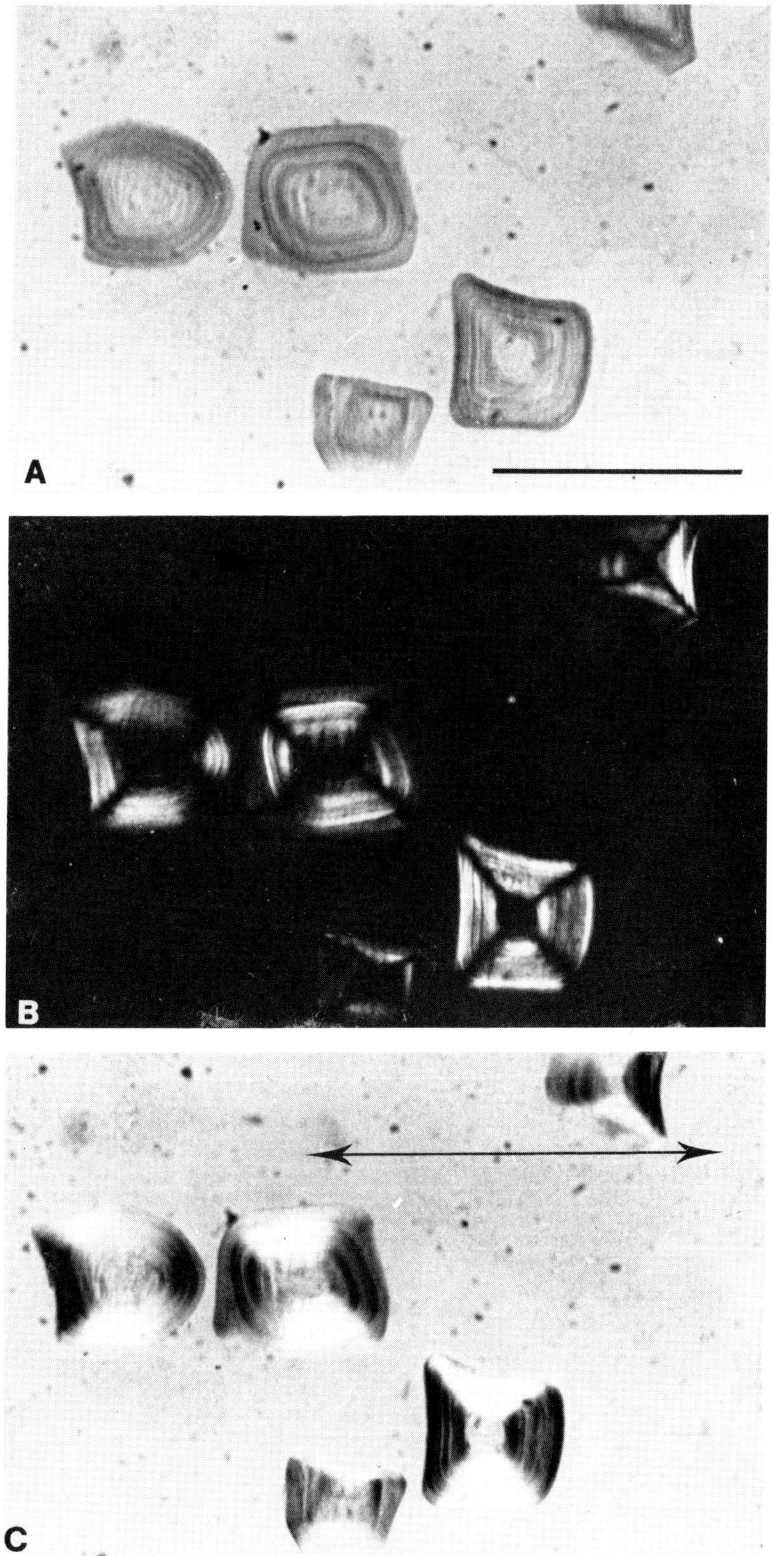

FIGURE 3. Prostatic corpora amylacea stained with congo red and taken in common (A) and polarized light without (B) and after compensation (C). The slow axis of the compensator is indicated by arrow. Note the concentric lamellar structure and the positive sign of birefringence with respect to the tangent of the lamellae. Scale: 100 μm.

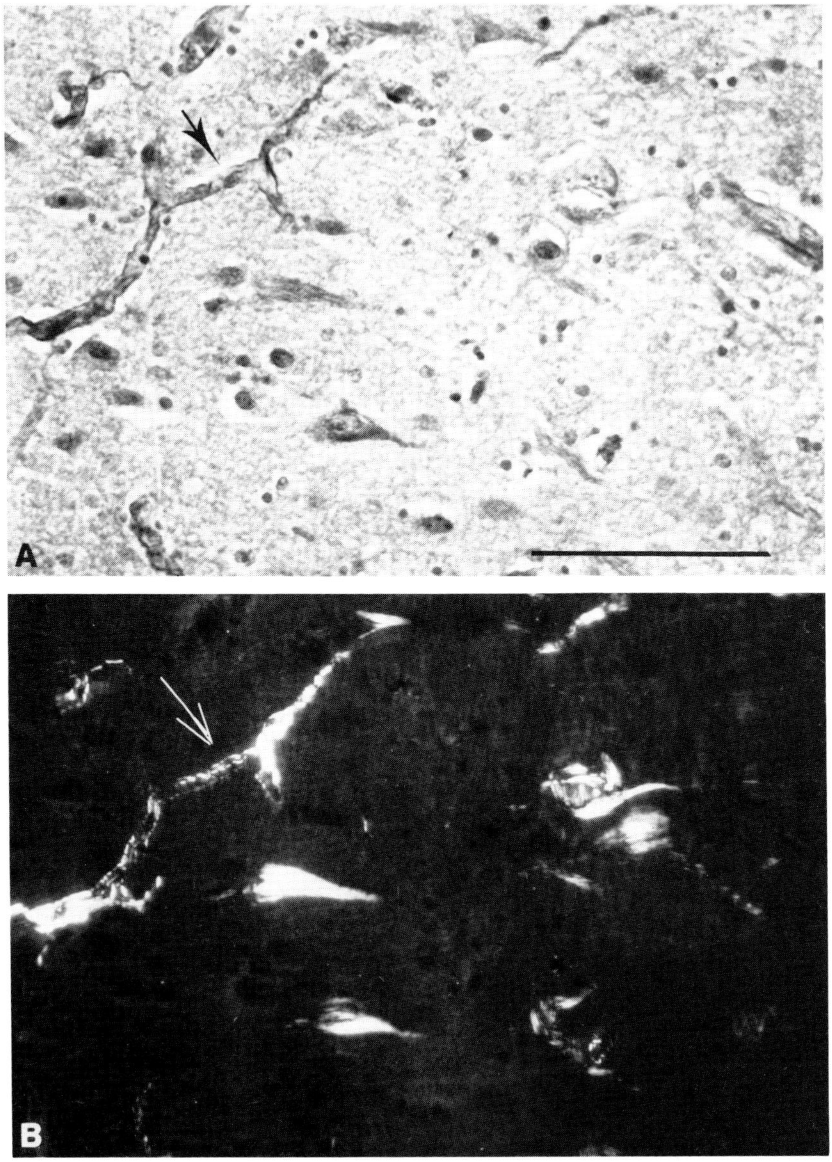

FIGURE 4. Neurofibrillary tangles in human brain suffering from Alzheimer's disease. Picrosirius red F3B staining. Normal (A) and polarized light (B) photomicrograph. The basement membranes of the blood capillaries are also stained and birefringent (arrows). Scale: 50 μm.

It has been found that uptake of cationic alcian blue molecules by amyloids was usually negligible at pH 2.6 but it was considerable at pH 5.7 in the presence of $MgCl_2$[67]. The high critical electrolyte concentration (CEC) registered in the same experiment, i.e., the concentration of $MgCl_2$ at which alcian blue ceased to stain amyloid indicated the presence of heparan sulfate molecules in amyloid deposits.

An interesting and specific although rather neglected topo-optical reaction of amyloid is the red birefringence induced by an orthochromatic staining with toluidine blue. It will be discussed under the next Section.

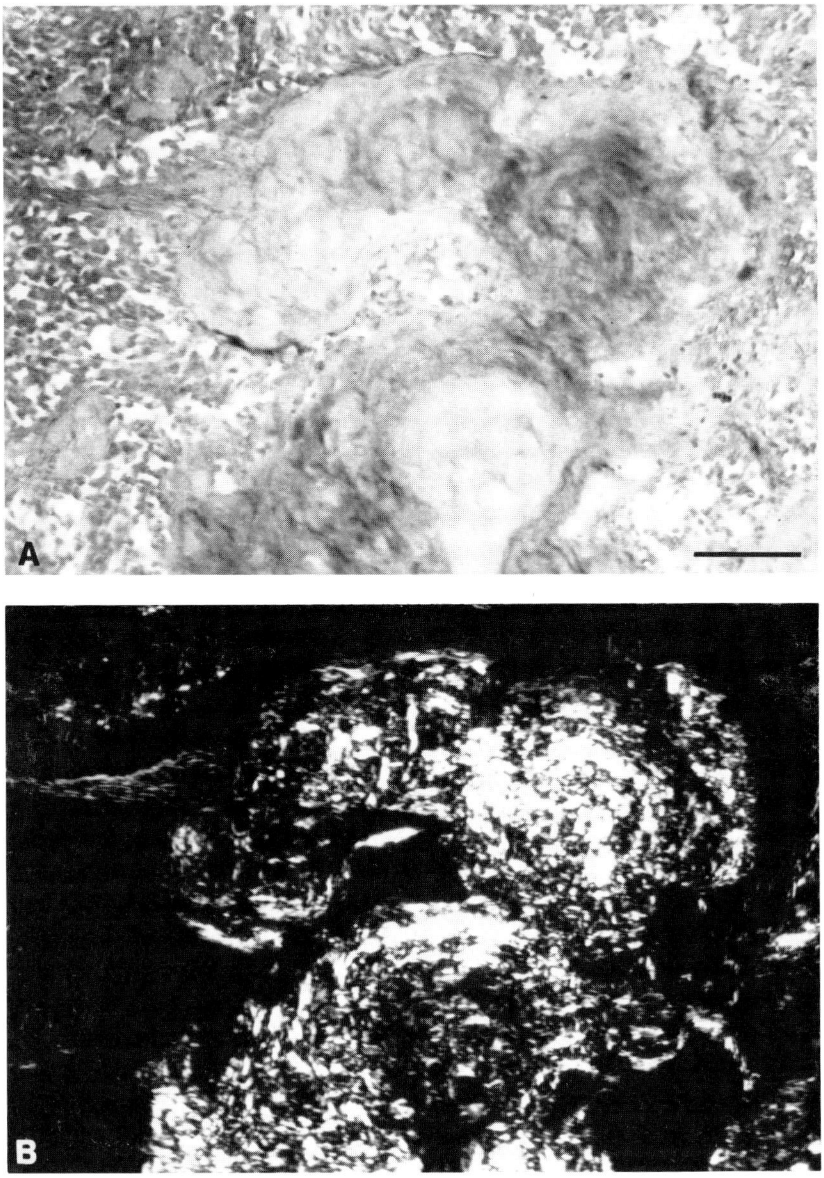

FIGURE 5. Iodine reaction of amyloid deposits in common blue (A) and polarized light (B). Scale: 100 μm.

IV. ANALYSIS OF TOLUIDINE BLUE TOPO-OPTICAL REACTION OF AMYLOID

As has been shown in Chapter 3, metachromatic staining is considered to be the best known topo-optical reaction. There are three optical phenomena induced by the ordered cationic dye aggregates formed on spatially ordered polyanions. These are metachromasia, dichroism, and birefringence. An orthochromatic staining produced by a metachromatic dye has been generally attributed to the unordered binding of dye molecules; it does not induce optical anisotropy.

This chapter demonstrates an exception to this rule. Despite its orthochromasia obtained with thiazine dye solutions (e.g., toluidine blue) at low pH values, amyloid exhibits an intense induced birefringence appearing in red color in polarized light.[67,68] This reaction has remained a histochemical curiosity. The purpose of our present study was to investigate the optical parameters and staining mechanism of the orthochromatic reaction of amyloid. Our results warrant a general interpretation of the theory of the topo-optical reactions. Therefore, the experimental procedures are described in detail. There is another reason why, in contrast to the other chapters, technical details are also given here. We used a wide range of optical analytical and histochemical methods in these experiments. It is believed that it will be instructive for those who are interested also in technical aspects of polarization microscopic analyses of the topo-optical reactions.

A. EXPERIMENTAL PROTOCAL

Investigations were performed on human medullary carcinoma of the thyroid containing high quantities of apudamyloid.[13,69] The specimens were fixed in a 1:3 mixture of 25% glutaraldehyde and saturated picric acid containing 1% acetic acid (4°C, 24 h), embedded in paraffin and sectioned at 6 to 8 μm. We wanted to compare the optical parameters of othochromatically stained amyloid with a typical metachromatic reaction. For this purpose, proximal epiphyseal cartilage matrix of tibias taken from young Wistar rats was chosen. The proximal portions of tibias were fixed in a mixture of ethanol-formaldehyde (4:1), decalcified in a 1:1 mixture of formic acid and 70% ethanol, embedded in paraffin, and cut at 6 to 8 μm. Before staining, the birefringent paraffin was carefully removed from the sections in xylene at 37°C for 24 h.

The sections were stained in 3 mmol/l toluidine blue (Chroma) for 10 min. The dye was dissolved in 0.1 mol/l Walpole's sodium acetate-HCl or in 0.1 mol/l Michaelis's sodium acetate-sodium barbiturate buffers. The pH of the staining solution varied from 1.0 to 7.0 with 0.5 intervals. The dye solution was blotted off with filter paper and then the staining was stabilized with a 6:1 mixture of 2% aqueous KI and 2% aqueous $K_3(Fe(CN)_6)$. It is important to note that any alcoholic post-treatment was avoided. The sections were mounted with a thin gum arabic layer which was allowed to dry 1 to 2 d, and the specimens were covered with Canada balsam. According to our observations, amyloid exhibits orthochromasia and intensive induced birefringence with toluidine blue below pH 3.5. Therefore, a staining solution with a pH of 2.5 to 3.0 was used in the majority of the investigations. Cartilage sections were also stained at this pH.

The following optical analytical procedures were used. Spectral absorption curves of stained structures were recorded in ordinary light between 500 and 700 nm using a cytophotometer (Carl Zeiss, Jena). For the investigation of linear dichroism (=direction dependent absorption of polarized light), the same cytophotometer equipped with a polarizer was used. Measurements of extinction were carried out on orthochromatic amyloid fibers as well as on metachromatic cartilage matrix trabecules oriented parallel (E II) and perpendicular (E ⊥) to the vibration plane of polarized light between wavelengths 480 and 660 nm. The polarized light microscopic investigations were made with a Zetopan-pol (Reichert, Wien) microscope equipped with metal interference filters and λ/10 Brace-Köhler compensator plate. Polarization colors and anomalous polarization colors appearing after additive and subtractive compensation were determined in mixed light. The optical path differences (γ) were determined in monochromatic light from 500 to 680 nm. From these data, curves of anomalous dispersion of birefringence were plotted showing the sign of induced birefringence and the magnitude of retardation values in the function of the wavelength.

Histochemical analysis. The method of the critical electrolyte concentration (CEC)[70] was applied to toluidine blue staining.[71] Alternating sections were stained for 10 min in a series of 3 mmol/l toluidine blue dissolved in 0.025 mol/l Walpole buffer at pH 5.8 containing $MgCl_2$ at 0.1 mol/l increments from zero to 1.8 mol/l.

Prior to staining one of the following four pretreatments was performed: methylation (in methanol containing 0.1 *N* HCl at 37°C for 1, 2, 3, 4, and 5 d); acetylation (in a 2:3

mixture of acetic anhydride and anhydrous pyridine at room temperature for 48 h);[72] periodic acid oxidation (in 1% periodic acid in water at room temperature for 20 min and 1 h); alkylation (in a 12.5% solution of methyliodide in methanol saturated with sodium carbonate at 37°C for 18 h).[73] In the last case, toluidine blue was used at pH 6.0 for 10 min.

The following enzymatic digestions were made prior to staining: testicular hyaluronidase (Reanal, Hungary) (4 ml/ml 0.1 mol/l phosphate buffer pH 6.0 at 37°C for 24 h); pectinase (Serva) (10 mg/ml 0.1 mol/l acetate buffer pH 4.5 at 37°C for 3 h); trypsin (DIFCO) (2 ml/ml 0.1 mol/l phosphate buffer pH 8.0 at 37°C for 0.5, 1, 1.5, 2 and 2.5 h); pepsin (officinal) (1 mg/ml 0.1 *N* HCl at 37°C for 16 h). Control sections were incubated in the solvents of the enzymes.

The stability of orthochromatic staining of amyloid was tested by immersion of sections stained with toluidine blue at pH 2.0 in pure ethanol, pure isopropyl alcohol, 10% NaCl or into 8 mol/l urea in water at room temperature for 2, 4, and 8 h.

B. RESULTS

1. Optical Phenomena

The amyloid deposits were stained with toluidine blue below pH 3.5 in a pale orthochromatic greenish-blue color. The stained areas proved to be strongly birefringent under the polarized light microscope as indicated by vivid red polarization color (Figure 6).

According to our cytophotometric results, the absorption maximum of the orthochromatically stained amyloid is at 625 nm (Figure 7). This absorption maximum is also characteristic for dilute aqueous solutions of the toluidine blue containing only single, nonaggregated dye molecules.[64] It seems, therefore, reasonable to suppose that the toluidine blue molecules are bound to the amyloid in monomeric form. It is known that a salient chemical precondition for metachromatic staining is the distance between the neighboring dye-binding groups. According to Sylvén[74] a metachromatic dye aggregate can develop on the surface of the chromotrope molecule if this distance does not exceed 0.5 to 0.7 nm. The orthochromatic absorption maximum of the amyloid indicates a greater distance between the bound toluidine blue molecules.

The dichroism of the orthochromatic amyloid was investigated in linearly polarized light without analyzer. Deep blue color (i.e., greater light absorption) was observed with parallel vibration direction of polarizer to the long axis of amyloid fibers, while very faint blue color (i.e., a smaller light absorption) could be seen when the polarizer was rotated 90°. This phenomenon is known as a linear dichroism with positive sign, and it clearly indicates that the main axes of dye molecules are aligned parallel to the long axis of amyloid fibers (Figure 8). In contrast to this, the cartilage matrix which was stained metachromatically with toluidine blue exhibited linear dichroism with negative sign indicating perpendicular orientation of the dye molecules to the matrix trabecules (Figure 8).

The sign of birefringence of the orthochromatic amyloid was determined in polarized light microscope by a compensator. If the slow vibration direction of the compensator plate and the optic axis of amyloid fibers were parallel to one another, a purple anomalous polarization color appeared in mixed polarized light. The longer wavelength of this interference color, as compared to the original red polarization color, indicated that an additon of retardations both of the amyloid and the compensator occured in this position. This assumption was supported by the investigation of the same phenomenon in red monochromatic light (λ = 630 nm). Amyloid fibers showing purple anomalous color in mixed light were bright in monochromatic red light. When the specimen was turned 90°, i.e., the slow vibration axis of compensator and the optical axis of amyloid fibers were perpendicular to one another, a subtraction of the retardation was indicated by the appearance of a greenish-yellow anomalous color in mixed light, and by the darkness of fibers compensated in monochromatic red light. Consequently, the orthochromatic staining of amyloid with tolui-

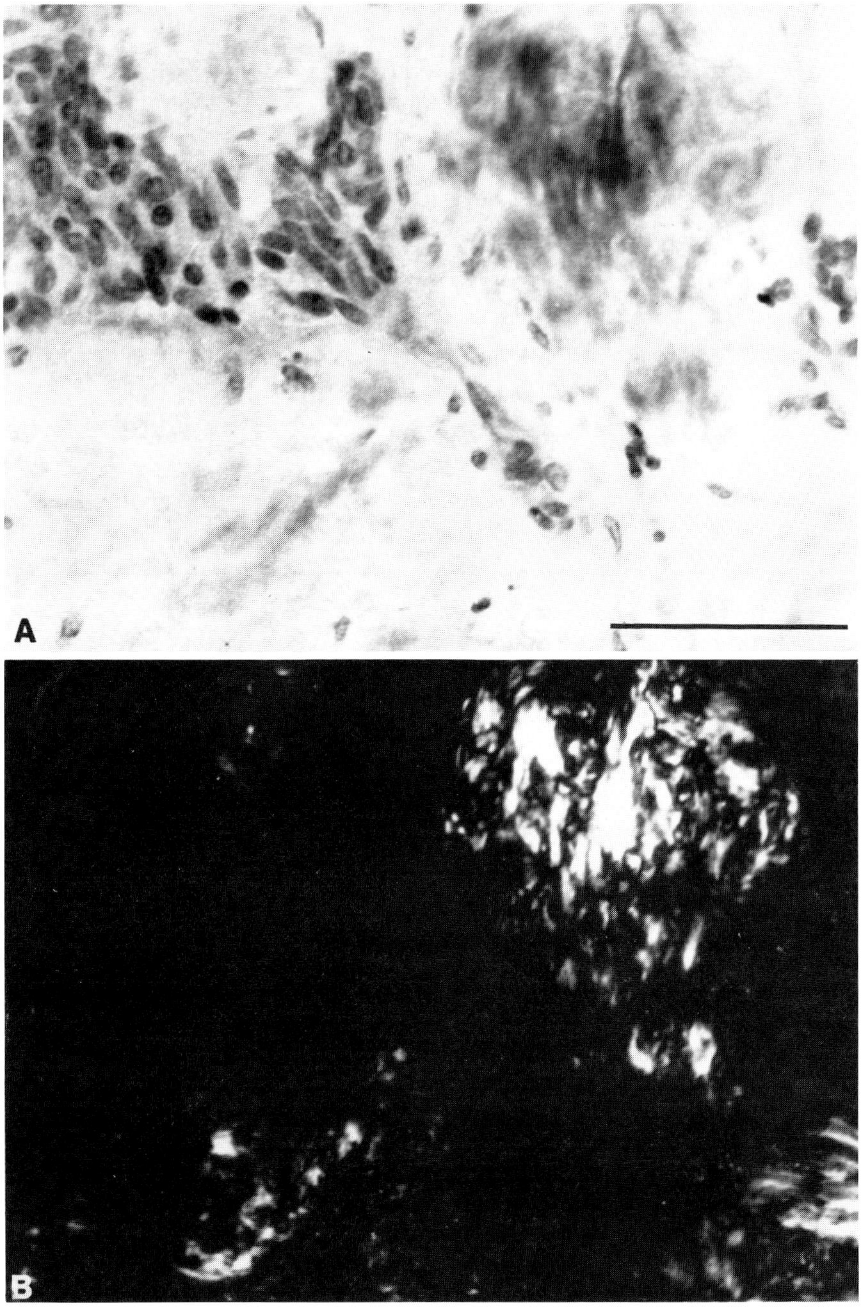

FIGURE 6. Human medullary carcinoma section after staining with toluidine blue at pH 3.0. (A) Amyloid deposits are blue in normal light. Nuclei are also stained. (B) The same visual field in the polarization microscope. The orthochromatically stained amyloid deposits are strongly birefringent. In reality, they appear as a red polarization color. Scale: 50 μm.

dine blue induced an optical anisotropy with positive sign: the planar dye molecules were arranged parallel to the long axis of the amyloid fibers. In other words, the orthochromatic toluidine blue staining induced an additive topo-optical reaction on the amyloid. One should be reminded here that a metachromatic toluidine blue staining induces in cartilage matrix, which contain polyanionic glycosaminoglycans, an inversive topo-optical reaction (i.e., a

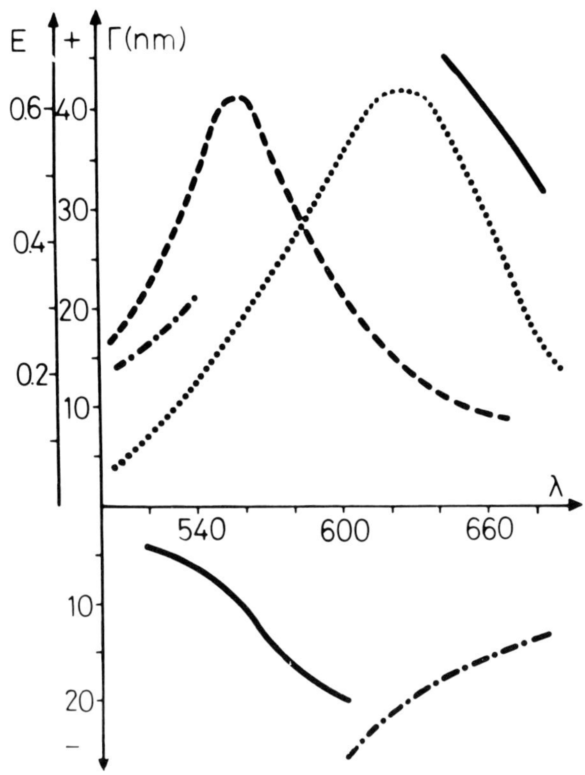

FIGURE 7. Comparison of optical characteristics of the orthochromatic amyloid and metachromatic cartilage matrix both stained with toluidine blue at pH 3.0. Spectral absorption curves of the apudamyloid (dotted line) and the cartilage matrix (broken line) taken in a cytophotometer. The difference between the absorption maxima (548 and 625 nm) indicates that the cartilage matrix binds metachromatic dye aggregates, while amyloid binds ony single dye molecule and stains orthochromatically. Anomalous dispersion curves of birefringence of apudamyloid (continuous lines) and the cartilage matrix (lines with dashes and dots) taken in a polarization microscope after staining with toluidine blue at pH 3 are also shown. The different positions of the anomalous dispersion curves are due to the differenct orientation of dye molecules bound to the cartilage matrix and to the amyloid. E, extinction; Γ, retardation values; λ, wavelengths.

birefringence with negative sign). This is due to an axiperpendicular binding of dye molecules to the chromotrope. The polarization color of the red metachromatic reaction is green and its anomalous colors are orange (after additive compensation) and blue (after subtractive compensation).

Comparing the anomalous dispersion curves of the birefringence of the orthochromatic amyloid and of the metachromatic cartilage matrix, the reverse optical behavior of these two reactions is obvious (Figure 7). The positive sign of birefringence of the amyloid (observed in red light) is inverted near the absorption maximum of orthochromasia. The birefringence of the metachromatically stained cartilage matrix behaves differently. It has a negative sign in yellow and red lights, and its inversion occurs below the metachromatic absorption maximum in green and blue lights. These data suggest that the spatial ordering of toluidine blue molecules to chromotrope molecules is reversed in orthochromatic and metachromatic topo-optical reactions.

2. Elucidation of the Staining Mechanism

Effect of pH on the staining solution — While amyloid was stained orthochromatically with toluidine blue below pH 3.5, it exhibited a typical red metachromasia above this pH

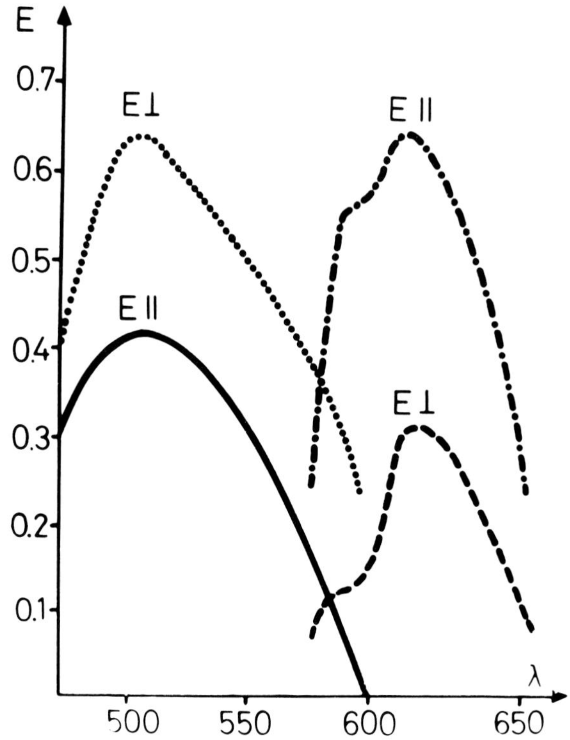

FIGURE 8. Absorption curves of the amyloid (lines with dashes and dots and broken line) and of the cartilage matrix (dotted and continuous lines) taken in a cytophotometer in polarized light showing the linear dichroism (direction-dependent light absorption) of chromotropes. Staining: toluidine blue at pH 3. The vibration direction of the polarized light is parallel at E II and perpendicular at E ⊥ to the amyloid and collagen fibers, respectively. The opposite character of dicroism of the two chromotropes indicates different dye orientations. The toluidine blue molecules are arranged parallel to the long axis of the amyloid fibers, and they are ordered perpendicularly to the long axis of the collagen fibers of the cartilage matrix.

value. As mentioned above, orthochromatic amyloid had a positive dichroism and a positive sign of induced birefringence (additive topo-optical reaction). The metachromatic amyloid, on the other hand, showed a negative dichroism, and its birefringence appearing in green polarization color, similar to the cartilage matrix, had a negative sign indicating the axiperpendicular orientaton of the dye molecules (inversive topo-optical reaction).

CEC of the amyloid-toluidine blue complex — The metachromatic reaction of amyloid was strongly diminished in the presence of 0.3 mol/l $MgCl_2$, and disappeared in the presence of 0.6 mol/l, or greater, $MgCl_2$ concentration. The metachromatic basophilia was gradually replaced by an orthochromatic staining effect between 0.3 to 0.6 mol/l $MgCl_2$. The orthochromasia and the associated red polarization color remained unchanged even at 1.8 mol/l $MgCl_2$. These results suggest that the orthochromatic staining must be due to a non-electrostatic interaction of toluidine blue and amyloid, while the metachromasia is the result of an ionic dye-binding to anions (probably to carboxyl and/or sulfate groups) of the amyloid.

Effects of the modification of different chemical groups of amyloid — An antecedent methylation for 1 and 2 d at 37°C did not change the orthochromatic staining and induced birefringence of the amyloid stained with toluidine blue at pH 2.0. It is known that this methylation technique completely abolishes the basophilia of sulfate and carboxyl groups.[75] It seems, therefore, reasonable to suppose that these groups are not involved in the orthochromatic staining of the amyloid. A prolonged methylation, however, caused a time-dependent gradual decrease of the orthochromatic staining, and a methylation for 5 d almost completely abolished the staining and birefringence. The prolonged methylation causes a conversion of hydroxyl radicals to methyl ether.[13] Thus, the hydroxyl groups may be responsible for the toluidine blue binding in orthochromatic staining of amyloid.

Both acetylation esterifying the hydroxyl groups and periodic acid pretreatment attacking the vicinal –OH groups of carbohydrates strongly reduced the orthochromatic staining and induced birefringence of amyloid. This finding supports our previous supposition that hydroxyl groups probably bound to proteins and glycoproteins of amyloid are involved in the orthochromatic dye-binding.

The alkylation technique does not modify the subsequent orthochromatic topo-optical reaction of amyloid. Methyl-iodide alkylates the carboxyl groups, but the sulfate and hydroxyl groups remain uneffected.[73] Considering that amyloid is not metachromatic below pH 3.5, it seems unlikely that the quantity of sulfate groups reaches a threshold concentration critical for this staining of amyloid. The unaltered staining of alkylated amyloid also warrants, therefore, the essential role of –OH groups played in orthochromatic dye-binding.

Effect of enzymatic digestion — No effect of antecedent enzymatic digestions could be observed on orthochromasia and birefringence of the amyloid when hyaluronidase, pectinase, and pepsin were used. The only exception was the trypsin digestion for 2 and 2.5 h. Although it caused no perceptible decrease in the stainability, it strongly reduced the intensity of birefringence. It may be, therefore, supposed that the spatial orientation of the dye-binding –OH groups is determined by a trypsin-sensitive structure-protein of the amyloid.

Effect of post-treatments with solvents — The orthochromatic staining and birefringence proved to be resistant to alcoholic post-treatments using either absolute ethanol or isopropyl-alcohol. However, the dye could be completely removed from the sections with 10% NaCl or 8 ml/l urea solutions used for splitting of H-bonds.[76] This suggests the possibility of a hydrogen bond that has formed between amyloid and toluidine blue.

C. DISCUSSION

The results described suggest that orthochromasia and optical anisotropy of the amyloid stained with toluidine blue below pH 3.5 can be regarded as a special kind of topo-optical reaction. This reaction differs from the metachromatic topo-optical reaction both in chemical mechanisms and optical behaviors (Table 1).

The orthochromatic reaction of the amyloid must also be distinguished from the generally known orthochromatic staining of any other biostructure. The latter type can be explained by one of the two alternative mechanisms: 1. The structure is isotropic (e.g., colloid of thyroid gland), randomly binding dye molecules which never induce optical anisotropy. 2. The structure is anisotropic and contains small amounts of oriented polyanions (i.e., extracellular matrix of tendon and bone tissues). Such structure exhibits orthochromatic basophilia which is due to electrostatic interaction of polyanions with cationic dyes at low pH. Increasing the pH above a threshold value (pH 3.5 to 4.5) results in the complete dissociation of the anionic binding sites. As a result of the increased charge density, the anions of the chromotrope become sufficiently close to each other to induce an ordered coplanar metachromatic dye aggregate which then shows birefringence in polarized light. Therefore, this type of orthochromasia is a "precursor" ("harbinger") to a metachromaticly oriented dye association, but it is clearly not due to an oriented dye binding.

The orthochromasia of the amyloid is neither a random dye-binding nor a "precursor"-type of a metachromatic topo-optical reaction. It is clearly an additive topo-optical reaction. The linearly arranged hydroxyl groups of amyloid fibrils bind the dye-molecules with hydrogen-bonds. The binding results in a parallel orientation of the dye molecules to the amyloid fibrils which are ordered with their long axes parallel to each other and to the long axis of the fiber[4,5,29] (Figure 9). The metachromasia of the amyloid induced above pH 3.5

TABLE 1
Summary of Optical Properties and Staining Mechanisms of Metachromatic and Orthochromatic Topo-Optical Reactions of Toluidine Blue (TB)

	Differences in molecular mechanisms and optical properties	Metachromatic TB-staining of cartilage matrix	Orthochromatic TB-staining of amyloid
1.	Types of dye-binding groups and macromolecules	$-OSO^-_3$, $-PO^-_4$ and $-CO^-_2$ groups of polyanions	$-OH$ groups of glycoproteins of amyloid
2.	Nature of dye-chromotrope interaction	Electrostatic force	Hydrogen bond
3.	Distance between dye molecules bound to the substrate	Smaller than 0.5—0.7 nm	Greater than 0.5—0.7 nm
4.	Interaction between dye molecules bound to the substrate	Coplanar aggregates are formed	Aggregates are not formed
	Optical manifestations	Absorption max.: $\lambda = 540$ nm polarization color: green	Absorption max.: $\lambda = 625$ nm polarization color: red
5.	Steric orientation of dye molecules	Axiperpendicular (inversive reaction)	Axiparallel (additive reaction)
6.	Optical manifestations	Dichroism with negative sign birefringence with negative sign at $\lambda = 650$ nm	Dichroism with positive sign birefringence with positive sign at $\lambda = 650$ nm

to 4.0 is, on the other hand, an inverse topo-optical reaction. This is the result of an electrostatic interaction of toluidine blue molecules and the anionic groups (carboxyls of dicarboxylic amino acid residues and sialic acids of the amyloid proteins as well as sulfates and carboxyls of the GAGs associated with the amyloid fibrils)[3,21,23,65,66] (Figure 10).

The staining mechanism does not account for the intensive optical anisotropy of the orthochromatic amyloid alone. There are a great number of other macromolecules containing hydroxyl groups, yet they never exhibit an orthochromatic topo-optical reaction. The only exception to this rule would be the observation of Scheuner and Hutschenreiter[77] that toluidine blue orthochromatically stains fibrin and induces dichroism and birefringence in polarized light. We have failed to reproduce this result on fibrin sponge, fibrin fibers isolated from human blood clot, and paraffin sections of fibrinous pericarditic heart using toluidine blue staining at different pH values, although we have obtained, at the same time, the same orthochromatic topo-optical reaction on different primary and secondary amyloids. The orthochromatic topo-optical reaction of different amyloids can be demonstrated not only with toluidine blue, but also with other cationic metachromatic dyes of planar character including dimethylmethylene blue, Janus red, and *N,N'*-diethylpseudoisocyanine. Thus, the optical behavior of the orthochromatic amyloid in polarized light seems to be a unique and specific phenomenon, and has a diagnostic value in pathology.

The observations of Missmahl[78] which we could reproduce, suggest a probable explantation for the birefringence induced by orthochromasia. In his experiments, mechanically made smears of toluidine blue stained substances were tested in polarization microscope, and the orthochromatic dye-stripes were dichroic and birefringent appearing in orange-red polarization colors. These phenomena were attributed to a mechanical axiparallel orientation of individual dye molecules. Single dye molecules are also ordered in the orthochromatic amyloid. In this structure, however, the twisted β-pleated configuration of amyloid fibrils with longitudinal grooves is the orienting factor for the dye molecules.[4,5,29] It should also be noted that according to our observations other orthochromatic structures having the same conformation (e.g., fibrin and keratin) have always failed to show induced optical anisotropy. Consequently, the conformation of the ordered macromolecule — although it is a crucial

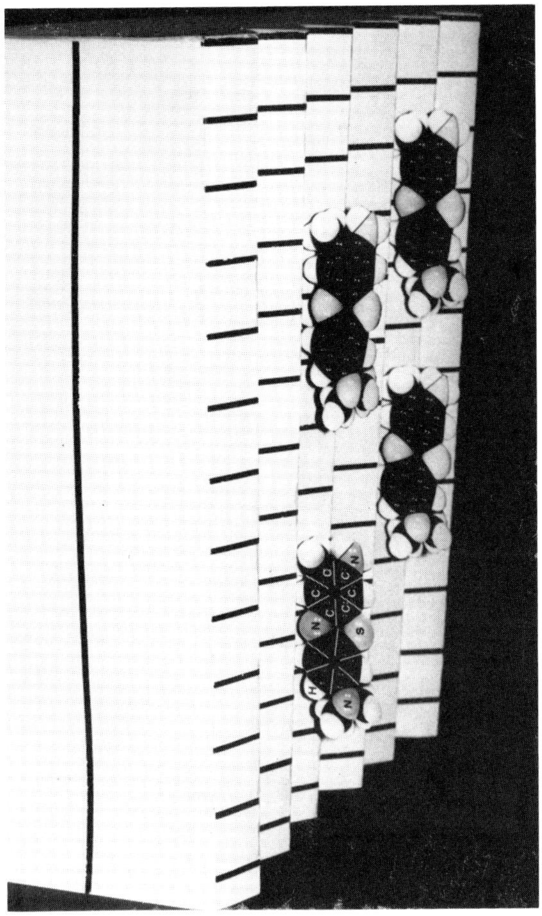

FIGURE 9. A scheme to show the binding of toluidine blue to the amyloid fibril below pH 3.5. The single dye molecules are ordered parallel to the long axis of the amyloid fibril composed of polypeptide chains with beta-pleated sheet conformation. The dye molecules are sitting in the grooves of the surface of amyloid filaments and they are bound to amyloid fibril proteins by H bonds. The long axis of filament is indicated by a straight line. Zigzag lines symbolyze the β-pleated sheet conformation.

factor — is not the only molecular basis of this topo-optical reaction. It is reasonable to suppose, therefore, that the hierarchic and ordered supramolecular organization of amyloid fibers is also involved in the ordering of the dye molecules even when they are bound in a greater distance from each other.

The curious orthochromatic topo-optical reaction suggests that the theory of the topo-optical reactions induced by metachromatic dyes must be reformulated. According to the original definition of Romhányi,[79] only metachromasia with induced birefringence refers to spatially oriented microstructure. On the basis of our findings, we propose that an orthochromatic staining can also induce optical anisotropy, if the stained structure is in a highly oriented state. It also follows that orthochromatic staining alone does not imply a structural isotropy without polarization microscopic analysis.

Using his "standard toluidine blue" method Wolman[68] proposed a macromolecular model of amyloid based on polarization microscopic analysis of orthochromatic staining. He believed that orthochromasia was due to the electrostatic interaction of dye molecules

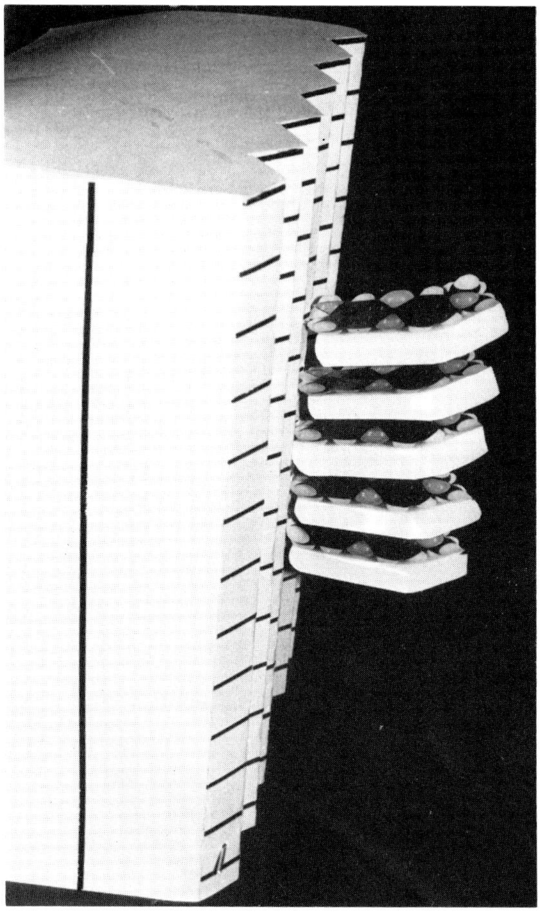

FIGURE 10. A schematic representation of the binding of toluidine blue to the amyloid fibril above pH 3.5. The dye molecules are in coplanar aggregates, and are regularly oriented with their long axes perpendicular to that of the fibril. In this case, the cationic dye molecules are bound by electrostatic forces to different anionic groups of amyloid fibril proteins or of glycoconjugates associated to the fibrils. Anionic groups are supposedly densely and regularly arranged along the major fibrillar axis.

and anionic groups of acid mucopolysaccharides associated with the amyloid deposits. Our results argue against this possibility. Furthermore, he also supposed that toluidine blue molecules are arranged perpendicular to the length of the amyloid fibrils. In our observations, however, the sign of dichroism and birefringence as well as the anomalous dispersion curves of birefringence clearly indicated an axiparallel binding of the dye molecules to the amyloid fibrils.

ACKNOWLEDGMENTS

This Chapter is dedicated to the memory of my friend, Dr. M. Szokol (former Associate Professor, Department of Pathology of this University, who left us suddenly in 1983) with whom several aspects of the amyloid histochemistry were discussed and whose advice helped us very much.

Warmest thanks are due to my wife and former co-worker, Dr. Róza Ádány (Department of Clinical Chemistry, University Medical School, Debrecen), who carried out the greater part of the optical analysis of the toluidine blue staining reaction. The courtesy of Professor Gy. Balázs (1st Department of Surgery) and Dr. M. Petkó (Department of Anatomy of this University) is acknowledged for providing samples of medullary carcinoma of the thyroid.

REFERENCES

1. **Glenner, G. G., Harbaugh, J., Ohms, J. I., Harada, M., and Cuatrecasas, P.,** An amyloid protein: the amino-terminal variable fragment of an immunoglobulin light chain, *Biochem. Biophys. Res. Commun.*, 41, 1287, 1970.
2. **Glenner, G. G., Terry, W., Harada, M., Isersky, C., and Page, D.,** Amyloid fibril proteins: proof of homology with immunoglobulin light chains by sequence analyses, *Science,* 172, 1150, 1971.
3. **Harada, M., Isersky, C., Cuatrecasas, P., Page, D., Bladen, H. A., Eanes, E. D., Keiser, H. R., and Glenner, G. G.,** Human amyloid protein: chemical variability and homogeneity, *J. Histochem. Cytochem.*, 19, 1, 1971.
4. **Glenner, G. G.,** Amyloid deposits and amyloidosis. The β-fibrilloses. (First of two parts), *N. Eng. J. Med.*, 302, 1283, 1980.
5. **Glenner, G. G.,** The bases of the staining of amyloid fibers: their physico-chemical nature and the mechanism of their dye-substrate interaction, *Prog. Histochem. Cytochem.*, 13/1, 1, 1981.
6. **Jaffe, E. S., Braylan, R. C., Nanba, K., Frank, M. M., and Berard, C. W.,** Functional markers: a new perspective on malignant lymphomas, *Cancer Treat. Rep.*, 61, 953, 1977.
7. **Benditt, E. P., Eriksen, N., Hermodson, M. A., and Ericsson, L. H.,** The major proteins of human and monkey amyloid substance: common properties including unusual N-terminal amino acid sequences, *FEBS Lett.*, 19, 169, 1971.
8. **Ein, D., Kimura, S., Terry, W. D., Magnotta, J., and Glenner, G. G.,** Amino acid sequence of an amyloid fibril protein of unknown origin, *J. Biol. Chem.*, 247, 5653, 1972.
9. **Levin, M., Franklin, E. C., Frangione, B., and Pras, M.,** The amino acid sequence of a major non-immunoglobulin component of some amyloid fibrils, *J. Clin. Invest.*, 51, 2773, 1972.
10. **Husby, G., Sletten, K., Michaelsen, T. E., and Natvig, J. B.,** Alternative, non-immunoglobulin origin of amyloid fibrils, *Nature,* 238, 187, 1972.
11. **Benditt, E. P. and Eriksen, N.,** Chemical classes of amyloid substance, *Am. J. Pathol.,* 65, 231, 1971.
12. **Glenner, G. G., Linke, R. P., and Pollock, P. S.,** Systemic amyloidosis and immunoglobulins, in *Symposium on Amyloidosis,* Wegelius, O. and Pasternack, A., Eds., Academic Press, New York, 1976, 233.
13. **Pearse, A. G. E., Ewen, S. W. B., and Polak, J. M.,** The genesis of apudamyloid in endocrine polypeptide tumors: histochemical distinction from immunamyloid, *Virchows Arch. (B),* 10, 93, 1972.
14. **Westermark, P.,** On the nature of amyloid in human islets of Langerhans, *Histochemistry,* 38, 27, 1974.
15. **Glenner, G. G., Eanes, E. D., Bladen, H. A., Linke, R. P., and Termine, J. D.,** β-pleated sheet fibrils; a comparison of native amyloid with synthetic protein fibrils, *J. Histochem. Cytochem.*, 22, 1141, 1974.
16. **Tashjian, A. H., Wolfe, H. J., and Voelkel, E. F.,** Human calcitonin. Immunologic assay, cytologic localization and studies on medullary thyroid carcinoma, *Am. J. Med.*, 56, 840, 1974.
17. **Sletten, K., Westermark, P., and Natvig, J. B.,** Characterization of amyloid fibril proteins from medullary carcinoma of the thyroid, *J. Exp. Med.*, 143, 993, 1976.
18. **Bladen, H. A., Nylen, M. U., and Glenner, G. G.,** The ultrastructure of human amyloid as revealed by the negative staining technique, *J. Ultrastruct. Res.*, 14, 449, 1966.
19. **Glenner, G. G. and Bladen, H. A.,** Purification and reconstitution of the periodic fibril and unit structure of human amyloid, *Science,* 154, 271, 1966.
20. **Cathcart, E. S., Skinner, M., and Cohen, A. S.,** Immunogenicity of amyloid, *Immunology,* 20, 945, 1971.
21. **Binette, P., Matsuzaki, M., Calkins, E., Alper, R., and Winzler, R. J.,** Carbohydrate composition of amyloid components, *Proc. Soc. Exp. Biol. Med.*, 137, 165, 1971.
22. **Pennock, C. A.,** Association of acid mucopolysaccharides with isolated amyloid fibrils, *Nature,* 217, 753, 1968.
23. **Pras, M., Nevo, Z., Schubert, M., Rotman, J., and Matolon, R.,** The significance of mucopolysaccharides in amyloid, *J. Histochem. Cytochem.*, 19, 443, 1971.

24. **Cohen, A. S. and Calkins, E.,** Electron microscopic observation on a fibrous component in amyloid of diverse origins, *Nature,* 183, 1202, 1959.
25. **Spiro, D.,** The structural basis of proteinuria in man. Electron microscopic studies of renal biopsy specimens from patients with lipid nephrosis, amyloidosis, and subacute and chronic glomerulonephritis, *Am. J. Pathol.,* 35, 47, 1959.
26. **Eanes, E. D. and Glenner, G. G.,** X-ray diffraction studies of amyloid filaments, *J. Histochem. Cytochem.,* 16, 673, 1968.
27. **Termine, J. D., Eanes, E. D., Ein, D., and Glenner, G. G.,** Infrared spectroscopy of human amyloid fibrils and immunoglobulin proteins, *Biopolymers,* 11, 1103, 1972.
28. **Glenner, G. G., Ignaczak, T. F., and Page, D. L.,** The inherited systemic amyloidoses and localized amyloid deposits, in *Metabolic Basis of Inherited Disease,* 4th ed., Stanbury, J. B., Wyngaarden, J. B., and Fredrickson, D. F., Eds., McGraw-Hill, New York, 1978, 1308.
29. **Cooper, J. H.,** Selective amyloid staining as a function of amyloid composition and structure: histochemical analysis of the alkaline congo red, standardized toluidine blue, and iodine methods, *Lab. Invest.,* 31, 232, 1974.
30. **Glenner, G. G.,** Amyloid deposits and amyloidosis. The β-fibrilloses. (Second of two parts), *N. Engl. J. Med.,* 302, 1333, 1980.
31. **Linder, E.,** Similarities between cytoskeletal "intermediate" 10 nm filaments and amyloid, in *Amyloid and Amyloidosis,* Glenner, G. G., Pinho e Costa, P., and Falcao de Freitas, A., Eds., Excerpta Medica, Amsterdam, Oxford, Princeton, 1980, 283.
32. **Selkoe, D. J.,** Molecular pathology of the aging human brain, *Trends Neurosci.,* 5, 332, 1982.
33. **Glenner, G. G.,** Alzheimer's disease, *Arch. Pathol. Lab. Med.,* 107, 281, 1983.
34. **Perry, E. K. and Perry, R. H.,** New insights into the nature of senile (Alzheimer-type) plaques, *Trends Neurosci.,* 8, 301, 1985.
35. **Masters, C. L., Multhaup, G., Simms, G., Pottgiesser, J., Martins, R. N., and Beyreuther, K.,** Neuronal origin of a cerebral amyloid: neurofibrillary tangles of Alzheimer's disease contain the same protein as the amyloid of plaque cores and blood vessels, *EMBO J.,* 4, 2757, 1985.
36. **Divry, P. and Florkin, M.,** Sur les propriėties optiques de l'amyloid, *C. R. Soc. Biol. (Paris),* 97, 1808, 1927.
37. **Marc, A. J., Gueft, B., and Moskal, J. F.,** Prostatic corpora amylacea: a study with the electron microscope and electron probe, *Arch. Pathol.,* 80, 487, 1965.
38. **Diezel, P. B. and Pfleiderer, A.,** Histochemische und polarisations-optische Untersuchungen am Amyloid, *Virchows Arch. (A),* 332, 552, 1959.
39. **Puchtler, H., Sweat, F., and Kuhn, J. G.,** On the binding of direct cotton dyes by amyloid, *J. Histochem. Cytochem.,* 12, 900, 1964.
40. **Beneke, G., Rakow, A. D., Rakow, L., and Schmitt, W.,** Nachweis von Fibrin und Amyloid mit morphologischen Methoden in Gewebeschnitten, *Beitr. Pathol.,* 141, 404, 1970.
41. **Lendrum, A. C., Slidders, W., and Fraser, D. S.,** Renal hyalin. A study of amyloidosis and diabetic fibrinous vasculosis with new staining methods, *J. Clin. Pathol.,* 25, 373, 1972.
42. **Stiller, D., Katenkamp, D., and Thoss, K.,** Der Färbemekanismus des Thioflavin T unter besonderer Berücksichtigung der Amyloiddarstellung, *Acta Histochem. (Jena),* 42, 234, 1972.
43. **Katenkamp, D. and Stiller, D.,** Comparisons of the texture of amyloid, collagen and Alzheimer cells, *Virchows Arch. (A),* 359, 213, 1973.
44. **Stiller, D. and Katenkamp, D.,** Ultrahistochemical demonstration of periodate reactive and alpha-amino groups in experimental amyloid, *Exp. Pathol.,* 9, 361, 1974.
45. **Puchtler, H., Sweat Waldrop, F., and Meloan, S. N.,** Application of thiazole dyes to amyloid under conditions of direct cotton dyeing: correlation of histochemical and chemical data, *Histochemistry,* 77, 431, 1983.
46. **Elghetany, M. T. and Saleem, A.,** Methods for staining amyloid in tissues: a review, *Stain Technol.,* 63, 201, 1988.
47. **Kelėnyi, G.,** Thioflavin S fluorescent and congo red anisotropic stainings in the histologic demonstration of amyloid, *Acta Neuropathol. (Berl.),* 7, 336, 1967.
48. **Benditt, E. P., Eriksen, N., and Berglund, C.,** Congo red dichroism with dispersed amyloid fibrils, and extrinsic cotton effect, *Proc. Natl. Acad. Sci. USA,* 66, 1044, 1970.
49. **Katenkamp, D. and Stiller, D.,** Polarisationsoptisch-histochemische Untersuchungen zur Kongorotfärbung des Amyloids, *Histochemie,* 29, 37, 1972.
50. **Mera, S. L. and Davies, J. D.,** Differential congo red staining: the effects of pH, non-aqueous solvents and the substrate, *Histochem. J.,* 16, 195, 1984.
51. **Wolman, M. and Bubis, J. J.,** The cause of the green polarization color of amyloid stained with congo red, *Histochemie,* 4, 351, 1965.
52. **Missmahl, H. P. and Hartwig, M.,** Polarizationsoptische Untersuchungen an der Amyloidsubstanz, *Virchows Arch. (A),* 324, 480, 1953.

53. **Puchtler, H., Sweat, F., and Levine, M.,** On the binding of congo red by amyloid, *J. Histochem. Cytochem.,* 10, 355, 1962.
54. **Carson, F. L. and Kingsley, W. B.,** Nonamyloid green birefringence following congo red staining, *Arch. Pathol. Lab. Med.,* 104, 333, 1980.
55. **Romhányi, G.,** Selective differentiation between amyloid and connective tissue structures based on the collagen specific topo-optical staining reaction with congo red, *Virchows Arch. (A),* 354, 209, 1971.
56. **Romhányi, G.,** Über die submikroskopische Struktur des Amyloids, *Zentralbl. Allg. Pathol.,* 80, 411, 1943.
57. **Romhányi, G.,** Über die submikroskopische Struktur des Amyloids, *Schweiz. Z. Pathol. (Pathol. Microbiol.),* 12, 253, 1949.
58. **Romhányi, G.,** Zur Frage der submikroskopischen Struktur des Amyloid, *Zentralbl. Allg. Pathol.,* 95, 130, 1956.
59. **Romhányi, G.,** Differences in ultrastructural organization of amyloid as revealed by sensitivity or resistance to induced proteolysis, *Virchows Arch. (A),* 357, 29, 1972.
60. **Delellis, R. L. and Bowling, M. C.,** The use of sirius red and congo red staining in routine histopathology, *Hum. Pathol.,* 1, 655, 1970.
61. **Sweat, F. and Puchtler, H.,** Demonstration of amyloid with direct cotton dyes. Experience with a new method for the selective staining of amyloid by sirius red F3BA and sirius supra scarlet GG-CF, *Arch. Pathol.,* 80, 613, 1965.
62. **Virchow, R.,** Über den Gang der amyloiden Degeneration, *Arch. Pathol. Anat. Physiol. Klin. Med.,* 8, 364, 1855.
63. **Virchow, R.,** Zur Cellulose-Frage, *Arch. Pathol. Anat. Physiol. Klin. Med.,* 8, 140, 1855.
64. **Carnes, H. and Forker, B. R.,** Metachromasy of amyloid. A spectrophotometric study with particular reference to the dye-chromotrope bond, *Lab. Invest.,* 5, 21, 1956.
65. **Dalferes, E. R., Jr., Radhakrishnamurthy, B., and Berenson, G. S.,** Acid mucopolysaccharides of amyloid tissue, *Arch. Biochem. Biophys.,* 118, 284, 1967.
66. **Pennock, C. A., Burns, J., and Massarella, G.,** Histochemical investigation of acid mucosubstances in secondary amyloidosis, *J. Clin. Pathol.,* 21, 578, 1968.
67. **Mowry, S. R. W. and Scott, J. E.,** Observations on the basophilia of amyloid, *Histochemie,* 10, 8, 1967.
68. **Wolman, M.,** Amyloid, its nature and molecular structure: comparison of a new toluidine blue procedure polarized light method with traditional procedures, *Lab. Invest.,* 25, 104, 1971.
69. **Powers, J. M. and Spicer, S. S.,** Histochemical similarity of senile plaque amyloid to apudamyloid, *Virchows Arch. (A),* 376, 107, 1977.
70. **Scott, J. E. and Dorling, J.,** Differential staining of acid glycosaminoglycans (mucopolysaccharides) by alcian blue in salt solutions, *Histochemie,* 5, 221, 1965.
71. **Módis, L.,** Topo-Optical Investigations of Mucopolysaccharides (Acid Glycosaminoglycans), in *Handbuch der Histochemie,* Vol. II. Part 4, Graumann, W. and Neumann, K., Eds., G. Fischer Verlag, Stuttgart, 1974.
72. **Lillie, R. D. and Fullmer, H. M.,** *Histopathologic Technic and Practical Histochemistry,* 4th. ed., McGraw-Hill, A Blackiston Publication, New York, 1976.
73. **Terner, J. Y.,** Histochemical alkylation: a study of methyliodide and its effect of tissues, *J. Histochem. Cytochem.,* 12, 504, 1964.
74. **Sylvén, B.,** Metachromatic dye-substrate interactions, *Quart. J. Microsc. Sci.,* 95, 327, 1954.
75. **Fischer, E. R. and Lillie, R. D.,** The effect of methylation on basophilia, *J. Histochem. Cytochem.,* 2, 81, 1954.
76. **Velican, C.,** Histochemical analysis of the structural stability of biopolymers (preliminary data), *Acta Histochem. (Jena),* 31, 261, 1968.
77. **Scheuner, G. and Hutschenreiter, J.,** Das Auftreten von Metachromasie, Doppelbrechung und Dichroismus durch die Toluidinblaureaktion, *Progr. Histochem. Cytochem.,* 7/2, 1, 1975.
78. **Missmahl, H. P.,** Metachromasie durch gerichtete Farbstoffzusammenlagerung von Farbstoffteilchen, erläutert am optischen Verhalten des Toluidinblaus, *Histochemie,* 3, 396, 1964.
79. **Romhányi, G.,** Über die submikroskopische strukturelle Grundlage der metachromatischen Reaktion, *Acta Histochem. (Jena),* 15, 201, 1963.

Chapter 11

CORRELATION BETWEEN POLARIZATION AND ELECTRON MICROSCOPIC FINDINGS IN ULTRASTRUCTURE RESEARCH OF PROTEOGLYCANS

TABLE OF CONTENTS

I. INTRODUCTION

Electron microscopy (EM) and polarization microscopy (PM) techniques reveal ultrastructural patterns in different ways. While EM directly visualizes the structures, PM is an indirect method, i.e., the ultrastructure can be inferred from the optical phenomena observed or measured in the microscope. In addition to this, the structures are differently altered during the different preparative methods of the specimens used in these two techniques, i.e., different artifacts are formed, observed, and compared. Furthermore, the minute section thickness and high magnification used in the transmission EM to image details of the ECM limits the analysis to a very restricted field. In contrast to these conditions, the PM permits analysis of much larger area of the tissue. EM visualizes bits and pieces of the trees in a forest, PM, on the other hand, provides high resolution of the whole forest.[1]

In spite of these considerations, many examples show that EM and PM reveal similar or identical ultrastructural patterns of the ECM. Generally, PM observations were made first which were supported and supplemented by subsequent electron microscopic reinvestigations. No polarization microscopic findings had to be essentially modified by EM.

Thus, PM and EM, especially if both of them are combined with histochemistry, can be regarded as two powerful and complementary methods in ultrastructure research. In certain fields, the application of the EM — due to its more precise attributes — proved to be more effective in description of orientation patterns in the ECM. For example, the PM revealed the spatial orientation pattern of the insoluble collagen in different ECMs, but only the EM investigations explored the fine structure of the collagen fibrils and the collagen-proteoglycan interactions. In other cases, however, the use of the PM has advantages over the EM, e.g., in showing the spatial orientation patterns of the GAGs and other carbohydrate molecules in the ECM and on the cell surface.

In this chapter an attempt will be made to show how EM and PM can complement each other in ultrastructure research of the GAG and PG molecules in the ECM of the cartilage tissue.

II. CRITICAL REMARKS TO THE ELECTRON MICROSCOPIC METHODOLOGY OF PROTEOGLYCANS

A. ELECTRON MICROSCOPY OF PROTEOGLYCANS WITHOUT SPECIFIC LABELINGS

In thin sections contrasted by conventional techniques (by salts of Os, U, Pb), differently sized electron dense structures may appear in the interfibrillary space of the extracellular matrix. Certainly, some of these granules and/or filaments may correspond to PGs. This approach, due to its aspecific character, cannot be recommended for ultrastructure research of the PGs.

A peculiar possibility has been suggested by Hunziker et al.[2] for electron microscopic examination of PG molecules not specifically contrasted. This is the high pressure freezing, freeze substitution (HPF/FS), and low temperature embedding method. This promising although sophisticated technique well preserves the PG molecules in an extended state.[3] PGs appear as a reticulum of fine filamentous threads throughout the matrix. Hunziker and Schenk[3] suggest that both core protein and GAG side chains of the PG molecules can be identified after this procedure in the electron micrographs. The advantages of this technique are apparent. The extraction of PGs caused by conventional fixation is avoided, and, furthermore, their collapse and dislocation caused by stainings and traditional embedding are prevented or decreased. However, further examinations seem to be necessary for more exact identification of the chemical nature of the differently sized matrix filaments obtained after the HPF/FS method. It also deserves more attention that the uranyl acetate used in the substitution medium does not really cause any modification in shape of the PG constituents. At present, we can conclude that highly different ultrastructural morphology of the ECM are revealed by using the conventional electron microscopic preparative techniques and the HPF/FS. There are, however, some similar findings. In cartilage ECM contacts between PGs and collagen fibrils have been demonstrated by both of these methods, although the pattern of PGs is shown differently. Using conventional techniques, PGs appear as regularly spaced structures along the collagen fibrils in different tissues contrasted with cationic reagents.[4-11] After HPF/FS, however, PGs of the cartilage matrix in contact with the fibrils fail to exhibit any characteristic and preferential orientation pattern.[3] Furthermore, contacts between PGs and the cell membrane of chondrocytes[3,12] as well as between PGs and extracellular matrix vesicles playing a role in mineralization[13,14] are shown by both of these techniques. It will be shown later in this chapter that according to our investigations PG molecules are bound to the cell membrane of chondrocytes in an oriented manner. HPF/FS reveals, however, a random network of filaments around the chondrocytes, although more thick and darkly stained strands, which are supposed to be the core proteins, are visible exhibiting a perpendicular orientation to the tangent of the cell membrane in such electron micrographs.[3]

We want to note here that isolated PG molecules can also be investigated by electron microscope. Spread PGs are investigated after either contrasting with cytochrome C,[15-25] or a negative staining[25] or by a rotary shadowing technique.[25] Immunohistochemical reactions can also be applied to show different parts of the isolated PG molecule.[26] Using this elegant technology, the "molecular anatomy" of PGs, i.e., shape and dimension both of the whole molecule and its different parts can be directly studied.

B. STAINS OR ANTIBODIES?

PGs can generally be shown by EM in the tissues if they are selectively contrasted. There are two categories of the histochemical methods which can be used to selectively stain PGs: the immunohistochemical and the cation binding (basophilic) reactions. No doubt antibodies are more specific than the simple stains or other cationic reagents. One must be reminded here, however, that — although a stain can never be regarded as a specific marker for a macromolecule — there are staining conditions which warrant the required specificity of the reaction.

In spite of the fact that antibodies are much larger than stains, the sensitivity of the immunohistochemical reactions is, at least at light microscopic level, higher than that of the basophilic reactions. It has been demonstrated that PGs could be shown by light microscopic immunohistochemical reaction in cartilage matrix where the cationic safranin O failed to stain these molecules.[27] The insensitivity of simple stainings has also been described by Poole.[28] He showed that no staining of cartilage matrix could be observed with toluidine blue after having lost 42% of its hexuronic acid content.

Immunohistochemical reactions, however, strongly modify the shape of the PG molecules using either polyclonal or monoclonal antibodies.[29-36] This is due to the large size of the antibodies and the electron dense markers (peroxidase, protein A-gold particles, ferritin, etc.).[9] Certain cation binding reactions, therefore, are more valuable tools for investigation of the spatial orientation patterns of the PG molecules because the relatively small dye molecules better preserve the native shape of the PGs. We have to keep in mind, however, that PGs without GAG side chains (which may occur under certain pathological conditions) can never be shown by cationic reagents.

It is obvious that both immunohistochemistry and "conventional" stainings have their specific values in ultrastructure research of the PGs. They should be regarded as complementary techniques, and the choice between them must be determined by the experimental goal. If ultrastructural localization of the core protein or a specific core protein region of a PG is needed, immunohistochemistry is the only appropriate tool. For the investigation of spatial orientation patterns of the PG molecules, however, their electron microscopic visualization by a staining of their GAG side chains with low molecular weight reagents ("mini-stains"[37]) seems to be a more successful way at this moment than immunohistochemistry.

C. PRE-EMBEDDING STAINING TECHNIQUES WITH CATIONIC REAGENTS

1. Fixation

The great majority of the methods recommended for electron-microscopic staining of the PGs are pre-embedding basophilic reactions (en bloc stainings). It means that the cationic reagents interact with the sample before dehydration, embedding, and cutting. The stain is either mixed with the fixative or it is applied separately after the fixation. The former variation has the obvious and frequently quoted advantage that the cationic dye precipitates the PGs hereby preventing their delocalization. The shortcoming of this staining protocol is that the dye which has to be present in relatively high concentration may modify the osmolarity of the fixative and it may result in a poor preservation of the ultrastructure of cells. In addition to this, the high pH of the fixative (pH 7.2 to 7.4) does not allow a specific cation binding reaction of PGs containing sulfated GAGs, because all of the carboxyl groups of different carbohydrates and proteins are also ionized and can bind cations at this pH value. Unfortunately, this obviously nonspecific staining condition is not always considered by some investigators, and this may result in erroneous interpretation of the electron micrographs. The problem of specificity of a cationic binding reaction will be separately discussed in this Chapter.

The other possibility is that the staining is carried out after fixation. In this case we have to count on a considerable loss of the hydrophilic and highly aqueous soluble PG molecules from the tissues. Engfeldt and Hjertquist[38] have demonstrated that approximately 70% of GAGs might be leached out of cartilage with conventional fixation techniques. A dislocation of PGs may also occur during the fixation using fixatives free of cationic reagents. Different alternative methods are suggested to overcome these difficulties.

Addition of cationic detergents such as cetylpyridinium chloride (CPC) or related quaternary ammonium compounds in the fixative may provide increased retention of PGs in the tissues.[39-41] These compounds are known to precipitate GAGs by binding to their anionic groups.[42] That is why these agents decrease or prevent the cationic dye-uptake of the GAG molecules. Washing the sections after a CPC-containing fixative in a 0.5 mol/l K-acetate in 20% ethanol restores the stainability of GAGs with cationic dyes because K^+ removes CPC and converts PGs to K-salt (personal communication of Dr. Margaret Aydelotte, Chicago). According to electron microscopic observations of Tuckett and Morriss-Kay,[43] CPC-supplementation of the fixative causes membrane disruption, evidenced by the enhanced intracellular staining and the appearance of strands and vesicles of basal laminae. According to these authors, the use of CPC is inadvisable if an ultrastructural distinction between surface-related and extracellular GAGs is desirable. They recommend the use of either the Sainte-Marie's (99 ml 95% ethanol, 1 ml acetic acid) or the Karnovsky's (2% paraformaldehyde, 2.5% glutaraldehyde in 0.1 mol/l cacodylate buffer) solutions at 4°C as good fixatives for GAG preservation and subsequent staining with alcian blue. Our laboratory has also had good results with the cold Karnovsky's fixative, but, similarly to the previously quoted authors, we also find a lack of consensus on which is the best fixative for the PG molecules.

It has been shown that isotonic buffer salt solutions and commercial tissue-culture media, mainly in the first 30 min of incubation, extract PGs from the cartilage matrix, and PGs within the different matrix compartments exhibit various susceptibility to extraction by different buffer salts.[44]

Chemical decalcification of mineralized tissues may also cause a considerable loss of PGs. To decrease this, the use of an alcoholic EDTA solution is recommended (0.2 mol/l trimethylammonium EDTA in 80% (v/v) ethanol/water with 4% (w/v) formaldehyde).[45,46]

2. Selection of Dyes

A high number of cationic reagents have been proposed for electron microscopic demonstration of the molecules. They electrostatically interact with anionic groups of the tissues, and under certain staining conditions (see below) allow a selective dye-uptake of the sulfate groups, they can selectively detect the sulfated GAG side chains of PG molecules.

Besides their cationic nature, the other common chemical feature of these reagents that they have one or more elements with an atomic number high enough to cause electron density. In some cation-binding reactions, however, the electron density is provided by a subsequent amplification using an electron-dense compound which can be bound to the dye-PG complex.

The most frequently used cationic reagents are the metal containing dyes. In addition, a number of other cations has also been recommended. The major groups of cationic probes for EM of PGs are the following: metal salts (e.g., ianthanum nitrate,[47] bismuth nitrate,[48,49] zirconium oxichloride[50,51]), metal colloids (e.g., colloidal iron,[52-55] colloidal thorium dioxide[56]), metal containing organic complexes (e.g., high iron diamine (HID),[40,57-59] chitosan-colloidal gold,[60] poly-L-lysine-colloidal gold,[61] cationic ferritin[62]), cationic proteins (e.g., lysozyme with subsequent osmification[63]). Metal salts and colloids strongly precipitate the PG molecules causing their extreme collapse. PG molecules appear, therefore, as differently sized granules after such reactions. Consequently, these labelling techniques are not useful when investigation of spatial orientation patterns of the PGs is intended.

The cationic dyes used in electron microscopic histochemistry of PGs can be divided into the following two categories: dyes without electron-dense atoms and metal-containing dyes.

Using a cationic dye without a heavy atom, a poststaining treatment with a compound containing an element of sufficiently high atomic number is indispensable (e.g., osmification, contrasting with lead and uranyl salts). In this case, the agent used in poststaining treatment is responsible for the electron density of the PG-dye-contrasting agent-complex. The electron dense compound does not necessarily accompany the dye,[9] and the aspecific binding of the secondary reagent cannot be excluded without proper control experiments. The following non-electron dense cationic dyes have been used in electron-microscopic histochemistry of PGs: toluidine blue O,[6,64] acridine orange,[65-70] safranin O.[7,64,67,71,72] After staining with these dyes, PGs appear as branching filaments or rods or — exceptionally — as bottle-brush structures in the electron micrographs. These observations indicate that the native shape of PGs is much less modified by these dyes than by metal salts or colloids.

Yet, contrary to this favorable feature, their use cannot be recommended for the following two reasons. First, the specificity of such stainings is doubtful. The pH of the staining solution is high (the stain is mixed with the fixative) to provide a specific dye-uptake of the GAGs, and the poststaining contrasting procedure may also cause aspecific reactions. Furthermore, the metal contaminants of the dyes, which can reach a considerable quantity,[73] may also interact with different tissue constituents. Using spectrographic and atomic absorption spectrometric techniques, we could also demonstrate several metals in different dyes (Fe, Pb, Zn). Their concentrations are varied from dye to dye and they are also different in different charges of the same dye. A constant finding was, however, the presence of Zn in thiazine dyes (Figure 1). Atomic absorption spectrometry revealed a rather high concentration in some samples, e.g., the range of Zn concentration varied between 1 to 15% (w/w) in toluidine blue. It is known that Zn^{2+} interacts with polyanions and it precipitates the GAGs.[74] It cannot be excluded, therefore, that the moderate electron density observed intra- and extracellularly by Shepard and Mitchell[6] after staining tissue samples with toluidine blue (without any poststaining treatment) was simply due to an interaction of Zn^{2+} with anionic groups of different tissue constituents.

The second reason why toluidine blue, acridine orange, and safranin are not recom-

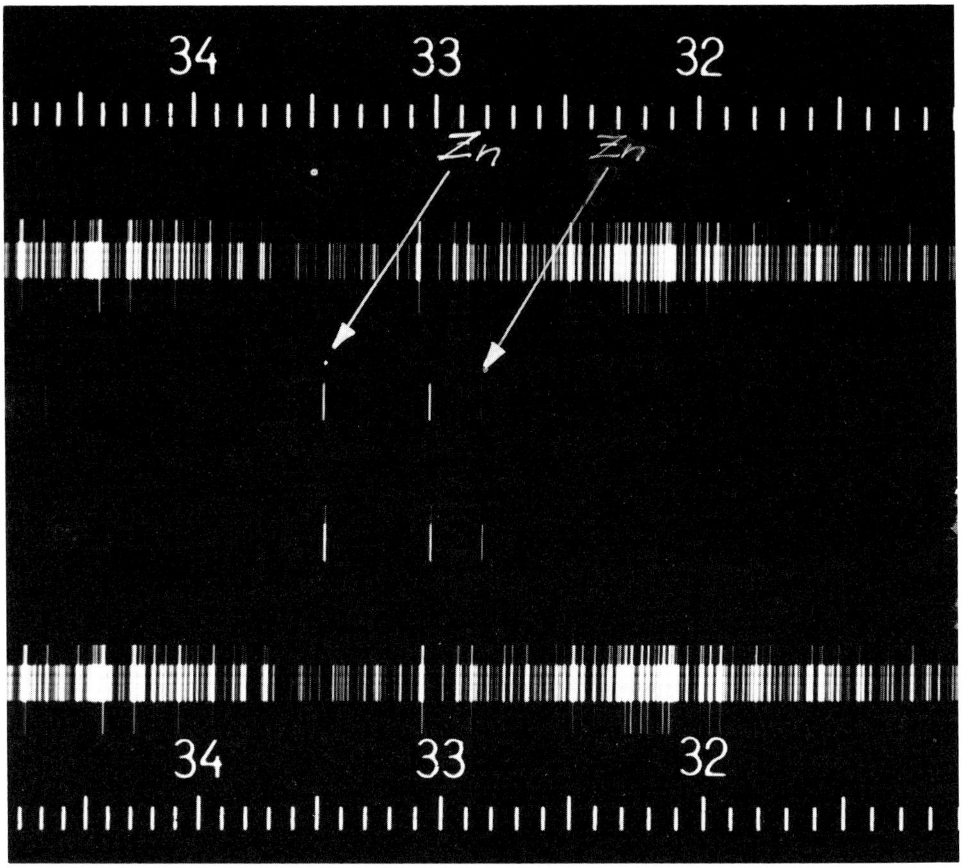

FIGURE 1. Detection of Zn in thiazine dyes by emission spectrography (Quartz spectrograph, Zeiss® Q24 type). Scales × 10 = wavelengths in nm. The following spectra are shown (in top-bottom direction): iron, toluidine blue O, dimethylmethylene blue, iron. Zn-lines are indicated by arrows.

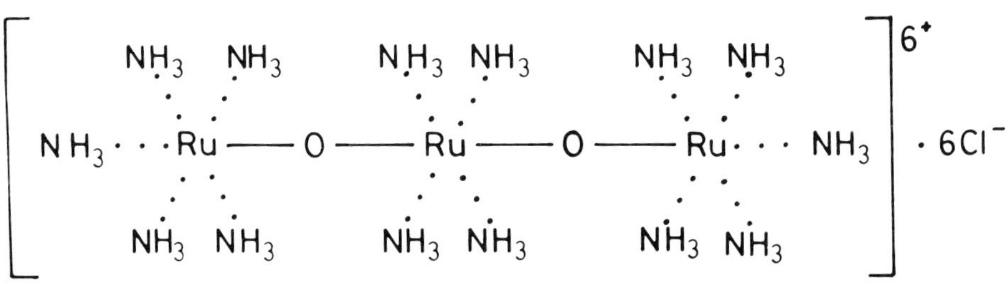

FIGURE 2. Structural formula of ruthenium red.

mended for block-staining of PG molecules will be separately discussed in this Chapter. This is the susceptibility of these staining reactions to the poststaining dehydration procedure.

The most popular cationic reagents used in EM of PGs are the metal-containing dyes. These can be again divided into two groups: the ruthenium compounds and the copper-phthalocyanins. Ruthenium red[15,16,19,67,71,72,75-79] (Figure 2) and ruthenium hexamine trichloride[3,12,80,81] belong to the first group. Contrary to the phthalocyanins, the ruthenium compounds, owing to their smaller size, strongly contrast and collapse PG molecules. PGs generally appear as differently sized granules in the electron microscope after these stainings

(Figure 3). We do not use them if we want to learn more about the spatial orientation of the PGs, but we can recommend them, especially the ruthenium hexamine trichloride, for investigaton of the cell coat.

Alcian blue 8GX is the most popular copper-containing phthalocyanin dye used in histochemistry of PGs. It is also applied as a stain for EM.[8-11,67,72,76,82-91] This relatively bulky dye (its diameter is about 2.6 to 2.8 nm) with its single Cu atom contrasts PGs moderately. The contrast can be amplified by a subsequent treatment with sodium tungstate.[8] After this staining, PGs appear as rod-like particles. The formation of the rods from the original bottlebrush structure of PG molecules is predominantly due to the collapse of their GAG side chains after electrostatic interaction with the dye (Figures 4 and 5). Other factors like dehydration may also contribute to the partial collapse. The longitudinal dimensions of the PGs are, however, relatively well preserved (Figure 6). Owing to the relatively large size of this dye, it brings about less collapse with consequently less contrast of PGs compared to other copper-phthalocyanins like cuprolinic blue and cupromeronic blue.[9,82,83] The bulky dye molecules being in interaction with numerous anionic groups of the GAGs may act as scaffolds for the GAG and PG structure.[37] At the same time, the large size of the dye explains its poor penetration of the tissues. To overcome this difficulty, thin tissue samples (possibly 100 to 300 μm sections made by a vibratome tissue sectioner) should be used before the staining is carried out with alcian blue or other phthalocyanins. An advantage of alcian blue and related dyes over other dyes is that they can be successfully utilized in critical electrolyte concentration (CEC) technique (for review see Scott[9,92]). Alcian blue staining with CEC guarantees a selective staining of sulfated GAGs and their PGs even when applying the dye at higher pH values. The same holds true for the other phthalocyanins.

The two other copper-containing phthalocyanins, cuprolinic blue[9,79,83,93-104] (Figures 6-8) and cupromeronic blue[83,105-112] have also been frequently used recently. Due to their smaller size compared to that of alcian blue, a staining with them results not only in a stronger electron density, but also a stronger collapse of the PG molecules.

It will be shown later that the binding of phthalocyanins to GAGs is not sensitive to subsequent alcoholic dehydration. Considering all of these advantageous features of the copper phthalocyanins, they are recommended for investigation of spatial orientation patterns of the PG molecules. Using these dyes, a predominant orthogonal orientation of PGs to collagen fibrils could be demonstrated electron microscopically in different soft tissues.[4-11,37,92,107-110] As will be shown later, these findings well fit the results of relevant polarization microscopic analyses.

3. Specificity

As has been discussed earlier, no cationic probes can be regarded as specific reagents for the GAGs and PGs. There are several possibilities for the histochemist, however, to make a simple staining specific for these molecules. The common premise of these additional histochemical procedures which achieve specificity is that the cationic probes electrostatically interact with the polyanionic GAGs. It is known that GAGs (except hyaluronic acid) are the only polysulfates in the normal ECM of vertebrates. If the staining conditions guarantee a selective dye-uptake of the ionized sulfates, the staining is specific for sulfated GAGs and hereby for PGs.

There are three alternative possibilities for this purpose: the staining must be carried out at low pH (below 2.5 to 3.0) where sulfates are readily dissociated, but the carboxyl groups of different carbohydrates and proteins are not ionized; the staining solution must contain salt in a certain concentration which blocks the dye-uptake of the carboxyl groups; a selective carboxymethylation must be carried out prior to the staining.

Staining with a cationic reagent at low pH cannot be recommended for the following two reasons: it is not optimal for a good preservation of the ultrastructure; and, although

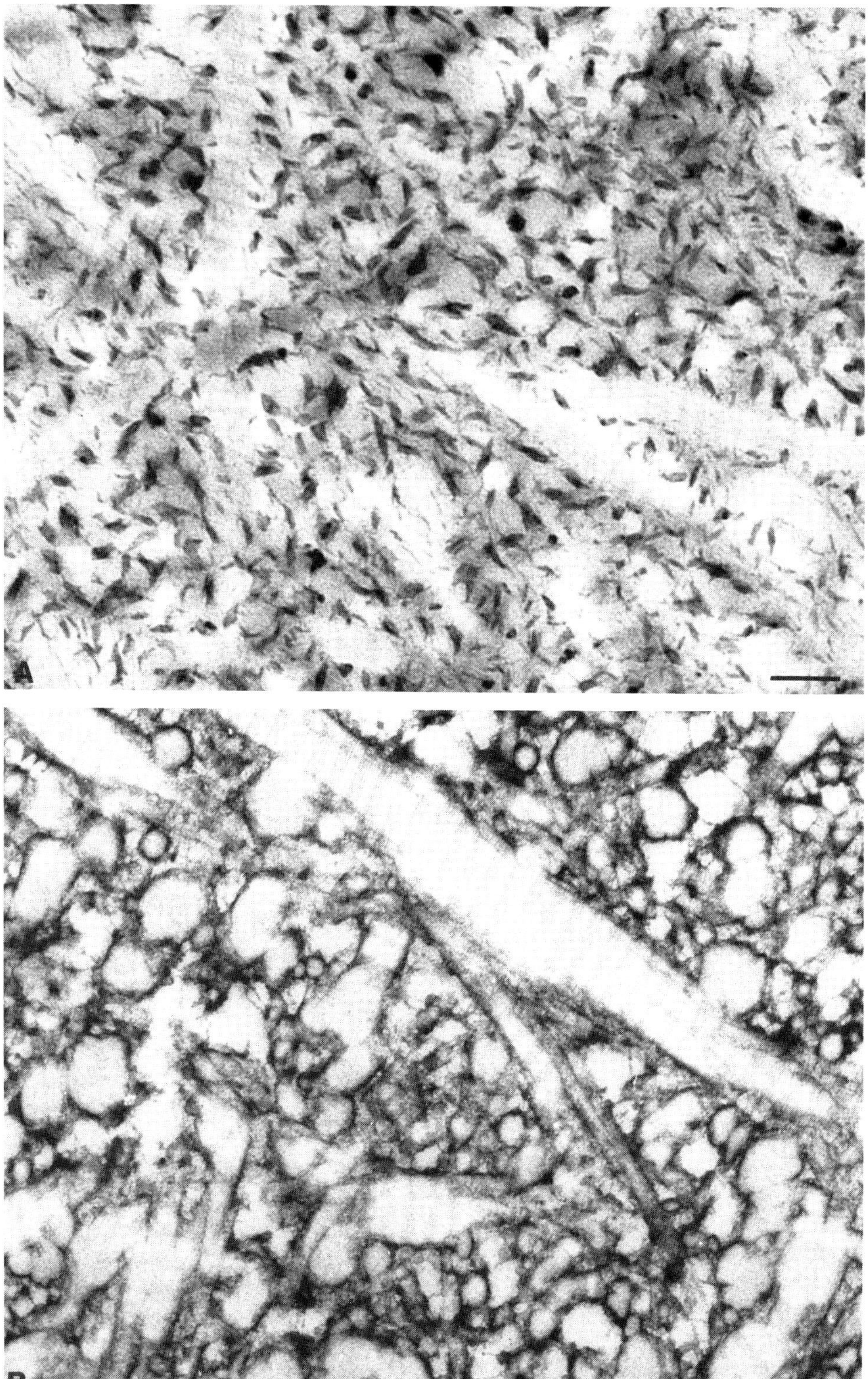

FIGURE 3. Electron micrographs taken from human articular cartilage matrix (15-year-old patient, autopsy material). Proteoglycans are stained by alcian blue 8GX (A) and ruthenium hexamine trichloride (B). Note the rod-like appearance of PGs in (A), and the total callapse of PGs in (B). Observe the relationship of PGs to collagen fibrils in (A); if collagen fibrils are longitudinally cut, a more or less perpendicular orientation of numerous PGs bound to these fibrils is seen. Scales: 200 nm.

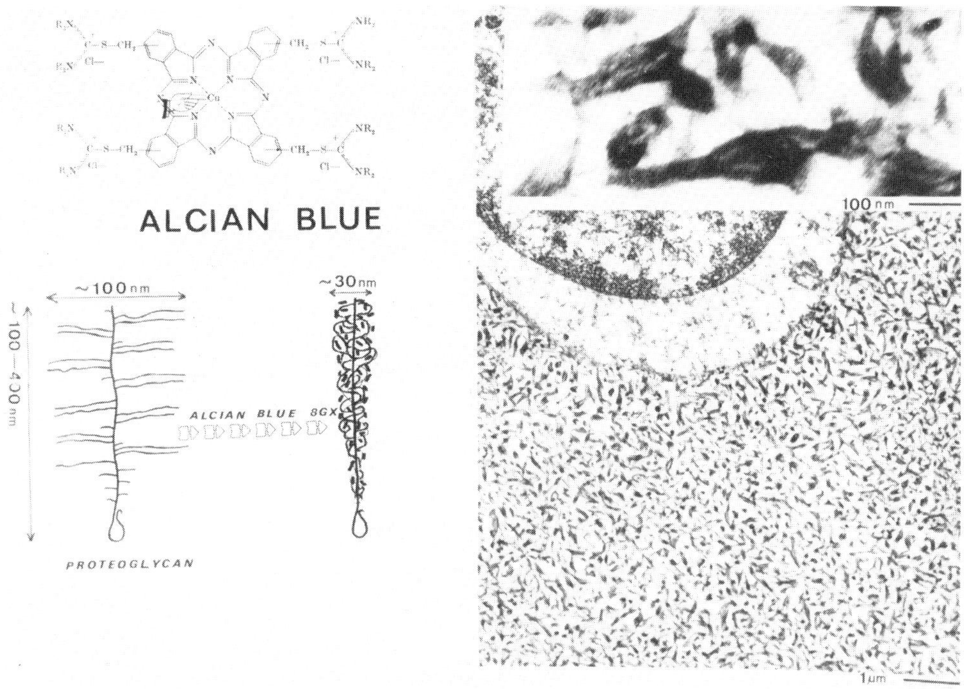

FIGURE 4. The diagram shows the structural formula of alcian blue 8GX, the collapsing effect of dye on GAG side chains of a PG molecule, and low- and high-power electron micrographs of cartilage PGs stained with this dye. Sample: human articular cartilage (4-year-old patient, autopsy material). Staining was carried out in block in presence of 0.3 mol/l $MgCl_2$. 0.5% sodium tungstate was used to amplify electron density. If not otherwise stated in this chapter, this procedure is used for staining of other tissue samples with phthalocyanins to show PGs electron microscopically. Note that PG molecules of the matrix appear as rod-like or leaflet-like particles.

the dissociation of carboxyls is suppressed at low pH, the tissue proteins are cationic (due to the proton uptake of the N atoms of their $-NH_2$ groups), and these cationic groups can electrostatically interact with the ionized sulfates decreasing or blocking the dye-uptake.[113]

The supplementation of the staining solution of phthalocyanin dyes with $MgCl_2$ is the generally accepted and widely used CEC-technique.[9,92] The dye-uptake of carboxyls is blocked by the competing Mg^{2+} at 0.2 to 0.3 mol/l $MgCl_2$ concentration, and the staining carried out even at high pH (generally at pH 5.7 to 5.8) will be specific for PGs in the ECM. A further advantage of this method is that the polycationic (tetravalent) phthalocyanins are supposedly transformed to monovalent cations, probably due to the interaction of some of their cationic charges with Cl^- ions of the staining solution.[9,83] Thus, more dye molecules can be bound to the chromotrope. This has two consequences: a higher intensity of the staining and a more complete "scaffolding" of the PGs by the increased number of bulky dye molecules bound to the GAG side chains of the PGs. It explains why PGs appear as rods (and not as granules) in the electron micrographs after staining with a phthalocyanin dye in the presence of 0.3 mol/l $MgCl_2$. Unfortunately, alcian blue 8GX frequently precipitates (at least partially) in the presence of $MgCl_2$, but other phthalocyanins do not. Despite this minor technical difficulty and some critical comments on the theory of the CEC

FIGURE 5. High-power electron micrograph taken from human articular cartilage matrix (4-year-old patient, autopsy material) showing the leaflet-like appearance of PGs after staining with alcian blue 8GX. Scale: 100 nm.

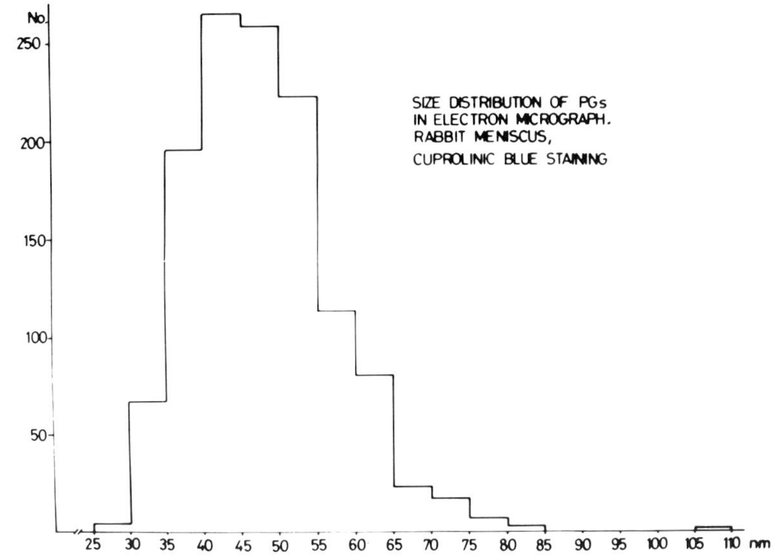

A

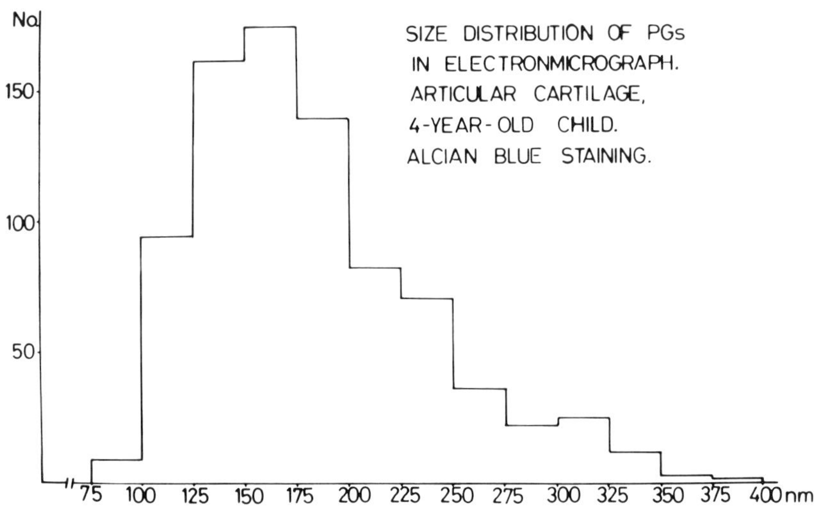

B

FIGURE 6. Size distribution of randomly selected proteoglycan particles appearing in the electron micrographs taken from two different cartilages stained by two different phthalocyanins. Only longitudinal dimensions of the particles were measured. PGs of rabbit meniscus stained with cuprolinic blue are smaller (A) than PG particles found in human articular cartilage matrix stained by alcian blue (B). Number of measurements = 1200 (A) and 800 (B).

technique,[114,115] the combination of CEC technique with phthalocyanins is one of the best histochemical reactions of PGs both for light and electron microscopy.

The third possibility to make sure that a basophilic reaction of the ECM will be specific for polysulfates is a specific carboxymethylation carried out prior to the staining with a cationic dye at high pH (pH 5 to 7.4).[116,117] In our experience, this is a good and reliable

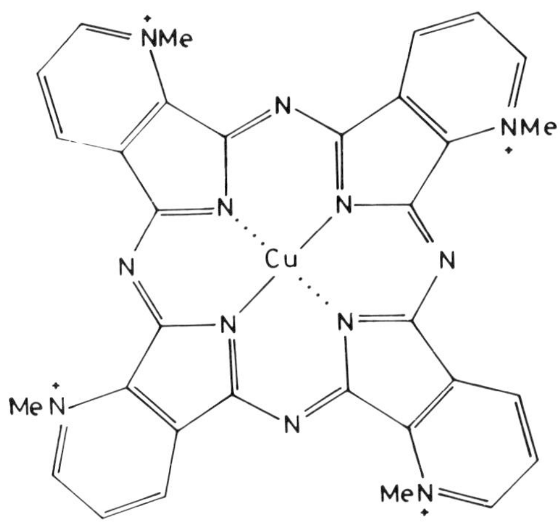

FIGURE 7. Structural formula of cuprolinic blue.

technique, when combined with a successive staining with phthalocyanins, both for transmission electron microscopy and energy dispersive X-ray microanalysis of PGs.[118]

Although one of the above techniques allows a specific staining of the PGs, the control of the staining effects cannot be neglected especially in pathological tissues or in any kind of investigaton when typing of GAGs is necessary. For this purpose, enzymatic extractions of GAGs with selectively acting hydrolyses are needed prior to the staining reaction (hyaluronidases, chondroitinases, keratanase, heparitinase).[9,119] It is worth mentioning here that a digestion with chondroitinase ABC combined with peanut or soybean-lectin-binding reactions has proved to be a selective light and electron microscopic technique to show oligosaccharides attached to the core protein of chondroitin sulfate PGs of the cartilage.[120]

4. Sensitivity Against Poststaining Alcoholic Treatments

Pre-embedding stainings must be followed by dehydration prior to the embedding. For those who are working with dye-binding reactions it is known that there are dyes which cannot be extracted from the tissues by alcohols used for dehydration but the binding of a great majority of cationic dyes to different tissue constituents is vulnerable to the dehydration procedure. Among others, the metachromatic dyes belong to this latter group. Several metachromatic dyes are commonly recommended for pre-embedding staining and electron-microscopic visualization of PGs. As has been described earlier, these dyes are the following: toluidine blue O, acridine orange, and safranin O. We investigated the sensitivity of these dyes to poststaining dehydration. Our quantitative data show a considerable loss of dye bound to tissue polyanions during dehydration with ethanol even using staining conditions rendering possible a sulfate-specific staining. The short summary of our experiments is included here just to convince electron microscopists that these staining techniques should be rejected.

Cryostate and vibratome sections were prepared from fresh, unfixed cartilages (bovine nasal cartilages, BNC, human fetal and postnatal articular cartilages from the femur condyles). Sections were freezed-dried at −20°C and divided into small groups. The weight of groups of lyophilized sections was measured with high accuracy and the groups of sections were put in staining solution. Staining, washing, and dehydration were carried out in a microtiter plate. The following dyes were used: toluidine blue O (SERVA), brominated toluidine blue (kindly provided by Nora Shepard, Montreal, this dye being synthesized in that laboratory and used for X-ray microanalysis of PGs[121]), acridine orange (Reanal, Hungary), safranin O (E. Gurr), ruthenium red (Fluka), cuprolinic blue (BDH), alcian blue 8GX

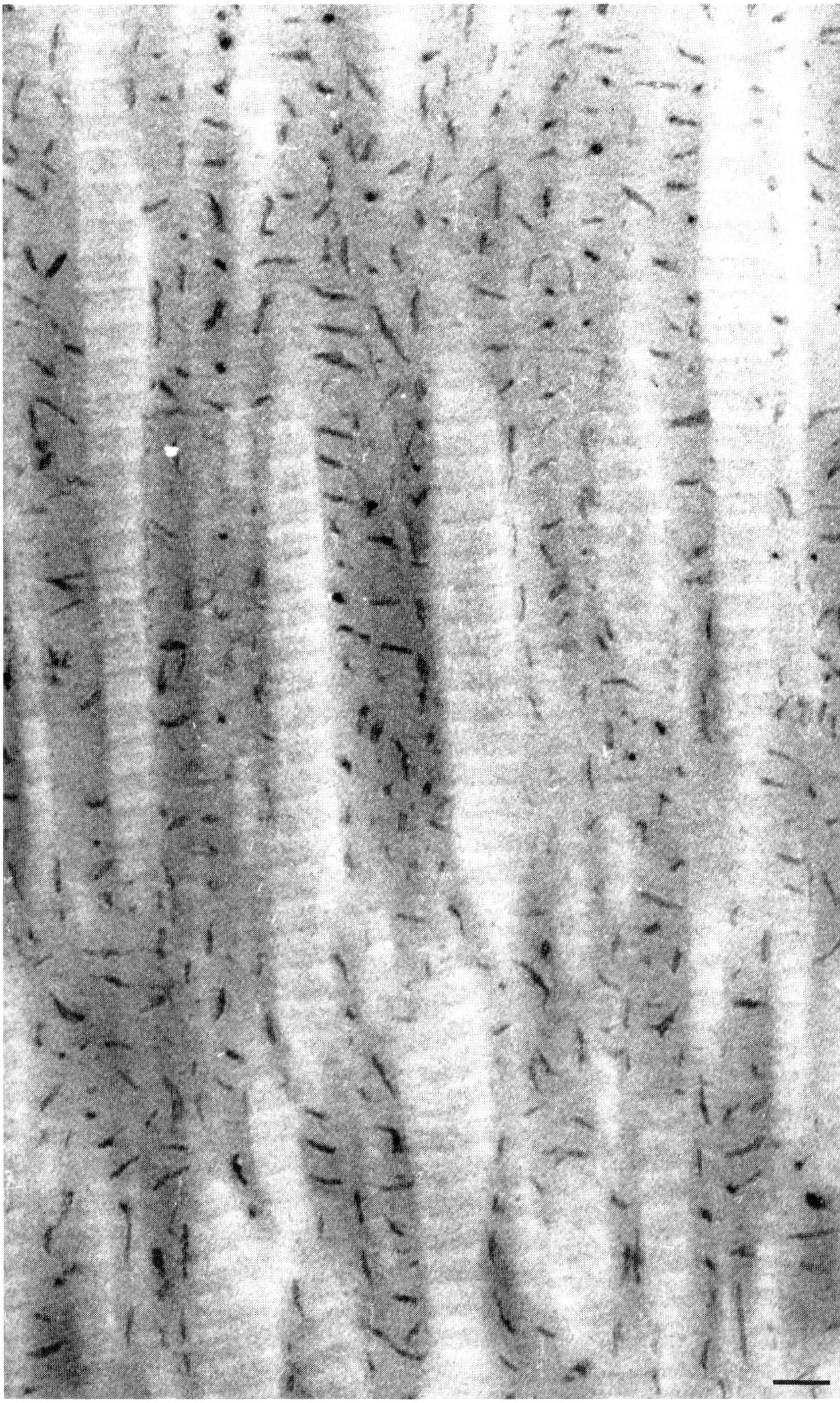

FIGURE 8. Electron micrographs taken from the meniscus of a rabbit knee joint after staining with cuprolinic blue. Note the numerous PG particles attached to the cross banded collagen fibrils. Many of these PGs are perpendicularly oriented to the long axis of the fibril. Scale: 100 nm.

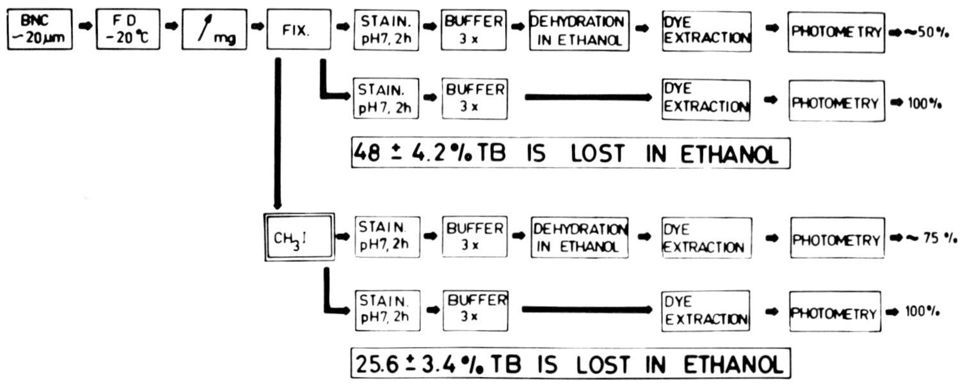

FIGURE 9. Sketch of our experiment devoted to investigation of the susceptibility of dye-binding reactions to poststaining alcoholic dehydration. Sections of cartilage tissue samples (here bovine nasal cartilage, BNC) were freezed and dried (FD), and their weights were measured. After fixation (FIX), stainings and poststaining treatments were carried out in microtitre plate. Stainings were performed without or after antecedent carboxymethylation using CH_3.[116] After staining, sections were washed in buffer, and were either dehydrated in a series of ethanol (from 30% up to 100%), or were subjected to a dye-extraction procedure (1ml conc. HCl, 99 ml 50% ethanol). Dye was also extracted from stained, washed, and dehydrated sections. Quantity of extracted dye was determined by spectrophotometry. Note that a considerable amount of toluidine blue is lost during dehydration even if the staining was carried out after carboxymethylation (after this treatment only tissue sulfate groups could bind the dye molecules). Acridine orange and safranin O behaved similarly.

(TAAB). Dyes were added either to the fixative (2.5% glutaraldehyde in 0.1 mol/l cacodylate, pH 7.2, or in 0.1 mol/l acetate buffer, pH 6) or they were applied after fixation and washing off the samples in buffer. The dye concentration was 6 mmol/l, and the staining time was 2 h at 22 to 24°C. After washing for 3 × 20 min in buffer, the groups of sections were treated either in a series of ethanol (from 30% to 100% with 10% increments) according to the protocol used in electron microscopic laboratories, or in a HCl-ethanol (1 ml conc. HCl, 99 ml 50% ethanol in water) to extract dye. Dilute ethanol in the presence of HCl readily extracts toluidine blue O, brominated toluidine blue, acridine orange, and safranin O. Alcian blue 8GX, cuprolinic blue, and ruthenium red cannot be extracted at all by this solution, and, similarly, the series of ethanol used for dehydration also failed to extract any trace of these dyes from the tissues. Therefore, further quantitative investigations were carried out only with toluidine blue O, brominated toluidine blue, acridine orange, and safranin O. The quantity of dye extracted from the stained and washed tissue samples was spectrophotometrically measured, and expressed in mg dye/mg tissue dry weight. These values were regarded as 100%, and they reflect the total dye binding capacity of the samples. Similar measurements and calculations were made after selective carboxymethylation + staining + washing. An example of the experimental set-up is shown in Figure 9. Other groups of sections were extracted after the complete dehydration, and the dye quantity remained in the tissue after dehydration was determined spectrophotometrically. Here again, the results were expressed in mg dye/mg tissue dry weight. These values indicate the amounts of dyes which cannot be dissolved by ethanol, and they are compared with values obtained in undehydrated samples. About 50% of the toluidine blue bound to the tissue is extracted during dehydration (Figure 9). Carboxymethylated and stained sections (in which only sulfate groups bound with toluidine blue) lost about 25% dye during dehydration in ethanol (Figure 9). This means that the sensitivity of toluidine blue reaction of PG is strongly diminished by the poststaining dehydration process. Similar findings were obtained for brominated toluidine blue, safranin O, and acridine orange. Somewhat more safranin O and somewhat less acridine orange was lost during dehydration than toluidine blue. These experiments were repeated with fixed, stained and osmificated (1% OsO_4 in buffer, 1 h) sections. It was

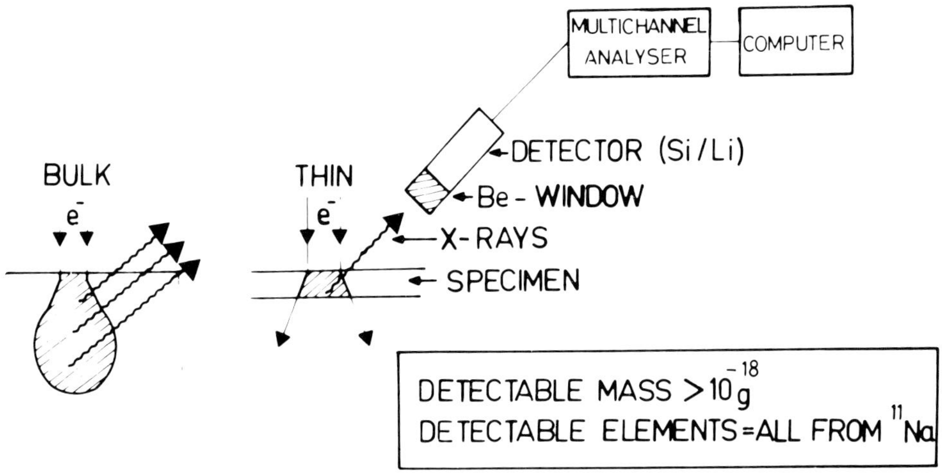

FIGURE 10. A highly simplified scheme of X-ray microanalysis. Bulk or thin tissue specimens are bombarded by electrons in the electron microscope. X-ray photons generated by interactions of bombarding electrons and different elements of the sample are collected by a detector, and analyzed by a X-ray spectrometer.

supposed that treatment of stained sections with osmium tetroxide may prevent loss of dye in ethanol. It was not the case, the amounts of dyes extracted by the dehydration process did not change at all.

These data suggest that dyes susceptible to poststaining dehydration cannot be optimal reagents in electron microscopy of the proteoglycan molecules, at least when using pre-embedding staining protocols. Taking together this and the facts that (1.) these stainings are carried out at high pH which does not allow a selective staining of GAGs, (2.) these dye molecules do not contain electron dense atoms and the electron density of the reaction product is provided by a poststaining amplification of uncontrolled specificity, we cannot consider toluidine blue, acridine orange, and safranin O as reagents which should be recommended for electron microscopic histochemistry of PGs even though in some cases nice, bottlebrush-shaped PGs are photographed after such stainings.[122]

D. X-RAY MICROANALYSIS

Energy-dispersive X-ray microanalysis has become a powerful tool of qualitative and quantitative histochemistry during the last two decades. The common version of this technique is suitable for the detection of elements of atomic number Z = 11 (sodium) or higher, and minimal quantities of 10^{-18} g of these elements can be detected in the volume of the tissue excited by the electron beam of the electron microscope (Figure 10).

1. Unstained or Stained Specimens?

X-ray microanalysis has also been used for identification and semiquantitative estimation of sulfated GAGs. Generally, unstained tissue samples are used for this purpose. In this case, the sulphur peak of the X-ray spectra is used for the estimation assuming that the X-ray photons are emitted only by the sulphur atoms of the sulfate groups of GAGs.[123-127]

It has been shown, however, in our laboratory that this assumption needs reconsideration.[118] X-ray spectra were taken and compared from unstained hyaline cartilage matrix without and after methylation (1ml conc. HCl, 99 ml absolute methanol, 56°C, 24 h). It is known that methylation removes sulfate groups.[128-130] Methylated sections could not be stained anymore by cationic dyes, but a considerable peak of sulfur was still registered in the X-ray spectrum taken from cartilage matrix after methylation (Figure 11). This finding suggests that other S-containing molecules other than sulfated GAGs may also be responsible

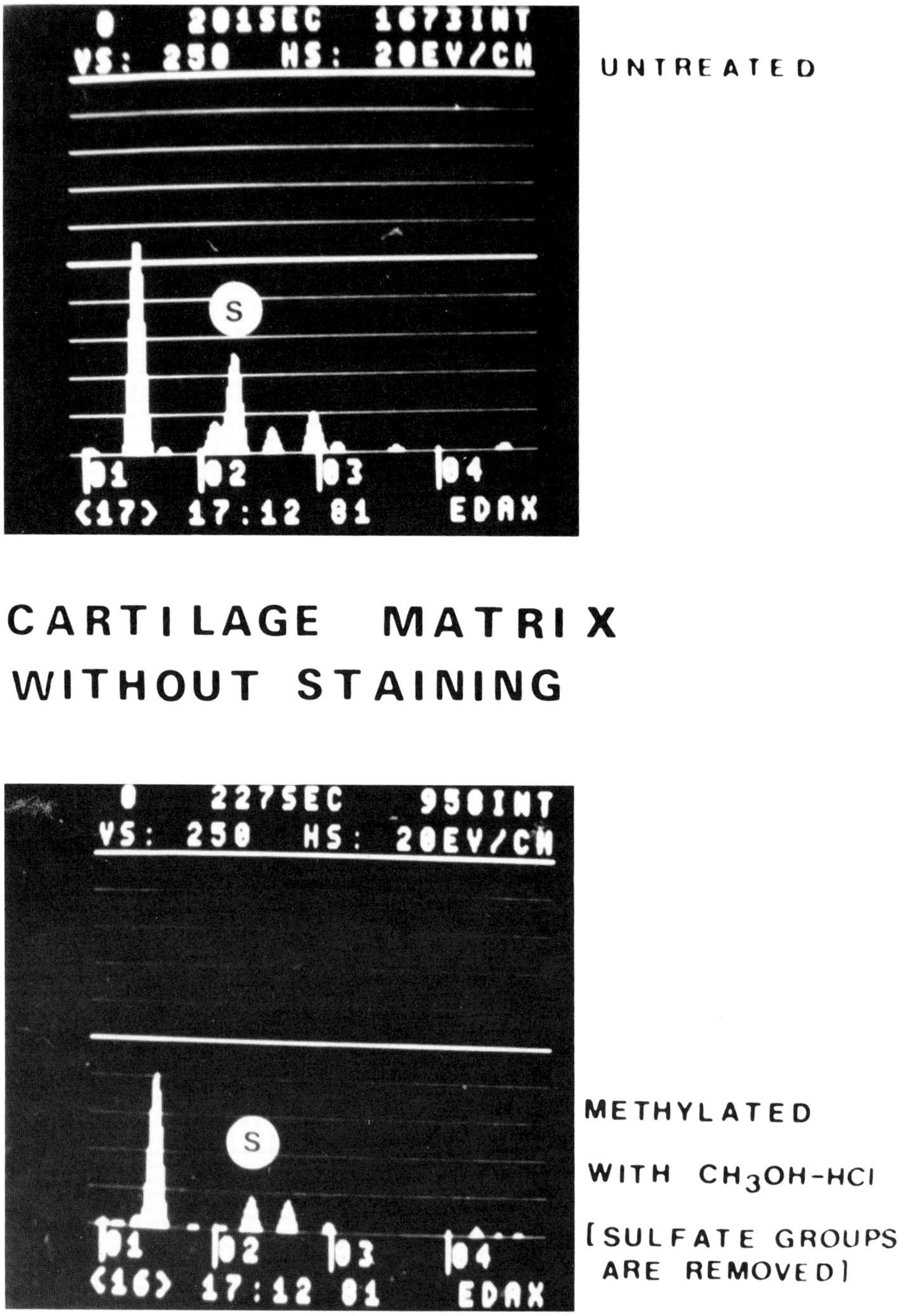

FIGURE 11. X-ray spectra recorded from unstained semifine sections of human articular cartilage (15-year-old patient, autopsy material). Samples were fixed and either untreated (upper spectrum) or methylated before the embedding (lower spectrum). Background of spectra is subtracted. Note the presence of a sulphur peak even after removal of sulphate groups by methylation. Thus, the peak of sulfur does not reflect exclusively the presence of sulfated GAGs in the tissue, S of sulfur-containing amino acids may also contribute to generation of X-ray photons of this peak. (From Módis, L. et al.,[118] *Bas. Appl. Histochem.*, 32, 415, 1988. With permission.)

for generating sulfur peaks in the X-ray spectra. X-ray microanalysis of sulfated GAGs in unstained specimens seems to be, therefore, an unreliable technique. For this reason, our attention was focused on the possibilities of X-ray microanalysis of PGs in tissue samples stained.

2. Energy-Dispersive X-Ray Microanalysis of Cartilage Matrix Stained with Cationic Probes

The use of X-ray microanalysis of sulfated GAGs in tissues stained with copper phthalocyanins had been suggested long ago, but the first bits of information available were only short, meeting abstracts without detailed description of the technique.[131,132] Butler and Heap[133,134] investigated GAGs with X-ray microanalysis in cat intervertebral discs stained with alcian blue before embedding. They failed to demonstrate any significant increase in copper content comparing stained to unstained specimens. They restained their specimens after embedding and cutting with the same dye, and they found copper both in the resin and in the tissue, but higher copper peaks were registered in the tissue than in the resin. Seemingly the aspecific absorption of the dye to the resin makes this technique unreliable. It is also to be questioned whether the large alcian blue molecules can readily penetrate the resin in which the sample is embedded. Mitchell et al.[121] and Mitchell and Shepard[135] used a pre-embedding staining with brominated toluidine blue and they proposed that the analysis of the bromine peak could be used for semiquantitative estimation of sulfated GAGs in the tissue. We showed in this chapter (II.C.4.), however, why toluidine blue is not an optimal reagent in electron-microscopic histochemistry of PGs.

We tried, therefore, the X-ray microanalysis of sulfated GAGs after a pre-embedding staining with alcian blue 8GX in cartilage matrix.[118] The staining was carried out after carboxymethylation,[116] and semifine sections placed on gold grids were examined with a JEOL JSM 35C electron microscope in scanning-transmission mode (STEM). X-ray spectra were recorded in selected area mode separately from minute territorial and interterritorial matrix areas by means of an EDAX system F consisting of an energy dispersive Si/Li detector, a 711A type multichannel analyzer and a Data General NOVA 3 computer with an FDEM program package of EDAX for mathematical handling of the spectra.

Stained and unstained specimens were compared. No copper peak could be observed in unstained samples. In contrast to this, depending on the matrix compartments, copper peaks of different size were found in the X-ray spectra taken in the cartilage matrix stained with alcian blue[118] (Figure 12). Always higher peaks were recorded from the territorial matrix than from the interterritorial matrix. It follows that the local concentration of the PG is higher in territorial matrix than in the interterritorial compartment. Similar observations were made after staining with cuprolinic blue (Figure 13). Maitland and Arsenault[91] studied growth plates of rats stained with alcian blue. Samples were slam-frozen, and freeze-substituted in 100% methanol containing alcian blue, embedded and cut. X-ray spectra were recorded from thin sections. They found much higher copper peaks in the pericellular matrix than in the interterritorial matrix. These data are in agreement with our X-ray microanalytical data. Furthermore, these results support our polarization-microscopic findings reported in Chapter 4 that GAGs form a more densely packed structure in pericellular and territorial matrix than in the interterritorial compartment.

An inherent limitation of the X-ray microanalysis of tissues stained with alcian blue or related dyes is connected with the size of this dye. This large molecule (2.6 to 2.8 nm) cannot equally well penetrate all portions of the matrix. GAGs of a dense matrix structure are less accessible for the dye than those of a loosely structured matrix area. This potential difficulty caused by the size of alcian blue may be decreased by application of a smaller reagent. We have promising preliminary results with Zr^{4+} (radius = 0.1454 nm) in showing GAGs with X-ray microanalysis (Figure 14). For this reaction $ZrOCl_2$ is used for a pre-

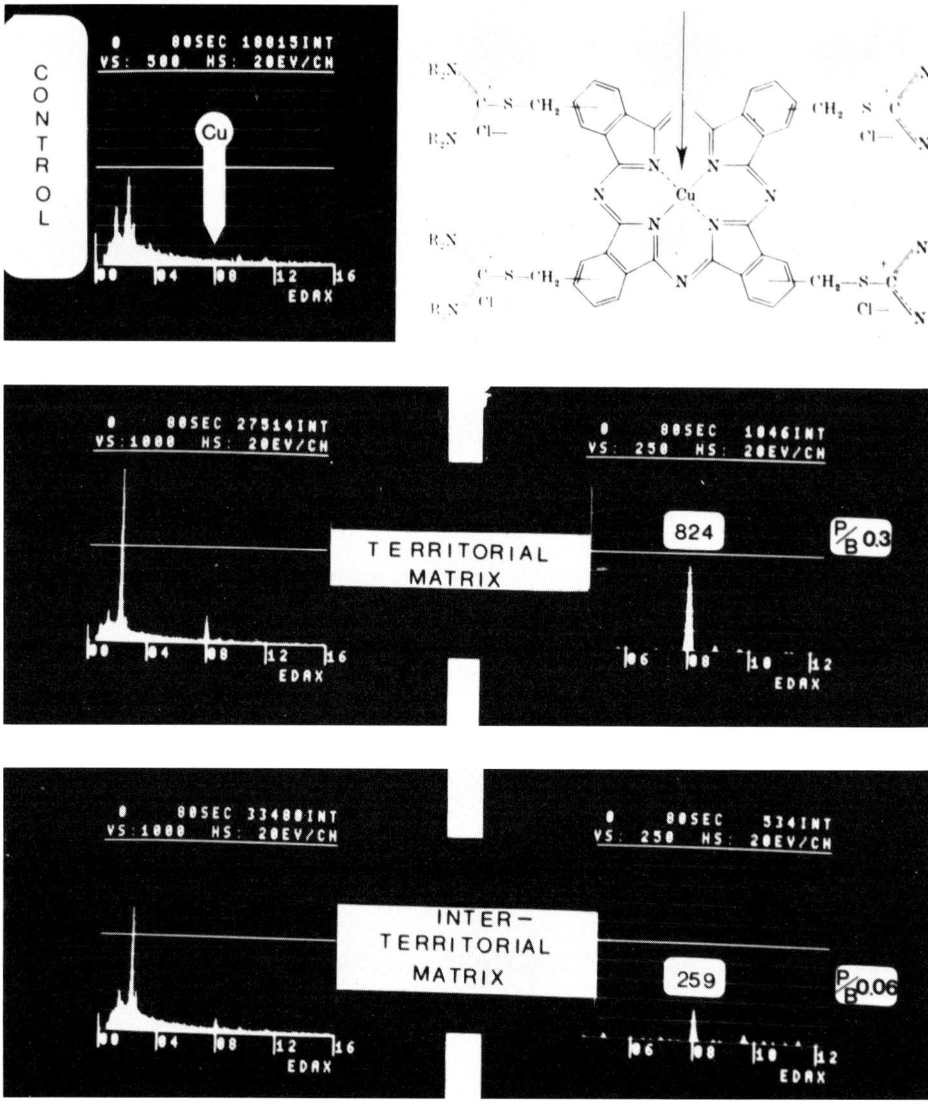

FIGURE 12. X-ray spectra recorded from the matrix of the intermediate zone of a 23-year-old human articular cartilage. Semifine sections of the specimen placed on gold grids were analysed. The upper left spectrum is taken from an untreated control section. Arrow indicates the place of K-alpha line of Cu. No copper is detected in unstained cartilage matrix. The lower four spectra are recorded after carboxymethylation[116] followed by staining with alcian blue 8GX. Complete spectra are shown on the left, and their parts with an other vertical scale after computerized background subtraction are seen on the right side. Cu peaks could be detected after staining both in territorial and interterritorial matrix compartments. Cu peaks and peak-to-background ratios (P/B) are higher in the territorial matrix than in the interterritorial matrix suggesting a higher local concentration of sulfated GAGs in the territorial matrix areas. (From Módis, L. et al.,[118] *Bas. Appl. Histochem.*, 32, 415, 1988. With permission.)

embedding ''staining''.[50,51] We must add, however, that these reaction strongly collapses PGs and, therefore, it cannot be used for transmission EM of PGs.

III. CORRELATIONS AND CONTRADICTIONS BETWEEN POLARIZATION AND ELECTRON MICROSCOPIC OBSERVATIONS IN CARTILAGE MATRIX

Observing transmission electron micrographs taken from cartilage samples processed with conventional and non-specific contrasting techniques, one would never say that the

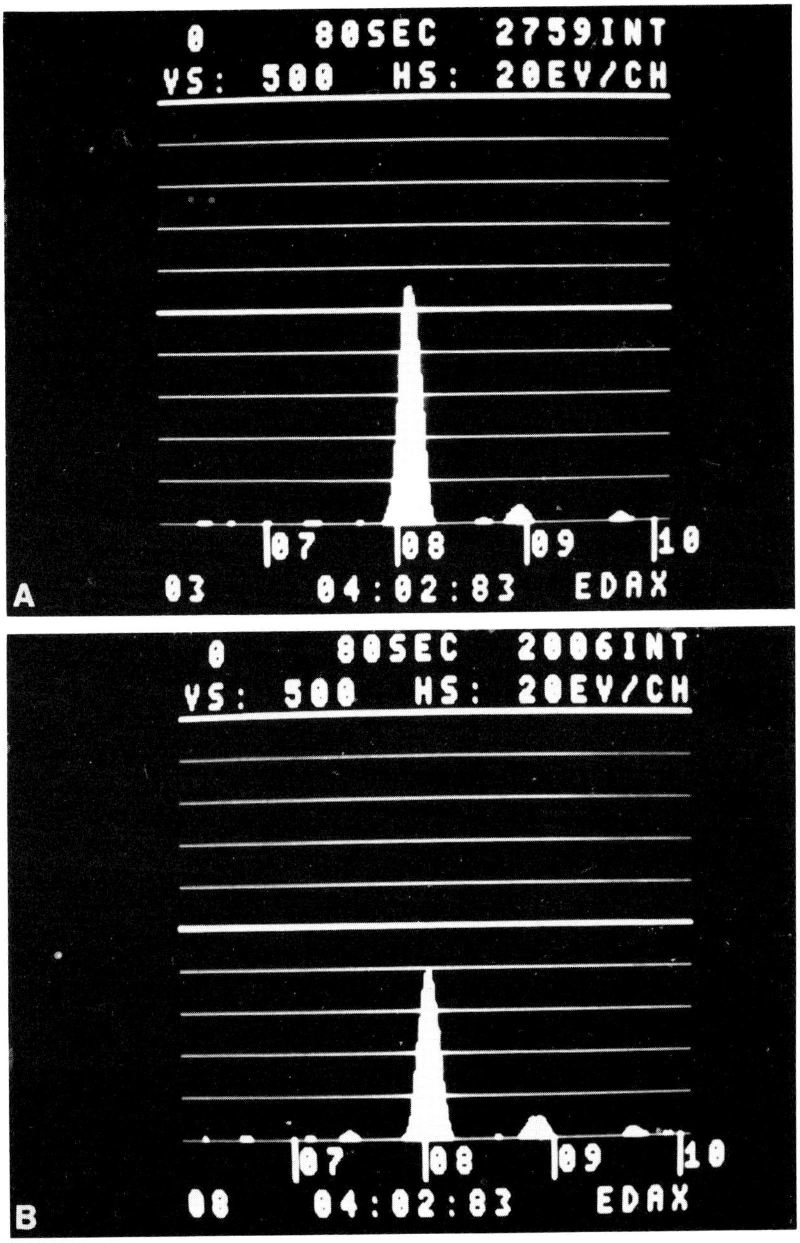

FIGURE 13. Parts of X-ray spectra taken from territorial (A) and interterritorial matrix (B) of human articular cartilage (15-year-old patient, autopsy material) stained with cuprolinic blue. Note the higher Cu peak in spectrum recorded from territorial compartment compared to the spectrum taken from interterritorial matrix.

interfibrillary matrix contains oriented constituents. The collagen fibrils are well visible in such pictures, and the presence of thicker fibrils in the interterritorial matrix and presence of finer fibrils in the territorial is frequently noted by electron microscopists.[33] A preferential collagen orientation was, however, only occasionally described after electron microscopic investigations in the different matrix compartments of hyaline cartilage (see Chapter 4). The cartilage ECM is, therefore, generally regarded as an amorphous system by electron microscopists.

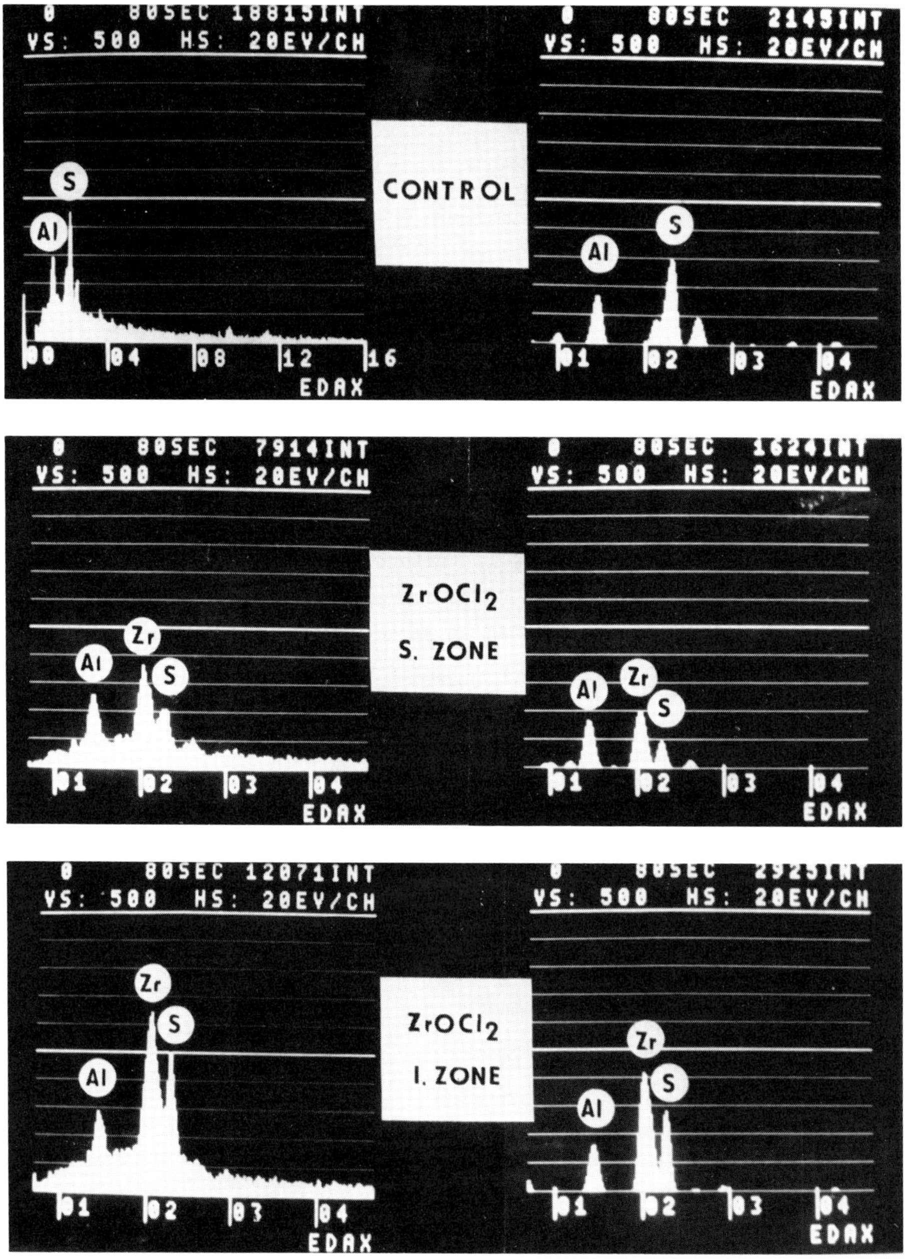

23-YEARS-OLD HUMAN ARTICULAR CARTILAGE

FIGURE 14. X-ray spectra recorded from semifine sections of human articular cartilage matrix (23-year-old patient, autopsy material). Complete spectra are shown on the left side, and their parts after computerized background subtraction are seen on the right side. The upper two spectra are recorded from unstained control samples. X-ray photons of Al peak are emitted by parts of the microscope stage, while that of S peak are emitted from the tissue. The lower four spectra are recorded from samples stained with $ZrOCl_2$ for sulfated GAGs. Note the appearance of Zr peaks in the spectra. Zr peak is higher in the spectrum taken from the intermediate (I) zone than in spectrum recorded from the matrix of superficial (S) zone.

After a staining for PGs, the electron micrographs reveal more details of the matrix. Depending on type of the reagent used for staining granules, filaments or rods appear in the interfibrillary space. Generally, no orientation pattern of PGs is inferred from such pictures.[9,76,84,105,136] Consequently, the old assumption that the cartilage matrix is an amorphous substance seems also to be supported by some electron microscopic histochemical findings. Polarization microscopy, on the other hand, has repeatedly revealed a preferential orientation both of collagen and GAGs in different matrix compartments. Do polarization microscopists see artifacts without any relevance to the biological structure or are the conclusions of their analyses correct? Numerous examples show that orientation patterns of biostructures were first inferred from polarization microscopical phenomena and later confirmed, and supplemented by electron microscopy (e.g., collagen and elastic fibers, striated muscle fibers, lipids of cell membrane, amyloid). We cannot quote polarization microscopic findings which had to be essentially modified or rejected by electron microscopic reinvestigations. Thus, we cannot conclude that polarization microscopy provides erroneous information even in analysis of hyaline cartilage matrix. Let us take some examples to support this view.

At the very beginning of cartilage differentiation we detected a circularly oriented GAG structure around the chondrocyte membrane (see Chapter 4) with polarization microscopic analysis of different topo-optical staining reactions. We concluded that GAGs of the pericellular matrix run parallel to the tangent of the cell membrane. No oriented collagen structure could be demonstrated at the same site and at the same stage of development in the cartilage matrix. It develops only at a later stage of differentiation. The orientation pattern of the subsequently deposited collagen was identical to that of the GAGs, the collagen fibrils assuming a circular orientation around the cells. Our conclusion was that the oriented sulfated GAG structure — which must be a part of an oriented PG structure — preceeded and supposedly facilitated the formation of the oriented collagen structure in vicinity of the cells. Meyer[137] suggested that mucopolysaccharides (= GAGs) ''are a code of extracellular information for the arrangement and organization of the fibrillar proteins, chiefly collagen fibrils''. We believe that the oriented PG microstructure deposited in the pericellular space, and which preceeds the formation of collagen fibrils, may serve with its orderly arranged anionic charge pattern — at least in cartilage tissue — as one of such extracellular ''codes''.

After staining with alcian blue or cuprolinic blue, electron microscopy revealed different localizations of PGs in extracellular space of chicken embryonic limb buds at early stages of development. At the stage of mesenchyme condensation, i.e., before the cartilage phenotype is expressed, a few scattered and small PG particles were found (Figure 15 A).

The cartilage phenotype is expressed at subsequent stages, and the number of PG molecules is considerably increased in the matrix. They show two different localizations: one group of PGs is attached to cell membrane, while other PGs are randomly distributed intercellularly (Figure 15 B). The staining reaction of these PGs could be abolished by antecedent digestions with testicular hyaluronidase or chondroitinase AC, but it remained unaffected by a digestion with heparitinase. It follows that they are chondroitin sulfate-PGs. PGs atached to the cell membrane show a preferential perpendicular orientation to the tangent of the cell membrane (Figure 16). This is the group of PGs which may be responsible for the pericellular birefringence seen after GAG-reactions just at the beginning of cartilage differentiation. The chondroitin sulfate side chains of these perpendicularly oriented PGs are in circular orientation in native state.

This circular orientation pattern of GAGs could be demonstrated by polarization optical investigaton but not by electron microscopy. It seems that polarization microscopic and electron microscopic histochemical reactions cause different modifications in the shape of PGs. Processing for EM results in stronger collapse of GAGs than PM techniques do.

According to our calculations, an embryonic cartilage cell is covered by approximately 4×10^4 PGs, and this PG layer contains about 4×10^8 anionic groups. These orderly arranged PGs may facilitate fibrillogenesis and their oriented charge density may provide a

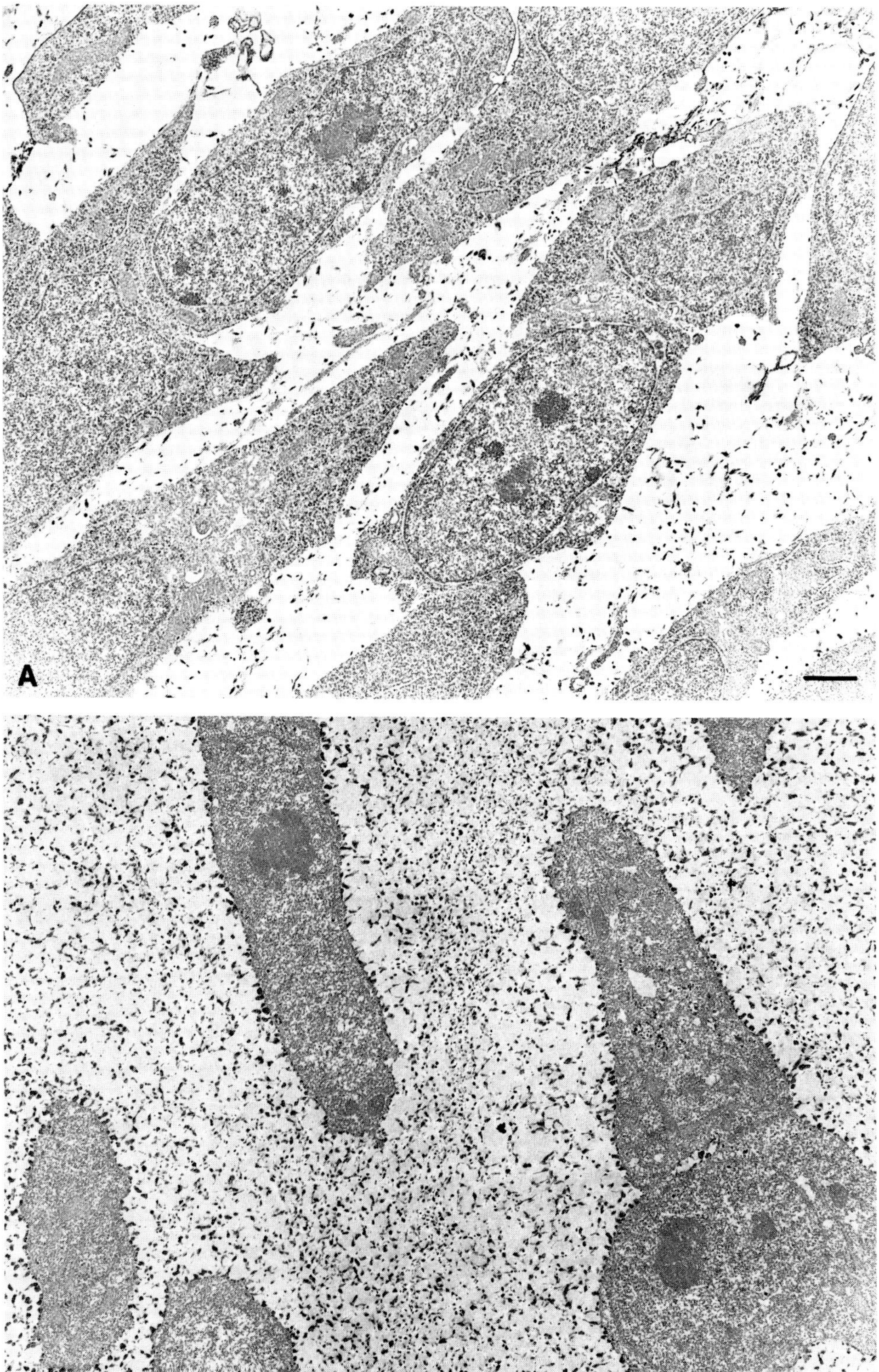

FIGURE 15. Electron micrographs taken from limb buds of chicken embryos after staining with alcian blue 8GX. (A) Four-day-old embryo (stage 24). Mesenchymal cells and narrow intercellular spaces containing scattered and small PG particles are visible. (B) Seven-day-old embyro (stage 30). Cartilage cells and abundant matrix with a high amount of PGs are seen. Note the PGs covering the outer surface of the chondrocytes. Scales: 1 μm.

guidance for the oriented deposition of the interacting collagen fibrils. It has been shown that cartilage PGs isolated from very young cartilages as well as aggregating cartilage PGs strongly stimulate collagen fibril formation under *in vitro* conditions.[138]

Pericellular PGs were demonstrated electron microscopically also by other authors in the cartilage matrix.[12,31,33,81,120,139] Ratcliffe et al.,[33] using quantitative immunoelectron microscopy, showed that the density of label for hyaluronic acid binding region of aggregating PG was higher in pericellular matrix than in the intercellular regions. Hunziker et al.[12] identified chondroitin sulfate-proteoglycans on the plasma membrane of chondrocytes and they suggested that these PGs are membrane intercalated molecules. We do not know how the pericellular PGs are fixed to the cell membrane. They can be bound to receptors, or they are fixed to a hyaluronic acid molecule (forming aggregate) and the hyaluronic acid may bind to the corresponding receptors of the cell membrane. Intercalation also cannot be excluded.

We want to note here that although both EM and PM revealed similar spatial orientation pattern of pericellular PGs at early stage of cartilage differentiation, the findings obtained by these two techniques were not identifical. Polarization-microscopic observations suggested that the GAGs are oriented in the matrix, but electron microscopy showed that there is a considerable number of PGs which are not bound to the cell membrane and they are randomly distributed. The presence of this population of the molecules could not be inferred from polarization optical phenomena.

The pericellular birefringence induced by GAG-specific, topo-optical reactions is increased after formation of collagen fibrils. According to polarization microscopic analyses, the collagen fibrils are also circularly arranged around the chondrocytes (see Chapter 4). Thus, GAGs and collagen fibrils run parallel to each other forming a basket-like oriented structure called cartilage capsule. The intensive birefringence of the pericellular cartilage capsule can be detected both after collagen- and GAG-specific topo-optical reactions from the early stages of cartilage differentiation going on both under *in vitro* or *in vivo* conditions (Figure 17). The circular orientation pattern of its collagen and GAG constituents remains unchanged until the normal cartilage phenotype is expressed. This microstructure seems to be one of the constant structural features of the cartilage ECMs.

After deposition of collagen fibrils in the pericellular capsule, electron microscopic investigation of embryonic cartilages stained for PGs revealed three different localizations of the PG molecules. One group of them remains in contact with the plasma membrane of chondrocytes preserving the previously discussed orientation pattern. Another population of PGs is in contact with the periphery of collagen fibrils (Figure 18). Many of these PGs are regularly spaced along the fibrils, and they assume a predominant axiperpendicular orientation. A third group comprises PGs randomly distributed in the interfibrillary space. The fine collagen fibrils are tangentially arranged. This arrangement has been shown by electron microscopic investigations of isolated chondrons[140-142] and different cartilage tissues.[81,143,144] It means that cartilage cells are surrounded by a three dimensional fibrillar structure in which the great majority of the fibrils are more or less parallelly arranged to the tangent of the cell membrane. The GAG side chains of PGs oriented perpendicularly the collagen fibrils must show the same orientation pattern (because in native state GAGs are perpendicular to the long axis, i.e., to the core protein of the PGs). These GAGs together with the similarly arranged GAG molecules of the membrane-bound PGs are responsible for the strong birefringence induced by metachromatic dyes in this matrix compartment.

At subsequent stages of differentiation, territorial matrix is formed in some zones of some hyaline cartilages. For example, territorial matrix appears in the early postnatal life in the middle and deep zones of articular cartilages of the human knee joint. The territorial matrix is a wide and basophilic zone around the chondrons. Electron microscopy reveals

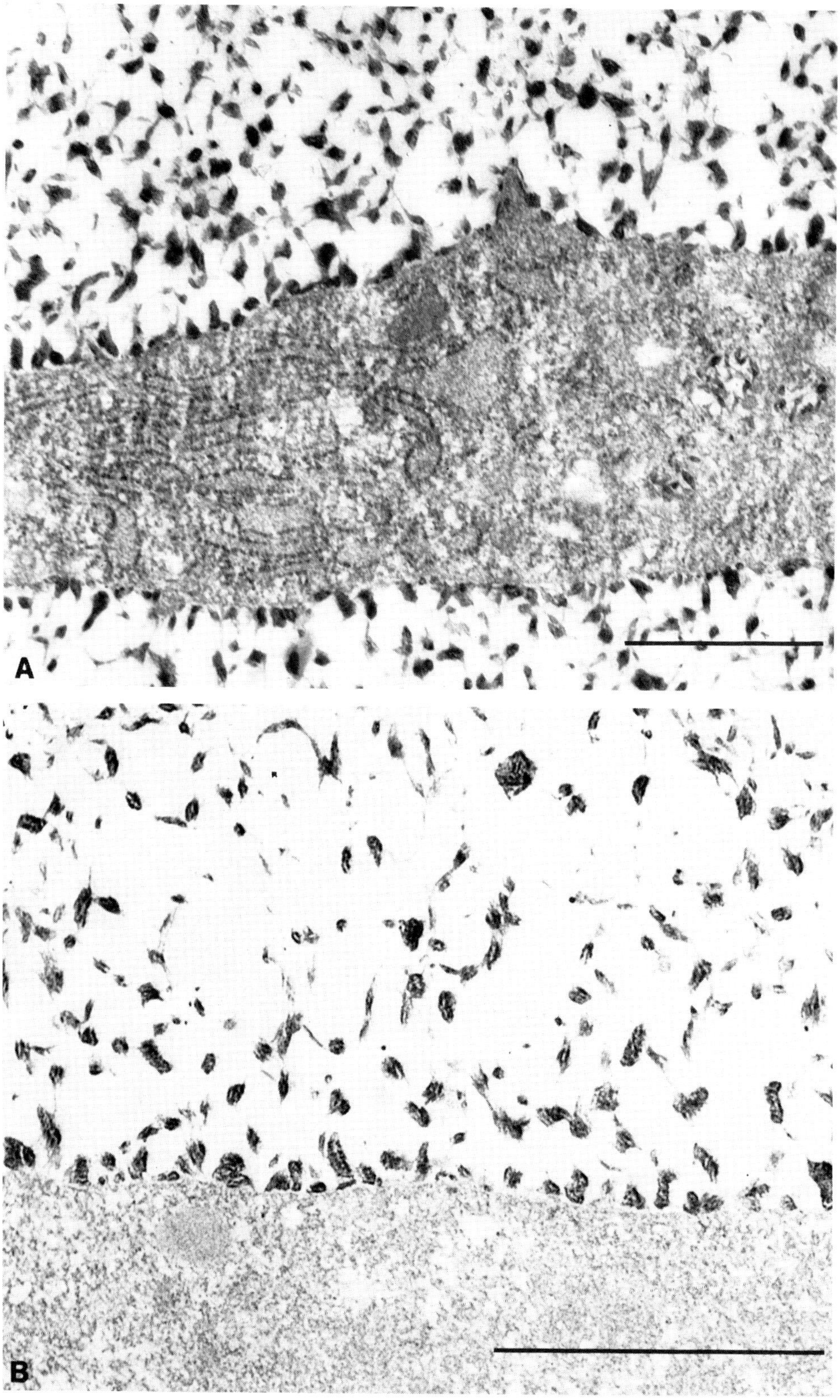

FIGURE 16. Electron micrographs taken from limb buds of 7-day-old chicken embryos after staining for PGs with alcian blue 8GX. Note the more or less perpendicularly oriented PGs which densely cover the outer surface of cell membrane. PGs without contact to the cell membrane are randomly distributed in the matrix. Scales: 1 μm.

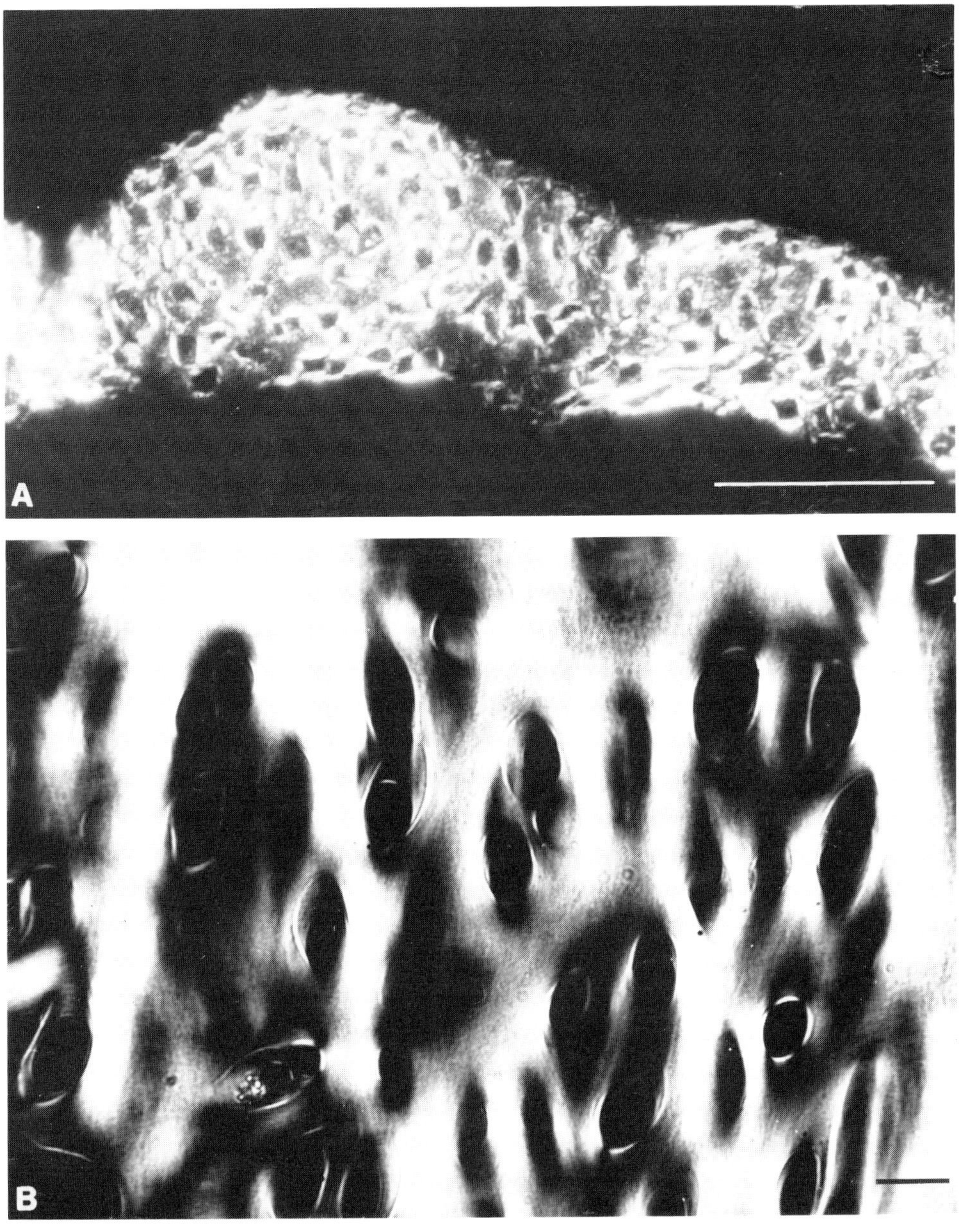

FIGURE 17. Microphotographs taken under polarization microscope after staining for sulfated GAGs with *N,N'*-diethylpseudoisocyanine. (A) Cross section of a 6-day-old chondrifying high density cell culture. (B) Section of human articular cartilage (7-month-old fetus, autopsy material). Note the strongly birefringent pericellular capsules in both of these specimens. The intercellular matrix is also anisotropic in the fetal cartilage (B). Scales: 100 μm (A), 10 μm (B).

here again tangentially arranged collagen fibrils[145] and proteoglycans periodically and perpendicularly bound to the fibrils (Figure 19). The membrane-bound and the randomly distributed interfibrillar PGs could also be detected. Polarization microscopic observations suggest a circular orientation pattern both of GAGs and collagen in the territorial matrix[146,147] (Figure 20, for details see Chapter 4). Quantitative polarization microscopic investigations reveal higher retardation values of birefringence induced both by GAG — and collagen — specific topo-optical reactions in territorial matrix than in the interterritorial matrix compartment. It follows that the territorial matrix contains more densely packed GAG and

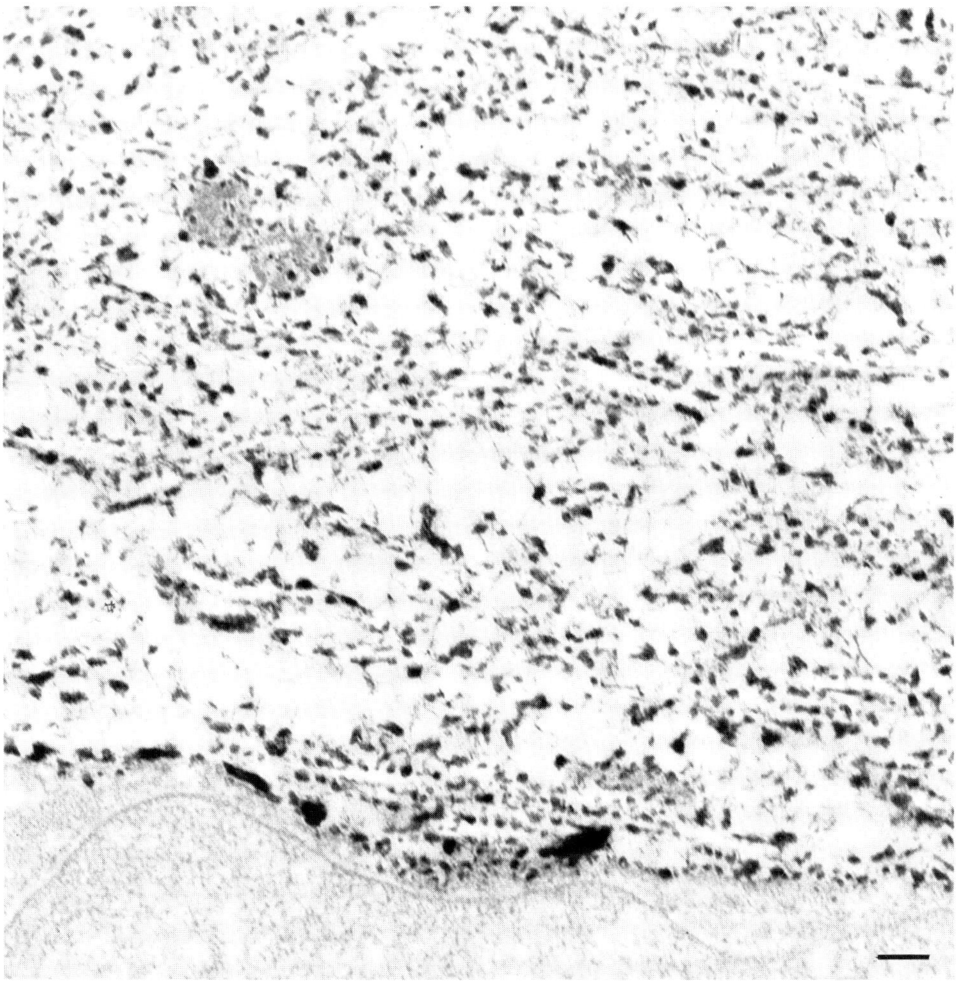

FIGURE 18. Electron micrograph taken from human articular cartilage (7-month-old fetus, autopsy material) after staining with cuprolinic blue. Three different localizations of PG can be distinguished: membrane-bound PGs, collagen fibril-bound PGs, and interfibrillary PGs. Molecules of the first and second groups are more or less orderly arranged in a monomolecular layer surrounding the cell membrane and the collagen fibrils. Note the tangential arrangement of collagen fibrils. Scale: 200 nm.

collagen structure than the interterritorial matrix. X-ray microanalytical data suggests a higher PG concentration in the territorial matrix than in the interterritorial areas.[118] We suppose that the territorial matrix with its dense and basket-like microstructure containing predominantly circularly oriented constitutents plays a role in protection of cartilage cells against compression and, furthermore, contributes to the resiliency of the cartilage.

The interterritorial matrix contains larger collagen fibrils, and they are longitudinally arranged to each other along preferential axes, e.g., tangentially in the superficial zone and perpendicularly to the surface in the deep zone of the articular cartilage,[144] parallel to the cell columns in the growth plate.[81,148] PG molecules are attached to these fibrils, and they show frequently a regular distribution and perpendicular orientation as shown by different electron microscopic techniques[84,86,87] (Figures 21 and 22). According to observations made on articular cartilage with the low angle, X-ray diffraction technique, PGs are regularly ordered on the type II collagen fibrils.[149] This oriented population of PGs explains the optical anisotropy of this matrix compartment after stainings for GAGs. Like the pericellular and

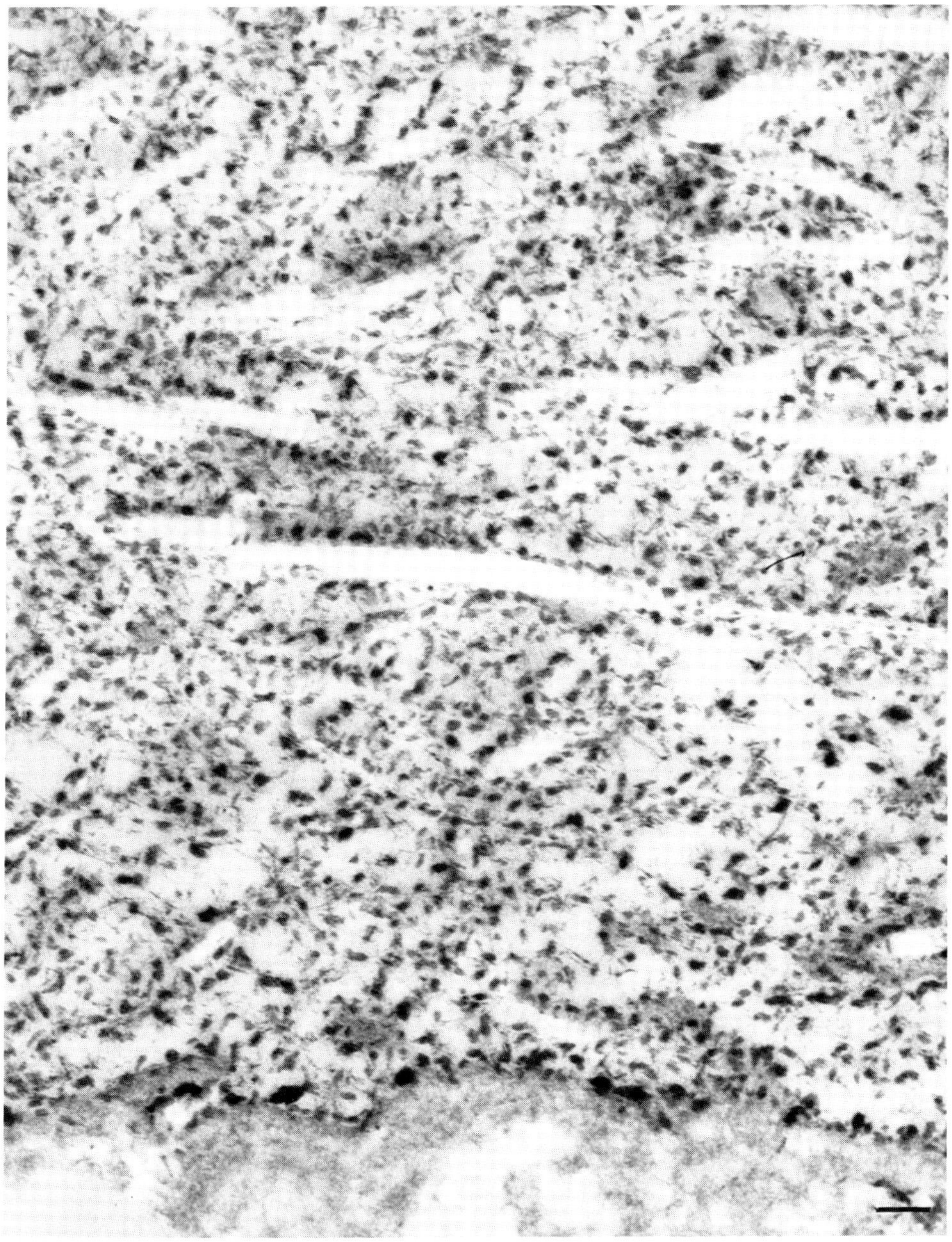

FIGURE 19. Electron micrograph of human articular cartilage (15-year-old patient, autopsy material) after staining with cuprolinic blue staining. The localization and orientation pattern of the PG particles is similar to that shown in fetal cartilage (compare with Figure 18). Scale: 200 nm.

territorial matrix compartments, interterritorial matrix areas also contain randomly distributed PGs (Figure 22).

Based on our polarization and electron microscopic observations, we can conclude that certain groups of PGs are spatially oriented in different matrix compartments of hyaline cartilage. In this sense, polarization-microscopic findings are supported by EM. On the other hand, EM revealed different localizations and orientations of PGs within matrix compartments. At this moment we can only speculate on the possible functions of the differently located PGs. The surface associated PGs are probably involved in fibrillogenesis. GAG side

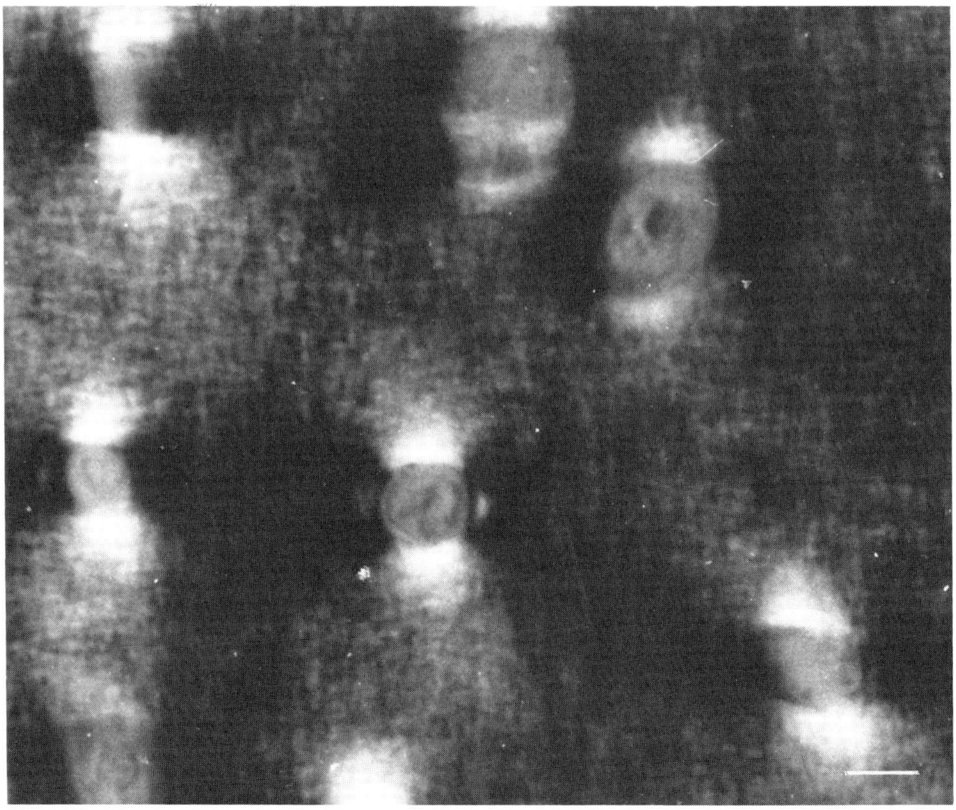

FIGURE 20. Polarized light microphotograph taken from the deep zone of a young human articular cartilage (1-year-old patient, autopsy material) after staining with *N,N'*-diethylpseudoisocyanine for sulfated GAGs. The strong birefringence of the territorial matrix is extinguished by a compensator plate in N-S direction. The spherulite-like appearance of birefringent territorial matrix compartments indicates the circular orientation of the GAG molecules in this matrix area. Similar optical findings were obtained in middle and deep zones of articular cartilages during entire postnatal life. Scale: 10 μm.

chains of PGs associated with the collagen fibrils are longitudinally arranged along the interacting fibrillar surface, these PGs may regulate the fiber diameter. The randomly distributed PGs are supposedly the major shock absorbers of the cartilage matrix. It is unlikely that this distribution of PGs is static. Keeping in mind that the major type of interaction between PGs and collagen is an electrostatic binding, which is relatively weak at physiological pH values, the existence of a dynamic PG distribution seems to be more probable.

We learned from polarization microscopy that the old terminology "hyaline cartilage matrix is an amorphous system" cannot be justified anymore. We learned from the electron microscopic studies that the hyaline cartilage matrix exhibits a structural — and maybe a functional — microheterogeneity. Besides the zones and major microscopic compartments (pericellular, territorial, interterritorial),[81,148] the existence of submicroscopic compartments (fibrillar, perifibrillar, interfibrillar, perimembrane) is suggested. GAGs, PGs and/or collagen fibrils exhibit a spatial orientation in all except the interfibrillar microcompartment.

The fibrillar compartment contains highly oriented and densely packed collagen. No GAGs or PGs could be found inside the fibrils. The perifibrillar compartment is rich in GAGs and other portions of PGs. GAGs ionically bind to the cationic charges of the fibrillar surface and are arranged parallel to the long axis of the fibril. Other, more specific protein-protein interactions may also occur between collagen and PGs.[46] The randomly distributed

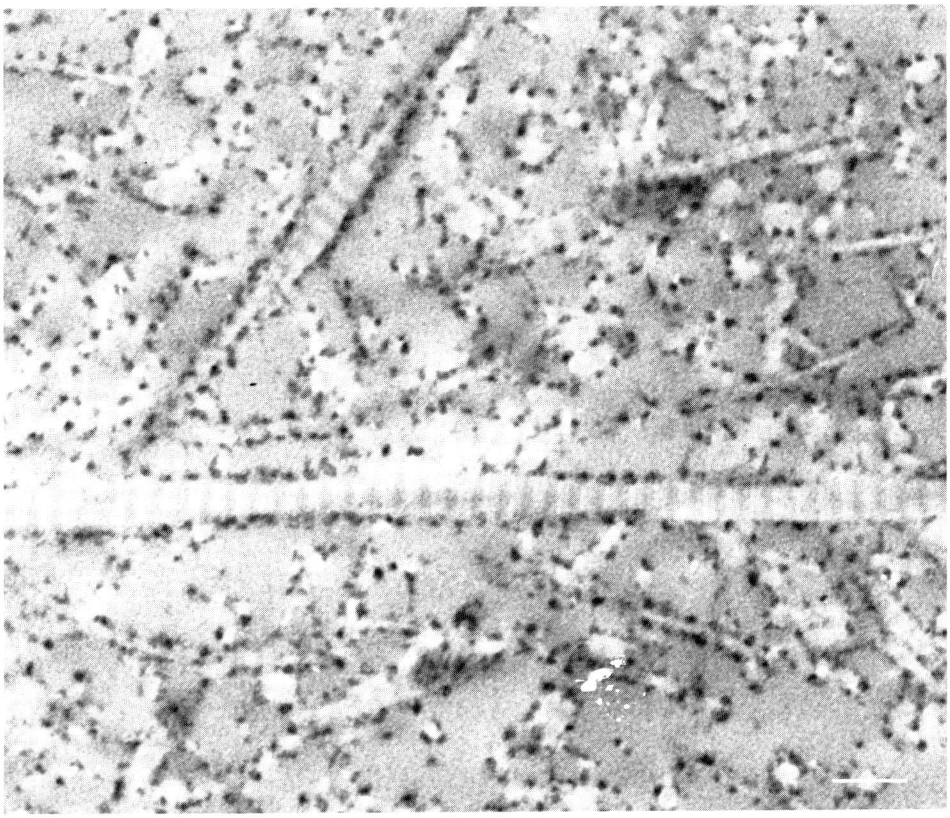

FIGURE 21. Electron micrograph showing the interterritorial matrix of a 15-year-old human articular cartilage. PGs stained with high iron diamine technique[40,57] exhibit a periodical arrangement along the collagen fibrils. This histochemical rection causes a stronger collapse of PGs than staining with phthalocyanine dyes. Scale: 200 nm.

PGs of the interfibrillary compartment are primarily space filling and shock absorbing molecules. They also prevent diffusion of larger molecules through the matrix.

The perimembrane compartment has an intricate structure. It contains perpendicularly oriented PGs and tangentially arranged GAGs, and is rich in anionic charges. Different minor collagens (type VI, IX, XI), and fine collagen type II fibrils are located here. They interact with each other and with other constituents (see Chapter 1 and 4). Some of the cartilage collagens (especially type XI) may form the "exocytoskeleton" of the cartilage cells. Collagens may bind to the cell membrane either via specific receptors, e.g., anchorin C II,[150,151] or chondronectin. Supposedly hyaluronic acid is also present and is in contact with its receptor located in the cell coat.[152-155] PGs may bind to the hyaluronic acid forming PG aggregates. Chondrocytes also bind other matrix macromolecules such as 36kDa and 58kDa matrix proteins.[153] Different hydrolytic enzymes may also be present in the pericellular matrix but in inactivated form.

Although the actually available data are sufficient to propose the existence of differently structured microcompartments in differentiated hyaline cartilage matrix, further efforts are needed to collect more evidence pro or contra this speculation.

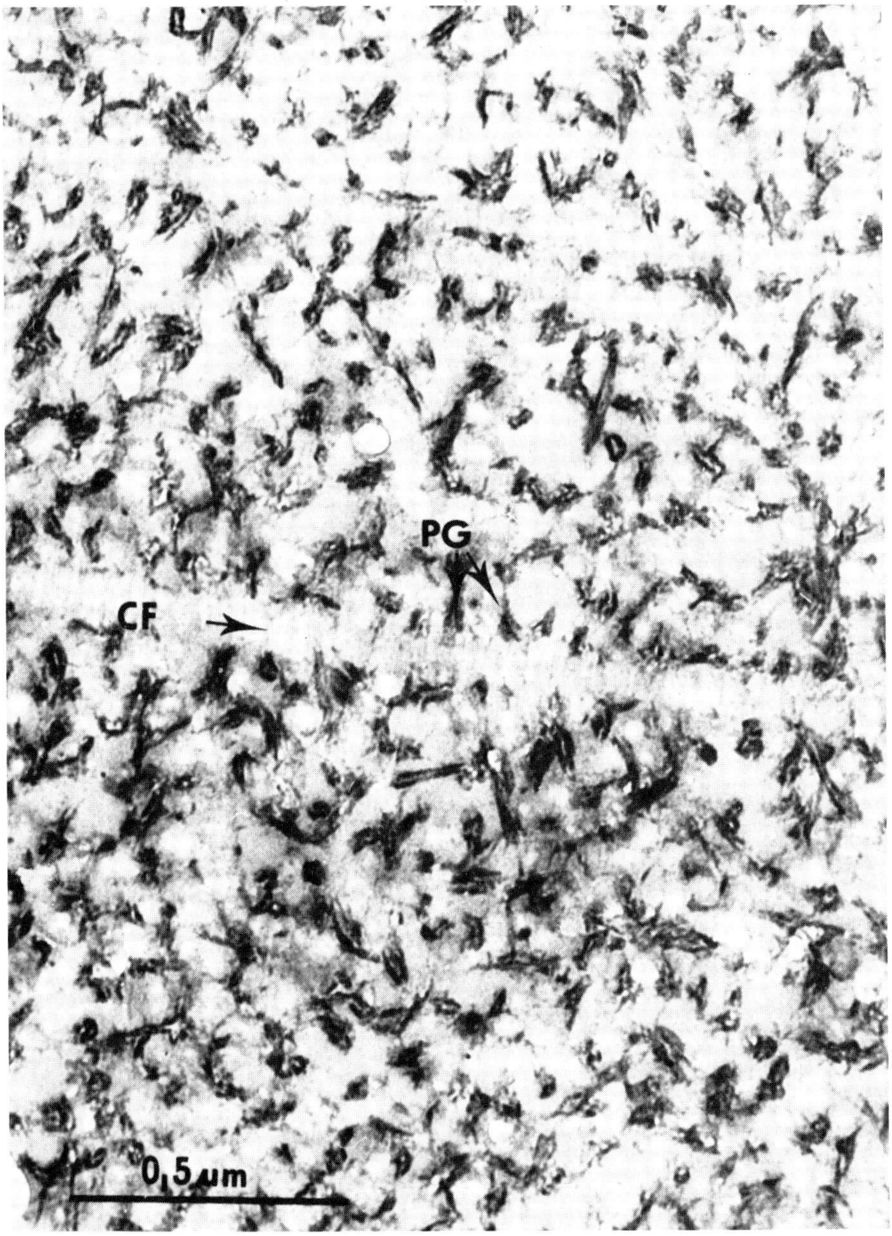

FIGURE 22. Electron micrographs taken from interterritorial matrix of human articular cartilage (23-year-old patient, autopsy material) stained with alcian blue 8GX. A more or less perpendicular arrangement of PGs bound to cross banded collagen fibrils (CF) is frequently seen.

ACKNOWLEDGMENTS

The author is indebted to Drs. I. Zs.-Nagy and G. Lustyik (F. Verzár International Laboratory for Experimental Gerontology, Hungarian Section, Debrecen) for their help in analytical electron microscopy. Thanks are due to Drs. Róza Ádány (Department of Clinical Chemistry, Debrecen) and Róza Zákány (Department of Anatomy, Debrecen) for carrying out the experiments on alcohol sensitivity of stainings. The kind help of Dr. Z. Nagy (deceased) and Florence Jeney (Central Research Laboratory, Debrecen) in spectrographic and atomic absorption investigations is highly appreciated. The author wishes to thank Dr. G. Lévai (Department of Anatomy, Debrecen) for critical reading of the manuscript. I am also grateful for the skillful technical help of Ms. Erzsébet M. Laczkó, (Department of Anatomy, Debrecen) and Mr. A. Katona (Central Research Laboratory, Debrecen).

REFERENCES

1. **Speer, D. P. and Dahners, L.,** The collagenous architecture of articular cartilage. Correlation of scanning electron microscopy and polarized light microscopy observations, *Clin. Orthop.*, 139, 267, 1979.
2. **Hunziker, E. B., Herrmann, W., Schenk, R. K., Mueller, M., and Moor, H.,** Cartilage ultrastructure after high pressure freezing, freeze substitution, and low temperature embedding. I. Chondrocyte ultrastructure — implications for the theories of mineralization and vascular invasion, *J. Cell Biol.*, 98, 267, 1984.
3. **Hunziker, E. B. and Schenk, R. K.,** Cartilage ultrastructure after high pressure freezing, freeze substitution, and low temperature embedding. II. Intercellular matrix ultrastructure-preservation of proteoglycans in their native state, *J. Cell Biol.*, 98, 277, 1984.
4. **Smith, J. W.,** The disposition of proteinpolyasaccharide in the epiphyseal plate cartilage of the young rabbit, *J. Cell Sci.*, 6, 843, 1970.
5. **Myers, D. B., Highton, T. C., and Rayns, D. G.,** Ruthenium red-positive filaments interconnecting collagen fibrils, *J. Ultrastruct. Res.*, 42, 87, 1973.
6. **Shepard, N. and Mitchell, N.,** Simultaneous localization of proteoglycan by light and electron microscopy using toluidine blue O. A study of epiphyseal cartilage, *J. Histochem. Cytochem.*, 24, 621, 1976.
7. **Shepard, N. and Mitchell, N.,** The localization of proteoglycan by light and electron microscopy using safranin O. A study of epiphyseal cartilage, *J. Ultrastruct. Res.*, 54, 451, 1976.
8. **Scott, J. E. and Orford, C. R.,** Dermatan sulphate-rich proteoglycan associates with rat tail-tendon collagen at the d band in the gap region, *Biochem. J.*, 197, 213, 1981.
9. **Scott, J. E.,** Proteoglycan histochemistry — a valuable tool for connective tissue biochemists, *Coll. Relat. Res.*, 5, 541, 1985.
10. **Scott, J. E. and Haigh, M.,** Keratan sulphate and the ultrastructure of cornea and cartilage: a 'stand-in' for chondroitin sulphate in conditions of oxygen lack?, *J. Anat.*, 158, 95, 1988.
11. **Scott, J. E. and Haigh, M.,** Proteoglycan-collagen interactions in intervertebral disc. A chondroitin sulfate proteoglycan associates with collagen fibrils in rabbit annulus fibrosus at the d-e bands, *Biosci. Rep.*, 6, 879, 1986.
12. **Hunziker, E. B., Herrmann, W., and Schenk, R. K.,** Ruthenium hexammine trichloride (RHT)-mediated interaction between plasmalemmal components and pericellular matrix proteoglycans is responsible for the preservation of chondrocytic plasma membranes *in situ* during cartilage fixation, *J. Histochem. Cytochem.*, 31, 717, 1983.
13. **Arsenault, A. L. and Hunziker, E. B.,** Electron microscopic analysis of mineral deposits in the calcifying epiphyseal growth plate, *Calcif. Tissue Int.*, 42, 119,1988.
14. **Arsenault, A. L., Ottensmayer, F. P., and Heath, I. B.,** An electron microscopic and spectroscopic study of murine epiphyseal cartilage: analysis of fine structure and matrix vesicles preserved by slam - freexing and freeze substitution, *J. Ultrastruct. Mol. Struct. Res.*, 98, 32, 1988.
15. **Thyberg, J.,** Electron microscopy of cartilage proteoglycans, *Histochem. J.*, 9, 259, 1977.
16. **Hascall, G. K.,** Cartilage proteoglycans: a comparison of sectioned and spread whole molecules, *J. Ultrastruct. Res.*, 70, 369, 1980.
17. **Iozzo, R. V. and Wight, T. N.,** Isolation and characterization of proteoglycans synthesized by human colon and colon carcinoma, *J. Biol. Chem.*, 257, 11135, 1982.

18. **Buckwalter, J. A. and Rosenberg, L.**, Structural changes during development in bovine fetal epiphyseal cartilage, *Coll. Relat. Res.*, 3, 489, 1983.
19. **Thyberg, C. J. O.**, Electron microscopy of proteoglycans, in *Ultrastructure of the Connective Tissue Matrix*, Ruggeri, A. and Motta, P. M., Eds., Martinus Nijhoff Boston, 1984, chap. 5.
20. **Buckwalter, J. A., Rosenberg, L. C., and Tang, L.**, The effect of link protein on proteoglycan aggregate structure, *J. Biol. Chem.*, 259, 5361, 1984.
21. **Buckwalter, J. A., Pedrini-Mille, A., Pedrini, V., and Tudisco, C.**, Proteoglycans of human infant intervertebral disc, *J. Bone Joint Surg. (Am.)*, 67, 284, 1985.
22. **Buckwalter, J. A., Kuettner, K. E., and Thonar, E. J-M.**, Age-related changes in articular cartilage proteoglycans: electron microscopic studies, *J. Orthop. Res.*, 3, 251, 1985.
23. **Pechak, D. G., Carrino, D. A., and Caplan, A. I.**, Electron microscopic characterization of chick embryonic skeletal muscle proteoglycans, *J. Cell Biol.*, 100, 1767, 1985.
24. **Thonar, E. J.-M. A., Buckwalter, J. A., and Kuettner, K. E.**, Maturation-related differences in the structure and composition of proteoglycans synthesized by chondrocytes from bovine articular cartilage, *J. Biol. Chem.*, 261, 2467, 1986.
25. **Laurie, G. W., Inoue, S., Bing, J. T., and Hassel, J. R.**, Visualization of the large heparan sulfate proteoglycan from basement membrane, *Am. J. Anat.*, 181, 320, 1988.
26. **Buckwalter, J. A., Poole, A. R., Reiner, A., and Rosenberg, L. C.**, Immunoferritin binding to proteoglycan monomers, *J. Biol. Chem.*, 257, 10529, 1982.
27. **Camplejohn, K. L. and Allard, S. A.**, Limitations of safranin O staining in proteoglycan-depleted cartilage demonstrated with monoclonal antibodies, *Histochemistry*, 89, 185, 1988.
28. **Poole, A. R.**, The relationship between toluidine blue staining and hexuronic acid content of cartilage matrix, *Histochem. J.*, 2, 425, 1970.
29. **Glant, T., Lévai, G., and Hadházy, Cs.**, The localization of proteoglycans and glycoproteins in the hyaline cartilage, *Histochemistry*, 53, 291, 1977.
30. **Glant, T. and Lévai, G.**, Localization of antigenic components of proteoglycan aggregate in bovine nasal cartilage, *Histochemistry*, 77, 217, 1983.
31. **Poole, A. R., Pidoux, I., Reiner, A., Tang, L-H., Choi, H., and Rosenberg, L.**, Localization of proteoglycan monomer and link protein in the matrix of bovine articular cartilage, *J. Histochem. Cytochem.*, 28, 621, 1980.
32. **Poole, A. R., Pidoux, I., Reiner, A., and Rosenberg., L.**, An immunoelectron microscope study of the organization of proteoglycan monomer, link protein, and collagen in the matrix of articular cartilage, *J. Cell Biol.*, 93, 921, 1982.
33. **Ratcliffe, A., Fryer, P. R., and Hardingham, T. E.**, The distribution of aggregating proteoglycans in articular cartilage, *J. Histochem. Cytochem.*, 32, 193, 1984.
34. **Glant, T. T., Mikecz, K., Roughley, P. J., Buzás, E., and Poole, A. R.**, Age-related changes in protein-related epitopes of human articular cartilage proteoglycans, *Biochem. J.*, 236, 71, 1986.
35. **Iozzo, R. V. and Clark, C. C.**, Chondroitin sulfate proteoglycan is a constituent of the basement membrane in the rat embryo parietal yolk sac, *Histochemistry*, 88, 23, 1987.
36. **Hunziker, E. B. and Herrmann, W.**, *In situ* localization of cartilage extracellular matrix components by immunoelectron microscopy after cryotechnical tissue processing, *J. Histochem. Cytochem.*, 35, 647, 1987.
37. **Scott, J. E.**, In praise of dyestuff histochemistry, *Bas. Appl. Histochem.*, 33, 9, 1989.
38. **Engfeldt, B. and Hjertquist, S. O.**, Studies on the epiphyseal growth zone. The preservation of acid glycosaminoglycans in tissues in some histotechnical procedures for electron microscopy, *Virchows Arch. (Zellpathol.)*, 1, 222, 1968.
39. **Radmehr, B. and Butler, W. F.**, Leaching of glycosaminoglycan from cow uterus and vagina, *Histochem. J.*, 10, 465, 1978.
40. **Albedi, F. M., Cassano, A. M., Ciaralli, F., Donelli, G., Giuliani, A., Mingazzini, P., and Marinozzi, V.**, Influence of cetylpyridinium chloride on the ultrastructural appearance of sulphated glycosaminoglycans in human colonic mucosa, *Histochemistry*, 89, 397, 1988.
41. **Young, H. E., Young, V. E., and Caplan, A. I.**, Comparison of fixatives for maximal retention of radiolabeled glycoconjugates for autoradiography, including use of sodium sulfate to release unincorporated (^{35}S)-sulfate, *J. Histochem. Cytochem.*, 37, 223, 1989.
42. **Scott, J. E.**, Aliphatic ammonium salts in the assay of acid mucopolysaccharides from tissues, *Meth. Biochem. Anal.*, 8, 145, 1960.
43. **Tuckett, F. and Morriss-Kay, G.**, Alcian blue staining of glycosaminoglycans in embryonic material: effect of different fixatives, *Histochem. J.*, 20, 174, 1988.
44. **Hunziker, E. B. and Graber, W.**, Differential extraction of proteoglycans from cartilage tissue matrix compartments in isotonic buffer salt solutions and commercial tissue-culture media, *J. Histochem. Cytochem.*, 34, 1149, 1986.
45. **Scott, J. E. and Haigh, M.**, Proteoglycan-type I collagen fibril interactions in bone and non-calcifying connective tissues, *Biosci. Rep.*, 5, 71, 1985.

46. **Scott, J. E.,** Proteoglycan-fibrillar collagen interactions, *Biochem. J.,* 252, 313, 1988.
47. **Shea, S. M.,** Lanthanum staining of the surface coat of cells, *J. Cell Biol.,* 51, 611, 1971.
48. **Serafini-Fracassini, A. and Smith, J. W.,** Observations on the morphology of the proteinpolysaccharide complex of bovine nasal cartilage and its relationship to collagen, *Proc. R. Soc. Lond. (Biol.),* 165, 440, 1966.
49. **Smith, J. W., Peters, T. J., and Serafini-Fracassini, A.,** Observations on the distribution of the proteinpolysaccharide complex and collagen in bovine articular cartilage, *J. Cell Sci.,* 2, 129, 1967.
50. **Módis, L. and Musy, J. P.,** Utilisation des sels de zirconium en histochimie. I. Etude au microscope photonique, *Acta Histochem. (Jena),* 41, 30, 1971.
51. **Sprumont, P., Musy, J. P., and de Blasi, V.,** Utilisation des sels de zirconium en histochimie. II., Etude au microscope électronique, *Acta Histochem. (Jena),* 42, 285, 1972.
52. **Curran, R. C., Clark, A. E., and Lovell, D.,** Acid mucopolysaccharides in electron microscopy. The use of the colloidal iron method, *J. Anat.,* 99, 427, 1965.
53. **Matukas, V. J., Panner, B. I., and Orbison, J. L.,** Studies on ultrastructural identification and distribution of protein-polysaccharide in cartilage matrix, *J. Cell Biol.,* 32, 365, 1967.
54. **Pihl, E., Gustavson, G. T. and Falkmer, S.,** Ultrastructural demonstration of cartilage acid glycosaminoglycans, *Histochem. J.,* 1, 26, 1968.
55. **Geyer G.,** Elektronenmikroskopische Histochemie, *Handbuch der Histochemie,* Vol. 1, Part 3, Graumann, W. and Neumann, K., Eds., G. Fischer Verlag, Stuttgart, 1977.
56. **Groot, C. G.,** Positive colloidal thorium dioxide as an electron microscopical contrasting agent for glycosaminoglycans, compared with ruthenium red and positive colloidal iron, *Histochemistry,* 71, 617, 1981.
57. **Spicer, S. S., Hardin, J. H., and Setzer, M. E.,** Ultrastructural visualization of sulphated complex carbohydrates in blood and epithelial cells with the high iron diamine procedure, *Histochem. J.,* 10, 435, 1978.
58. **Kogaya, Y. and Furuhashi, K.,** Ultrastructural distribution of sulfated glycosaminoglycans in epithelial-mesenchymal interface of developing rat tooth germs, *J. Histochem. Cytochem.,* 35, 585, 1987.
59. **Takagi, M., Parmley, R. T., Toda, Y., and Austin, R. L.,** Ultrastructural cytochemistry and immunohistochemistry of sulfated glycosaminoglycans in epiphyseal cartilage, *J. Histochem. Cytochem.,* 30, 1179, 1982.
60. **Horisberger, M. and Clerc, M.-F.,** Chitosan-colloidal gold complexes as polycationic probes for the detection of anionic sites by transmission and scanning electron microscopy, *Histochemistry,* 90, 165, 1988.
61. **Skultelsky, E. and Roth, J.,** Cationic colloidal gold — a new probe for the detection of anionic surface sites by electron microscopy, *J. Histochem. Cytochem.,* 34, 693, 1986.
62. **Danon, D., Goldstein, L., Marikovsky, Y., and Skultelsky, E.,** Use of cationized ferritin as a label of negative charges on cell surface, *J. Ultrastruct. Res.,* 38, 500, 1972.
63. **Eisenstein, R., Arsenis, C., and Kuettner, K. E.,** Electron microscopic studies of cartilage matrix using lysozyme as a vital stain, *J. Cell Biol.,* 46, 626, 1970.
64. **Stiller, D., Barenwald, G., and Schwanengel, H.,** Elektronenmikroskopische Darstellung von Proteoglykanen des Gelenkknorpels, *Beitr. Orthop. Traumatol.,* 28, 477, 1981.
65. **Shepard, N. and Mitchell, N.,** Acridin orange stabilization of glycosaminoglycans in beginning endochondral ossification. A comparative light and electron microscopic study, *Histochemistry,* 70, 107, 1981.
66. **Shepard, N. and Mitchell, N.,** Ultrastructural modifications of proteoglycans coincident with mineralization in local regions of rat growth plate, *J. Bone Joint Surg. (Am.),* 67, 455, 1985.
67. **Chen, K. and Wight, T. N.,** Proteoglycans in arterial smooth muscle cell cultures: an ultrastructural histochemical analysis, *J. Histochem. Cytochem.,* 32, 347, 1984.
68. **Mallinger, R.,** Long-spacing collagen: elektronenmikroskopische Beobachtungen an Mäuserippenknorpeln und *In-vitro*-Präzipitäten von gelösten Typ II Kollagen, *Verh. Anat. Ges.,* 78, 209, 1984.
69. **Mallinger, R. and Stockinger, L.,** Histochemistry of the extracellular matrix of aging hyaline cartilage, *Folia Histochem. Cytochem.,* 25, 129, 1987.
70. **Mallinger, R. and Stockinger, L.,** Amianthoid (asbestoid) transformation: electron microscopical studies on aging human costal cartilage, *Am. J. Anat.,* 181, 23, 1988.
71. **Reale, E., Luciano, L., and Kühn, K. W.,** Ultrastructural architecture of proteoglycans in the glomerular basement membrane, *J. Histochem. Cytochem.,* 31, 662, 1983.
72. **Reale, E., Luciano, L., and Kühn, K.,** Cationic dyes reveal proteoglycans on the surface of epithelial and endothelial kidney cells, *Histochemistry,* 82, 513, 1985.
73. **Marshall, P. N. and Lewis, S. M.,** Metal contaminants in commercial thiazine dyes, *Stain Technol.,* 50, 143, 1975.
74. **Sato, C. S. and Gyorkey, F.,** Differential precipitation of glycosaminoglycans by zinc at pH 7.0, *Curr. Sci.,* 46, 415, 1977.
75. **Luft, J. H.,** Ruthenium red and violet. II. Fine structural localization in animal tissues, *Anat. Rec.,* 171, 369, 1971.

76. **Thyberg, J., Lohmander, S., and Friberg, U.,** Electron microscopic demonstration of proteoglycans in guinea pig epiphyseal cartilage, *J. Ultrastruct. Res.*, 45, 407, 1973.
77. **Shepard, N. and Mitchell, N.,** The use of ruthenium red and *p*-phenylene diamine to stain cartilage simultanously for light and electron microscopy, *J. Histochem. Cytochem.*, 25, 1163, 1977.
78. **Ruggeri, A., Dell'Orbo, C., and Quacci, D.,** Electron microscopic visualization of proteoglycans with ruthenium red, *Histochem. J.*, 9, 249, 1977.
79. **Tammi, R. H., Hyyrylainen, A. M. H., Maibach, H. I., and Tammi, M. I.,** Ultrastructural localization of keratinocyte surface associated heparan sulphate proteoglycans in human epidermis, *Histochemistry*, 87, 243, 1987.
80. **Hunziker, E. B., Herrmann, W., and Schenk, R. K.,** Improved cartilage fixation by ruthenium hexammine trichloride (RHT), *J. Ultrastruct. Res.*, 81, 1, 1982.
81. **Eggli, P., Herrmann, W., Hunziker, E. B., and Schenk, R. K.,** Matrix compartments in the growth plate of the proximal tibia of the rat, *Anat. Rec.*, 211, 246, 1985.
82. **Scott, J. E.,** Histochemistry of alcian blue. III. The molecular biological basis of staining by alcian blue 8GX and analogous phthalocyanins, *Histochemie*, 32, 191, 1972.
83. **Scott, J. E.,** Collagen-proteoglycan interactions. Localization of proteoglycans in tendon by electron microscopy, *Biochem. J.*, 187, 887, 1980.
84. **Ruggeri, A., Dell'Orbo, C., and Quacci, D.,** Electron microscopic visualization of proteoglycans with alcian blue, *Histochem. J.*, 7, 187, 1975.
85. **Schofield, B. H., Williams, B. R., and Doty, S. B.,** Alcian blue staining of cartilage for electron microscopy. Application of the critical electrolyte concentration principle, *Histochem. J.*, 7, 139, 1975.
86. **Dell'Orbo, C., Quacci, D., and Ruggeri, A.,** Nouve tecniche istochimiche per la dimostrazione dei proteoglicani estracellulari al microscopio elettronico, *Riv. Istoch. Norm. Pat.*, 21, 193, 1977.
87. **Morocutti, M., Marchini, M., Colombo, B., and Ruggeri, A.,** Heterogeneity of proteoglycan particles in thin sections and replicas of human articular cartilage, *Acta Biol. Hung.*, 35, 233, 1984.
88. **Meek, K. M., Scott, J. E., and Nave, C.,** An X-ray diffraction analysis of rat tail tendon treated with cupromeronic blue, *J. Microsc.*, 139, 205, 1985.
89. **Chan, L. and Wong, Y. C.,** Cytochemical characterization of cuprolinic blue-stained proteoglycans in the epithelial-stromal interface of the guinea pig lateral prostate, *The Prostate*, 14, 133, 1989.
90. **Crawford, R.,** Ultrastructure of the basal lamina and its relationship to extracellular matrix of embryos of the starfish *Pisaster ochraceus* as revealed by anionic dyes, *J. Morphol.*, 199, 349, 1989.
91. **Maitland, M. E. and Arsenault, A. L.,** Freeze-substitution staining of rat growth plate cartilage with alcian blue for electron microscopic study of proteoglycans, *J. Histochem. Cytochem.*, 37, 383, 1989.
92. **Scott, J. E.,** Affinity, competition and specific interactions in the biochemistry and histochemistry of polyelectrolytes, *Biochem. Soc. Trans.*, 1, 787, 1973.
93. **Scott, J. E. and Orford, C. R.,** Dermatan sulphate-rich proteoglycan associates with rat tail tendon collagen at the d band in the gap region, *Biochem. J.*, 197, 213, 1981.
94. **Van Kuppevelt, T. H. M. S. M., Domen, J. G. W., Cremers, F. P. M., and Kuyper, C. M. A.,** Staining of proteoglycans in mouse lung alveoli. I. Ultrastructural localization of anionic sites, *Histochem. J.*, 16, 657, 1984.
95. **Van Kuppevelt, T. H. M. S. M., Cremers, F. P. M., Domen, J. G. W., and Kuyper, C. M. A.,** Staining of proteoglycans in mouse lung alveoli. II. Characterization of the cuprolinic blue-positive, anionic sites, *Histochem. J.*, 16, 671, 1984.
96. **Van Kuppevelt, T. H. M. S. M., Cremers, F. P. M., Domen, J. G. W., Van den Brule, A. J. C., and Kuyper, C. M. A.,** Ultrastructural localization and characterization of proteoglycans in human lung alveoli, *Eur. J. Cell Biol.*, 36, 74, 1985.
97. **Van Kuppevelt, T. H. M. S. M., Rutten, T. L. M., and Kuyper, C. M. A.,** Ultrastructural localization of proteoglycans in tissue using cuprolinic blue according to the critical electrolyte concentration method: comparison with biochemical data from the literature, *Histochem. J.*, 19, 520, 1987.
98. **Van Kuppevelt, T. H. M. S. M., Janssen, H. M. J., van Beuningen, H. M., Cheung, K.-S., Schijen, M. M. A., Kuyper, C. M. A., and Veerkamp, J. H.,** Isolation and characterization of a collagen fibril-associated dermatan sulphate proteoglycan from bovine lung, *Biochim. Biophys. Acta*, 926, 296, 1987.
99. **Young, R. D.,** The ultrastructural organization of proteoglycans and collagen in human and rabbit scleral matrix, *J. Cell Sci.*, 74, 95, 1985.
100. **Rutten, T. L. M., van Kuppevelt, T. H. M. S. M., Janssen H. M. J., and Kuyper, C. M. A.,** Ultrastructural localization of a chondroitinase-sensitive, cuprolinic blue-positive filament in developing mouse lung, *Eur. J. Cell Biol.*, 45, 256, 1987.
101. **Völker, W., Schmidt, A. and Buddecke, E.,** Mapping of proteoglycans in human arterial tissue, *Eur. J. Cell Biol.*, 45, 72, 1987.
102. **Lormée, P., Lécolle, S., Septier, D., Le Denmat, D., and Goldberg, M.,** Autometallography for histochemical visualization of rat incisor polyanions with cuprolinic blue, *J. Histochem. Cytochem.*, 37, 203, 1989.

103. **Nefussi, J.-R., Septier, D., Collin, P., Goldberg, M., and Forest, N.,** A comparative ultrahistochemical study of glycosaminoglycans with cuprolinic blue in bone formed *in vivo* and *in vitro*, *Calcif. Tissue Int.*, 44, 11, 1989.
104. **Chan, L. and Wong, Y. C.,** Ultrastructural localization of proteoglycans by cationic dyes in the epithelial-stromal interface of the guinea pig lateral prostate, *The Prostate*, 14, 147, 1989.
105. **Scott, J. E.,** Contributions of light and electron histochemical techniques to the study of proteoglycan function in cartilage, *Acta Biol. Hung.*, 35, 227, 1984.
106. **Scott, J. E.,** Proteoglycan-collagen interactions, *CIBA Symposia*, 124, 104, 1986.
107. **Scott, J. E.,** Molecules for strength and shape: our fibre-reinforced composite bodies, *Trends Biochem. Sci.*, 12, 318, 1987.
108. **Scott, J. E.,** Proteoglycan-collagen interactions. Structure in the amorphous ground substance, *Collection Czechoslovak Chem. Commun.*, 53, 202, 1988.
109. **Scott, J. E. and Haigh, M.,** 'Small'-proteoglycan: collagen interactions: keratan sulphate proteoglycan associates with rabbit corneal collagen fibrils at the 'a' and 'c' bands, *Biosci. Rep.*, 5, 765, 1985.
110. **Scott, J. E. and Haigh, M.,** Proteoglycan-type I collagen fibril interactions in calcifying and non-calcifying connective tissues, *Biochem. Soc. Trans.*, 13, 933, 1985.
111. **Meek, K. M., Elliott, G. F., and Have, C.,** A synchrotron X-ray diffraction study of bovine cornea stained with cupromeronic blue, *Coll. Relat. Res.*, 6, 203, 1986.
112. **Quantock, A. J. and Meek, K. M.,** Axial electron-density of human scleral collagen-location of proteoglycans by X-ray diffraction, *Biophys. J.*, 54, 159, 1988.
113. **Szirmai, J. A.,** Quantitative approaches in the histochemistry of mucopolysaccharides, *J. Histochem. Cytochem*, 11, 24, 1963.
114. **Goldstein, D. J. and Horobin, R. W.,** Rate factors in staining by alcian blue, *Histochem. J.*, 6, 157, 1974.
115. **Horobin, R. W. and Goldstein, D. J.,** The influence of salt on the staining of tissue sections with basic dyes: an investigation into the general applicability of the critical electrolyte concentration theory, *Histochem. J.*, 6, 599, 1974.
116. **Terner, J. Y.,** Histochemical alkylation: a study of methyliodide and its effect on tissues, *J. Histochem. Cytochem.*, 12, 504, 1964.
117. **Ádány, R.,** A selective topo-optical reaction of sulfated glycosaminoglycans, in *VIIth European Symposium on Connective Tissue Research*, Prague, (Abstr.), 1980, 262.
118. **Módis, L., Lustyik, Gy., Ádány, R., and Zs.-Nagy, I.,** Energy dispersive X-ray microanalysis of sulfated glycosaminoglycans in cartilage matrix stained with alcian blue 8GX, *Bas. Appl. Histochem.*, 32, 415, 1988.
119. **Avasthi, P. S. and Koshy, V.,** The anionic matrix at the rat glomerular endothelial surface, *Anat. Rec.*, 220, 258, 1988.
120. **Takagi, M., Saito, I., Kuwata, F., and Otsuka, K.,** Specific binding of peanut agglutinin and soybean agglutinin to chondroitinase ABC-digested cartilage proteoglycans: histochemical, ultrastructural cytochemical, and biochemical characterization, *Histochem. J.*, 20, 88, 1988.
121. **Mitchell, N., Shepard, N., and Harrod, J.,** The use of brominated toluidine blue O in X-ray microanalysis for proteoglycan, *Histochemistry*, 68, 245, 1980.
122. **Wight, T. N. and Hascall, V. C.,** Proteoglycans in primate arteries. III. Characterization of the proteoglycans synthesized by arterial smooth muscle cells in culture, *J. Cell Biol.*, 96, 167, 1983.
123. **Maroudas, A.,** X-ray microprobe analysis of articular cartilage, *Connect. Tissue Res.*, 1, 153, 1972.
124. **Boyde, A. and Shapiro, I. M.,** Energy dispersive X-ray elemental analysis of isolated epiphyseal growth plate chondrocyte fragments, *Histochemistry*, 69, 85, 1980.
125. **Middleton, J. F. S., Oates, K., O'Connor, P., Orford, C. R., and Gardner, D. L.,** Demonstration by X-ray microprobe analysis of relationship between chondrocytes and tertiary surface-structure of hyaline articular cartilage, *Connect. Tissue Res.*, 13, 1, 1984.
126. **Wroblewski, J.,** Elemental changes associated with chondrocyte differentiation in rat rib growth plate, *Histochemistry*, 87, 145, 1987.
127. **Mitchell, N. and Shepard, N.,** The sulphur content of chondrocyte nuclei, *Histochemistry*, 80, 73, 1984.
128. **Kantor, T. G. and Schubert, M.,** A method for the desulfation of chondroitin sulfate, *J. Am. Chem. Soc.*, 79, 152, 1957.
129. **Spicer, S. S. and Lillie, R. D.,** Saponification as a means of selectively reversing the methylation blockade of tissue basophilia, *J. Histochem. Cytochem.*, 7, 123, 1959.
130. **Scudder, P., Tang, P. W., Hounsell, E. F., Lawson, A. M., Mehmet, H., and Feizi, T.,** Isolation and characterization of sulphated oligosaccharides released from bovine corneal keratan sulphate by the action of endo-β-galactosidase, *Eur. J. Biochem.*, 157, 365, 1986.
131. **Scott, J. E., Dorling, J., and Chandler, J. A.,** Applications of electron-histochemical technique to cartilage structure, in *7th European Rheumatology Congress, Brighton, Abstracts*, Faraday House, London, 1971, 24.9.

132. **Ohkura, T., Takashio, M., and Watanabe, T.,** Ultrahistochemical analysis of acid mucopolysaccharides using an energy dispersive X-ray analyzer, *J. Electron Microsc. (Tokyo),* 21, 240, 1972.
133. **Butler, W. F. and Heap, P.,** Correlation between alcian blue staining of glycosaminoglycans of cat nucleus pulposus and TEM X-ray probe microanalysis, *Histochem. J.,* 11, 137, 1979.
134. **Butler, W. F. and Heap, P.,** An ultrastructural study of glycosaminoglycans associated wtih collagen and other constituents of the cat anulus fibrosus, *Histochem. J.,* 14, 113, 1982.
135. **Mitchell, N. and Shepard, N.,** Pericellular proteoglycan concentrations in early degenerative arthritis, *Arthritis Rheum.,* 24, 958, 1981.
136. **Orford, C. R. and Gardner, D.,** Proteoglycan association with collagen d band in hyaline cartilage articular cartilage, *Connect. Tissue Res.,* 12, 345, 1984.
137. **Meyer, K.,** Reflectons on ''mucopolysaccharides'' and their protein complexes, in *Chemistry and Molecular Biology of the Intercellular Matrix,* Vol. 1, Balazs, E. A., Ed., Academic Press, London, 1970, 5.
138. **Kuijer, R., van de Stadt, R. J., de Koning, M. H. M. T., van Kampen, G. P. J., and van der Korst, J. K.,** Influence of cartilage proteoglycans on type II collagen fibrillogenesis, *Connect. Tissue Res.,* 17, 83, 1988.
139. **Iozzo, R. V. and Pacifici, M.,** Ultrastructural localization of the major proteoglycan and type II procollagen in organelles and extracellular matrix of cultured chondroblasts, *Histochemistry,* 86, 113, 1986.
140. **Poole, C. A., Flint, M. H., and Beaumont, B. W.,** Morphology of the pericellular capsule in articular cartilage revealed by hyaluronidase digestion, *J. Ultrastruct. Res.,* 91, 13, 1985.
141. **Poole, C. A., Flint, M. H., and Beaumont, B. W.,** Chondrons in cartilage: ultrastructural analysis of the pericellular microenvironment in adult human articular cartilages, *J. Orthop. Res.,* 5, 509, 1987.
142. **Poole, C. A., Flint, M. H., and Beaumont, B. W.,** Chondrons extracted from canine tibial cartilage: preliminary report on their isolation and structure, *J. Orthop-Res.,* 6, 408, 1988.
143. **Weiss, C., Rosenberg, L., and Helfet, A. J.,** An ultrastructural study of normal young adult human articular cartilage, *J. Bone Jt. Surg. (Am.),* 50, 663, 1968.
144. **Clark, J. M.,** The organization of collagen in cryofractured rabbit articular cartilage: a scanning electron microscopic study, *J. Orthop. Res.,* 3, 17, 1985.
145. **Clarke, I. C.,** Articular cartilage: a review and scanning electron microscope study. II. The territorial fibrillar architecture, *J. Anat.,* 118, 261, 1974.
146. **Módis, L., Ádány, R., and Lakatos, I.,** Polarisationsoptische Analyse der menschlichen embryonalen Knorpelmatrix, *Acta Histochem.* (Jena) Suppl., 26, 305, 1982.
147. **Módis, L. and Ádány, R.,** Sphaerit-Struktur der territorialen Knorpelmatrix, *Verh. Anat. Ges.,* 75, 217, 1981.
148. **Schenk, R. K., Eggli, P. S., and Hunziker, E. B.,** Articular cartilage morphology, in *Articular Cartilage Biochemistry,* Kuettner, K. E., Schleyerbach, R., and Hascall, V. C., Eds., Raven Press, New York, 1986, 3.
149. **Ronziere, M.-C., Berthet-Colominas, C., and Herbage, D.,** Low-angle X-ray diffraction analysis of the collagen-proteoglycan interaction in articular cartilage, *Biochim. Biophys. Acta,* 842, 170, 1985.
150. **Mollenhauer, J. and von der Mark, K.,** Isolation and characterization of a collagen-binding glycoprotein from chondrocyte membranes, *EMBO J.,* 2, 45, 1983.
151. **Von der Mark, K., Mollenhauer, J., Kühl, U., Bee, J., and Lesot, H.,** Anchorins: a new class of membrane proteins involved in cell-matrix interactions, in *The Role of Extracellular Matrix in Development,* Trelstad, R. L., Ed., Alan R. Liss, New York, 1984, 67.
152. **Lacy, B. E. and Underhill, C. B.,** The hyaluronate receptor is associated with actin filaments, *J. Cell Biol.,* 105, 1395, 1987.
153. **Heinegard, D. and Sommarin, Y.,** Macromolecules and interactions at the chondrocyte cell surface, *J. Clin. Chem. Clin. Biochem.,* 24, 937, 1986.
154. **Knudson, C. B. and Toole, B. P.,** Changes in the pericellular matrix during differentiation of limb bud mesoderm, *Dev. Biol.,* 112, 308, 1985.
155. **Roughley, P. J., White, R. J., and Poole, A. R.,** Identification of a hyaluronic acid-binding protein that interferes with the preparation of high-buoyant-density proteoglycan aggregates from adult human articular cartilage, *Biochem. J.,* 231, 129, 1985.

Chapter 12

METHODOLOGICAL APPENDIX

TABLE OF CONTENTS

I. INTRODUCTION

Realizing that the polarization microscopic techniques are generally not well known, it was decided that some selected methods will be collected and described in detail. Two criteria were used in the selection. First, methods for polarization-microscopic investigations of constituents of the extracellular matrix were chosen. Second, this short collection includes predominantly those techniques with which we have personal experience. Therefore, this selection is a little bit subjective.

II. SOME REMARKS ON PREPARATION OF TISSUE SAMPLES

Generally, materials fixed and embedded in wax are used. Cryostat sections, smears, and cell cultures grown on glass coverslips can also be used. The fixation is not discussed here. Generally, for carbohydrates alcoholic fixatives are recommended.[1] Some reactions can also be carried out on semifine sections made from materials embedded in resin. In this case, Epon or metacrylates should be used for embedding.

Using paraffin embedding, the complete removal of paraffin before polarization microscopical analysis is essential because paraffin is strongly birefringent. Depending on the type of paraffin used, sections must be treated in xylene at 37°C or 56°C for 2 to 24 h. An alternative solvent to paraffin is a 1:1 mixture of methanol and chloroform, which is used at room temperature for 2 to 12 h, generally combined with an antecedent treatment in xylene. The presence of paraffin can easily be checked by gentle warming of the sections until the paraffin melting point is reached (e.g., by a lighter). If warmed slides are placed again under the polarization microscope, the reappearance of birefringent paraffin can be followed as the specimen is cooled. Generally, after conventionally used dewaxing procedures, paraffin is left in the sections especially in cell nuclei.[2-4]

III. IMBIBITION ANALYSIS

This technique is a well established tool of ultrastructure research.[5-13] It can be carried out both with frozen or embedded and unstained sections. It is generally used to study the structure of collagen. Analyzing the imbibition curves obtained after measurements of retardation values under the polarization microscope, form and intrinsic birefringence of the structure can be distinguished. Form birefringence arises from the difference in refractive index between the structure under investigation and the surrounding medium which fills the spaces between the oriented, generally supramolecular but submicroscopic constituents (''micellae'') of the structure. Intrinsic birefringence is related to the molecular organization of the structure, and cannot be extinguished by imbibing fluids of different refractive indices, because these fluids cannot penetrate intermolecular spaces. Intrinsic birefringence is a characteristic property of a structure.

A series of imbibing fluid is prepared just before the experiments. The use of compounds as imbibing media which chemically bind to the structure must be avoided. Generally, mixtures of glycerol and water are used.[10-12] Instead of water, physiological saline can also be used.[14] The refractive index of each solution is measured by a refractometer. Using mixtures of glycerol and water, the refractive index (n) changes as a function of glycerol concentration as follows: water (n: 1.33); 20% (v/v) glycerol (n: 1.36); 40% glycerol (n: 1.39); 60% glycerol (n: 1.42); 80% glycerol (n: 1.44); 100% glycerol (n: 1.46). If imbibing media of higher refractive indices are needed, different oils or other fluids are chosen, e.g., monobromnaphtaline (n: 1.65) and its mixtures with isopropanol (n: 1.52 of a 1:1 v/v mixture, n: 1.57 of a 4:1 mixture, n: 1.6 of a 8:1 mixture).

Procedure:

1. The section is carefully cleared of wax, gradually hydrated in a series of ethanol of decreasing concentration and soaked in water for 30 min to 1 h. The section is covered by a coverslip and put under the microscope. The areas where measurements will be performed are selected and drawn to facilitate their refinding in the next imbibition media. Generally, small areas are chosen for measurements, e.g., a collagen fiber bundle, or a cartilage matrix trabecula. Sign of birefringence and retardation values are determined by using compensator plates in monochromatic green light. The required minimum number of measurements is determined by calculation of the coefficient of variation (CV). The CV should be less than 5%. If the structure is relatively homogeneous (e.g., a tendon), 10 to 20 measurements are sufficient to obtain the required accuracy.
2. The coverslip is removed, and the section is immersed and soaked in the next medium which is changed two or three times. After 1 h, the specimen is covered and the previously selected areas are measured again. This step is repeated until the last medium. If different compounds are used to produce media of different refractive indices (e.g., 100% glycerol for n: 1.46 and paraffin oil for n: 1.48), the complete imbibition of the section by medium containing the new compound can take considerably more time than 1 h. Pilot measurements of retardation are necessary to establish the appropriate imbibition period. After a complete imbibition the retardation value remains constant.
3. After finishing the measurements in all media, retardation values expressed in narometers are plotted against refractive indices. The paraboloid curves obtained are called form birefringence curves. If the curve touches the abscissa, the structure does not exhibit intrinsic birefringence. If there is a distance between the site of minimum retardation value of the curve and the abscissa, the structure has intrinsic birefringence, and the retardation at the minimum value of the graphs is proportional to the intrinsic birefringence. Curves taken from different structures can be compared by different mathematical methods. The simplest way is to use the *t*-test to compare corresponding retardation values measured at a given refractive index, e.g., to compare intrinsic birefringence of different structures. If form birefringence curves recorded from different sections are compared, care must be taken that the thickness of the sections be identical, because retardation values are also dependent upon the thickness of the structure. Comparison of imbibition curves taken from untreated tissues and from those subjected to enzymatic extraction of a matrix constitutent may provide valuable information on the ultrastructure of the extracted macromolecule. It has been shown by such investigations that GAGs are in parallel orientation to collagen in different connective tissues[7,9] (see also Chapters 2, 3, and 4).

IV. TOPO-OPTICAL REACTIONS

A. REACTIONS FOR COLLAGEN

1. Phenol[15,16]

Procedure: Sections of embedded material are carefully dewaxed in xylene. Cryostate sections are dried. Sections are mounted in a 1:1 (w/w) mixture of phenol and Canada balsam.

Comments: The mixture is prepared at 56°C and kept in dark glass. An alternative possibility is to use a series of phenol diluted in Canada balsam at concentrations ranging from 1 to 10% (w/w).[17] In this case, parallel sections are mounted in different media. Depending on the GAG-content of the tissue, 4 to 8% phenol concentration is sufficient to invert the positive sign of collagen to a negative sign. Phenol is supposed to bind to cationic groups of the collagen.[16] The advantage of this "most classical" topo-optical reaction is its specificity. No other birefringent structure than collagen changes the sign of its original birefringence with phenol. The shortcoming of this simple method is its insensitivity. Structures which exhibit birefringence with other topo-optical reactions frequently fail to give a phenol reaction (e.g., glomerular basement membrane, young embryonic cartilage matrix). Another, although minor, drawback is its colorless nature which makes it sometimes difficult to recognize different structures. To overcome this difficulty, it can be combined with a simple staining of nuclei with hematoxylin. In this case, nuclear staining is carried out before mounting the sections in a mixture of phenol and Canada balsam.

2. Congo Red[18]

Procedure: For fixation, Carnoy's or Zenker's fluids are preferred. Frozen or dewaxed paraffin sections are stained with a 2% aqueous congo red for 20 min. After short rinsing in distilled water, sections are mounted in 25% (w/v) aqueous gum arabic, or in 50% (w/v) fructose in egg-white. The mounting media are allowed to dry on the slides without using a coverslip at room temperature or in an exsiccator. After drying, preparations are mounted in Canada balsam with a coverslip.

Comments: Collagen fibers exhibit a strong birefringence of negative sign. Birefringence of amyloid and collagen stained by congo red can be distinguished using hydophilic mounting media, the sign of birefringence of amyloid always being positive. Antecedent acetylation + digestion with trypsin (2 mg enzyme in 1 ml phosphate buffer at pH 8.2, 37°C, 4 h) prevents the congo red topo-optical reaction of collagen by extraction of collagen from the tissue.[19]

3. Xylidine-Ponceau 2R[20]

Procedure: Dewaxed and hydrated sections are stained in a 0.1% xylidine-ponceau 2R dissolved in 0.1 mol/l acetate buffer at pH 1.7 for 15 min. After washing in the same buffer for 24 h, sections are dehydrated and mounted in Canada balsam.

Comments: Collagen fibers exhibit a positive dichroism and birefringence of positive sign after this staining. The staining mechanism is an electrostatic interaction between the dye and collagen. Antecedent digestion with testicular hyaluronidase enhances the staining intensity in cartilage tissue.

4. Sulfation[21]

Procedure: Deparaffinized sections are treated in 1% aqueous periodic acid at 37°C for 2 h, washed, and dehydrated in a series of ethanol of increasing concentration. Sections are brought through two changes of absolute ethanol into acetic anhydride for 10 min and then placed into the following freshly made sulfation mixture for 10 min: 1 ml concentrated sulfuric acid, 100 ml acetic anhydride. Sulfated sections are treated with acetic anhydride for 10 min and are hydrated in a series of ethanol of descending concentration and washed in water. Staining is carried out in 0.1% toluidine blue at pH 1 to 1.5 for 5 to 10 min. The staining effect is stabilized after blotting off the dye solution by dropping 2% aqueous potassium ferricyanide solution on the sections. After 2 to 3 min, the precipitant is removed and the section is mounted in gum arabic which is allowed to dry without coverslip. The dried layer of gum arabic is then overlaid with Canada balsam and covered with a coverslip.

Comments: Collagen fibers are metachromatically stained and they exhibit a strong birefringence with negative sign with respect to the long axis of the fibers. This is due to the sulfation of –OH groups of the collagen and the dye-uptake of sulfates bound to these –OH groups. Vicinal –OH groups of carbohydrates are oxidized by periodic acid and they cannot be sulfated. Nonvicinal –OH groups of carbohydrates are sulfated, but they are too far from each other to induce metachromasia and birefringence. It is possible that the presence of sulfated GAGs in the tissues causes aspecific staining. To avoid this, the untreated section must be stained by toluidine blue at pH 1 to 1.5 to check the dye uptake of the tissue before the sulfation is carried out. If the tissue is stained, removal of sulfate groups by a methylation must be the first step of the sulfation reaction (for methylation technique see Section VI.). Sulfation followed by a toluidine blue staining is a sensitive topo-optical reaction of collagen, its sensitivity being about four times higher than that of the phenol reaction. Sulfation, without subsequent staining, induces an inversive topo-optical reaction because of the axi-perpendicularly oriented sulfate groups bound to hydroxyls of the collagen.

5. Picrosirius Red F3B[22,23]

Procedure: Dewaxed and hydrated sections are stained for 30 min or 1 h in a 0.1% sirius red F3B or F3BA dissolved in saturated aqueous picric acid. Sections are dehydrated in absolute ethanol for 3 × 3 min, cleared and mounted in Canada balsam. An antecedent staining of cell nuclei with Weigert's haematoxylin is recommended.

Comments: Collagen is stained in red color and birefringent. The sign of birefringence is positive with respect to the long axis of the collagen fibers. The staining mechanism is an electrostatic interaction between the elongated polyanionic dye and collagen. Other structures than collagen can also bind sirius red. Therefore, the staining is not specific for collagen. As it has been shown in Chapters 3 and 10, intermediate filaments of the cytoskeleton of different cells and amyloid fibrils deposited extracellularly under pathological conditions are also stained and birefringent after picrosirius red reaction. The specificity of this topo-optical reaction can, therefore, be formulated as follows: a birefringence of positive sign appearing after picrosirius red staining in the extracellular matrix of normal tissues is a collagen-specific phenomenon. In tissues with high local concentration of sulfated GAGs in the matrix, an antecedent digestion with papain[23] or testicular hyaluronidase[24] significantly increases the sensitivity of the reaction. This is due to the removal of GAGs which mask the dye-binding groups of collagen. Antecedent methylation also proved to be successful in increasing the sensitivity.[24] Methylation blocks and removes fixed anionic charges of the matrix constituent hereby decreasing the repulsion of the polyanionic dye molecules. In our experience, staining with sirius supra blue[25,26] also results in an additive topo-optical reaction of collagen. It is important to note that the use of picrosirius red topo-optical reaction, or any other staining except immunohistochemistry, is not recommended for collagen typing.[27]

B. REACTIONS FOR ELASTIC FIBERS

1. Aniline[28,29]

Procedure: After dewaxing and clearing in xylene and terpineol, sections are mounted in aniline oil. To prevent drying, the edges of the coverslips are surrounded by wax or varnish.

Comments: Aniline is a colorless topo-optical reaction resulting in a birefringence with negative sign of the elastic fibers. Collagen fibers are isotropic, muscle elements retain their original positive sign of birefringence. Longitudinal sections of mature elastic fibers (e.g., ligamentum nuchae fibers) exhibit a three banded anisotropic pattern suggesting a complex helicoid micellar texture in the fibers.[30,31]

2. Permanganate-Bisulfite-Toluidine Blue (PBT)[32]

Procedure: Sections are oxidized in freshly prepared 0.5% $KMnO_4$ dissolved in 0.1 N HCl for 10 to 30 min. After rinsing in distilled water and decolorization in 1%

oxalic acid, sections are treated with freshly made and saturated aqueous potassium pyrosulfite (potassium metabisulfite) for 30 min. After a short washing in distilled water, staining is carried out with 0.05 to 0.1% toluidine blue at pH 1.0 to 1.5 for 5 to 10 min with a poststaining stabilization (see under C.1.), and sections are mounted in gum arabic, and after drying, in Canada balsam with a coverslip.

3. Bisulfite-Toluidine Blue (BT)[32,33]

Procedure: The sections are treated with freshly prepared saturated aqueous potassium pyrosulfite for 30 min without any antecedent oxidation and then stained with toluidine blue as described previously.

Comments to 2 and 3: Elastic fibers are metachromatic in normal light and they exhibit a birefringence of negative sign in polarized light. Antecedent digestion with elastase abolishes the PBT reaction of elastic fibers. If bisulfite-toluidine blue is used without antecedent oxidation with permanganate, the naturally occuring free aldehydes of the tissues can be stained, e.g., in young elastic fibers.[32,33] If the oxidation period is shorter (5 to 10 min), RNA is stained and birefringent in cells rich in densely packed rough endoplasmic reticulum, e.g., in plasma cells.[34] During the PBT reaction of elastic fibers, potassium permanganate is supposed to oxidize the crosslinks of elastin (desmosine, isodesmosine) yielding aldehyde groups. By the addition of bisulfite, aldehydes are transformed to hydroxysulphonic acids[35] which can be selectively stained by a cationic dye (e.g., toluidine blue) at low pH. If elastic fibers are detected by PBT reaction in a tissue containing higher amounts of sulfated GAGs, a methylation must be carried out prior to the PBT reaction. Methylation removes the sulfate groups (see VI). The necessity of an antecedent methylation can easily be checked by staining a section of the specimen with toluidine blue used for development of the PBT reaction.

C. REACTIONS FOR POLYANIONIC SUBSTANCES INCLUDING GLYCOSAMINOGLYCANS

1. Toluidine Blue[36]

Procedure: The specimens are stained with 0.05 to 0.1% toluidine blue dissolved in acetate or phosphate buffers at different pH values for 5 to 10 min. For selective staining of sulfated GAGs a pH below 2.5 to 3.0 is chosen. If staining of constituents rich in carboxyls (e.g., cell coat) is required the pH of the staining solution must be between 4.5 to 7.2. Excess dye is blotted off the section by filter paper or removed by a rinsing in buffer used to dissolve the dye, and then staining is stabilized by a precipitant solution for 2 to 3 min. The precipitant is either a 2% aqueous potassium ferricyanide or a 6:1 mixture of 2% aqueous potassium iodide and 2% aqueous potassium ferricyanide. The former one well preserves the matachromatic color and latter one has good preservation effects on both on the original orthochromatic and metachromatic colors. After precipitation, sections are mounted in 25% aqueous gum arabic containing 2% fructose (to prevent cracking of dried gum arabic) and 0.1% potassium ferricyanide without coverslip. Preparations are kept in the horizontal position and dust - free atmosphere until the gum arabic layer is dryed. It takes generally 12 to 24 h except at high humidity. In this case, the use of an exsiccator with dry phosphoric oxide is recommended. After complete drying of gum arabic, its refractive index will be about 1.50 to 1.53, and the preparations can be mounted in Canada balsam or in any other synthetic hydrophobic medium and with coverslip. Any kind of alcohol treatments should be avoided during the poststaining procedure.

Comments: Structures containing polyanions are metachromatic. Metachromasia appears in red. If the chromotrope is arranged in a preferential direction, in addition to the staining, it exhibits an optical anisotropy. In the great majority of cases, toluidine blue molecules are bound orthogonally to the chromotrope. Therefore, the relative sign of birefringence induced by this dye is generally negative. Tissue constituents stained metachromatically are not necessarily birefringent. Presence of metachromasia indicates the high anionic charge density of the chromotrope. In this case the dye-binding groups are not at a distance from each other greater than 0.5 to 0.7 nm, otherwise dye-dye interactions could not be developed and the metachromatic dye aggregate could not be formed. Generally, polysulfated GAGs fit this requirement. If the chromotrope molecules are arranged in an oriented manner, an optical anisotropy can also be observed. Birefringence appears in green or greenish yellow polarization color. Red polarization color indicates either a low degree of orientation or an overstaining. For distinction, more parallel sections are stained using different staining times. If the appearance of green polarization color cannot be observed at all, the metachromatically stained structure must have a low level of orientation. For measurement of retardation value of birefringence, generally red transilluminating light is used. Toluidine blue staining is frequently used to enhance the birefringence of sulfated GAGs. DNA of cell nuclei or chromosomes are also birefringent if histones are removed prior to the staining.[37-39] Cell coats of isolated cells exhibit birefringence if the staining is carried out at pH 4.5 to 7.2, this is due to the oriented dye binding by the neuraminic acid residues of different cell surface glycoproteins.[40-41]

2. Rivanol[42-48]

Procedure: The sections are stained in 0.35% rivanol dissolved in McIlvaine's Na_2HPO_4 - citric acid buffer for 10 min. If sulfated GAGs are stained, the pH must be less than 2.5 to 3.0. When DNA is stained, 3.5% aqueous rivanol solution is used. After rinsing in the diluent of the dye, staining is stabilized with 2% aqueous potassium ferricyanide for 2 to 3 min. The sections are mounted in gum arabic and dried afterwards in Canada balsam.

Comments: Rivanol can induce a birefringence of negative relative sign. Depending on the pH of the staining solutions, different polyanions can be stained for polarization microscopic investigation. Furthermore, it can be used for fluorescence microscopic visualization of sulfated GAGs,[49] for precipitation of GAGs from solutions[50] and turbidometric determination of chondroitin sulfate.[51]

3. Dimethylmethylene Blue[1]

Procedure: The sections are stained with 0.1% dimethylmethylene blue dissolved in McIlvaine's buffer for 5 to 10 min. After washing in the solvent of the dye, the sections are mounted without poststaining stabilization in gum arabic and dried afterwards in Canada balsam.

Comments: The staining effects and optical behavior are similar to those of toluidine blue. The metachromasia induced by this dye is, however, stronger. Orthochromatic and metachromatic tones are more distinct, and the selective appearance of metachromasia in tissue areas containing sulfated GAGs can be observed even at higher pH values (about pH 5) where toluidine blue staining induces generally an overall metachromatic staining. The birefringence of structures stained with dimethylemethylene blue is more intense as compared to toluidine blue. This dye has been recommended for biochemical assay of sulfated GAGs.[52,53]

4. *N,N'*-Diethylpseudoisocyanine[1,54]

Procedure: The specimens are covered by drops of 0.1% aqueous dye and stained for 5 min. A 3% acetic acid can also be used as a diluent. After blotting with filter paper, the sections are mounted in gum arabic and dried afterwards in Canada balsam. We carry out this staining without any poststaining stabilization, but the use of dilute (0.1 to 0.2%) aqueous potassium ferricyanide as precipitant can also be recommended.

Comments: This dye has a purple metachromatic color, and its orthochromatic staining effect appears in red. The main advantage of this staining over other basophilic topo-optical reactions is the extreme high intensity of birefringence induced.[1] It can also be used to study the ultrastructure of the cell coat.[55] The staining solution must be kept in the dark and its use over 1 month is not recommended.

5. Alcian Blue[56-60]

Procedure: The sections are stained for 2 h in 0.05% alcian blue 8GX dissolved in 0.25 mol/l acetate buffer at pH 5.8 containing 0.3 mol/l $MgCl_2$. The staining solution is freshly prepared. After washing in buffer, the sections can be either dehydrated in graded ethanol, cleared in xylene, and mounted in Canada balsam, or mounted in gum arabic with a coverslip without any dehydration.

Comments: Sulfated GAGs are stained in the presence of 0.3 mol/l $MgCl_2$ in the staining solution.[56] By increasing the concentration of the $MgCl_2$, the staining of different sulfated GAGs can be abolished. For example, chondroitin sulfates are not stained at 0.6 mol/l but keratan sulfate and heparin are.[56] We observed that the alcian blue staining can also induce birefringence.[1] The sign of birefringence is negative indicating an axiparallel binding of the dye molecules to the GAGs. The birefringence appears in red, while the anomalous polarizations are purple (after additive compensation) and greenish yellow (after subtractive compensation). Dichroism can also be observed.[61] Due to its relative large size, alcian blue is not an optimal topo-optical reagent of sulfated GAGs, the birefringence induced by this dye being weaker compared to that generated by the metachromatic dyes (see Chapter 3). Its advantage is, however, that it can be used in the critical electrolyte concentration technique which is a powerful histochemical method to show different GAGs,[56] although some alcian blue 8GX samples, unfortunately, easily precipitate in the presence of $MgCl_2$.

6. Chemically Intensified Basophilic Reaction (CIBR)

Procedure: The dewaxed sections are hydrated and then oxidized with 2% aqueous periodic acid at 37°C for 2 h. After washing in distilled water, they are treated in freshly prepared 1% aqueous $NaBH_4$ for 4 min at room temperature. After washing, the specimens are subjected to a second treatment in periodic acid (1%, room temperature, 24 h). The sections are washed again, and they are treated with freshly prepared, saturated aqueous potassium pyrosulfite (metabisulfite) for 30 min, and, after a short rinsing in distilled water they are stained with 0.1% dimethylmethylene blue or toluidine blue at pH 2.5 for 5 to 10 min or with 0.5% cuprolinic blue at pH 2.5 for 30 min. Excess dye is removed by blotting with filter paper or rinsing in buffer, and dimethylmethylene blue and cuprolinic blue staining are followed by mounting in gum arabic while toluidine blue metachromasia is stabilized by a poststaining precipitation before mounting.

Comments: Our CIBR is a combination of several histochemical reactions. Oxidation with periodic acid is the first step of the classical PAS reaction, and it yields aldehydes at the sites of the vicinal alcoholic −OH groups of different carbohydrates. The aldehydes are blocked by the subsequent treatment with sodium borohydride. The prolonged treatment with periodic acid results in the oxidation of vicinal hydroxyls of the uronic acid residues of different GAGs,[62-64] e.g., chondroitin sulfates, and the aldehyde groups formed in the uronic acids are transformed to hydroxysulfonic acids by the addition of potassium pyrosulfite[35] which can be stained by the cationic dyes. If uronic acid residue is present in the sulfated GAGs, it will bind two dye molecules after this reaction which will be added to the dye bound by the sulfate group of the hexosamine residue. Altogether, three cationic dye molecules are bound to one disaccharide unit of the chondroitin sulfate molecules after CIBR reaction. Using a simple staining at pH 2.5, only one dye molecule is taken up by the sulfate groups of one disaccharide unit. Thus, CIBR is a sensitive topo-optical reaction of uronic-acid containing sulfated GAGs. We observed in hyaline cartilage matrix that staining with cuprolinic blue — like toluidine blue and dimethylmethylene blue — results in an optical anisotropy with negative sign.

D. REACTION FOR PERIODATE-REACTIVE CARBOHYDRATES

1. Aldehyde-Bisulphite-Toludine Blue (ABT)[65]

Procedure: The sections are treated with 1% aqueous periodic acid for 30 min then, after rinsing in distilled water, with freshly saturated aqueous sodium pyrosulfite for 30 min. After rinsing with distilled water for a short time, the sections are stained with toluidine blue at pH 1 to 1.5 for 5 to 10 min. The poststaining treatment is described in Section C.1.

Comments: This is a modified PAS reaction. The result of this reaction is a metachromatic staining and a birefringence with negative sign if the stained constituent is spatially oriented. The advantage of the ABT reaction over the PAS technique is the ability to induce birefringence if the stained constituents are spatially oriented. The aldehydes generated by the periodic acid are transformed to hydroxysulfonic acids[35] and they are detected by the toluidine blue. Before carrying out this reaction, a section must be stained by the toluidine blue used for the development. If metachromasia is observed in the section, the ABT reaction must be preceded by a methylation. Otherwise the sulfate-containing constituents (GAGs, some epithelial mucins) will interfere with the ABT reaction. ABT reaction can be combined with an antecedent mild alkaline hydrolysis (0.5% KOH in 70% ethanol, room temperature, 30 min). This treatment removes the O–acyl groups of some carbohydrates,[66] and results in the formation of –OH groups in these sites. If these hydroxyls are in the vicinity of alcoholic –OHs, the intensity of the ABT reaction will be increased.[67] The ABT reaction, when evaluated in polarized light, proved to be a sensitive technique to show microorganisms (bacteria, fungi) in tissue sections.[67,68] The interpretation of the ABT reaction (similarly to the PAS reaction) is difficult. The only thing that can be said after this reaction is that periodate-reactive monosaccharides are present in the structure. Such monosaccharides are rather common in different glycoproteins. The ABT reaction, therefore, cannot be used for selective localization of a matrix macromolecule. If metachromatic intracellular granules are seen after ABT reaction, they are probably glycogen granules. Glycogen generally exhibits a moderate birefringence appearing in red polarization color after this reaction. In this case, a digestion with amylase must be performed prior to the ABT reaction, and the disappearance of the reaction will verify the diagnosis.

E. REACTIONS FOR AMYLOID

1. Iodine[69]

Procedure: The dewaxed sections are stained with Lugol's iodine solution (2 g KI and 1 g I in 100 ml distilled water containing 0.1 *N* HCl) for 10 s. The specimens are mounted in a 1:9 mixture of iodine solution and glycerol.

Comments: As it has been discussed in Chapter 10, this reaction has rather historical than diagnostic value. Amyloid is birefringent, and supposedly its P-component is responsible for the oriented binding of the iodine.[69] The relative sign of birefringence is positive.

2. Congo Red

Procedure: 1. Alkaline Congo red;[70] Mayer's acid hemalum staining for 10 min, washing in distilled water three times, and staining in Congo red for 20 min. After rapid dehydration in absolute ethanol 3 ×, the sections are cleared in xylene and mounted in Canada balsam or Permount. Stock solution: 80% ethanol saturated with NaCl. Working solution (it must be prepared freshly and used within 15 min): 0.5 ml 1% aqueous NaOH is added to 50 ml of the stock solution. 2. Romhányi's technique:[71] 1 to 2% aqueous congo red for 10 to 20 min. After rinsing in distilled water, the sections are mounted in gum arabic.

Comments: Amyloid is stained in orange-red colors, and it exhibits a birefringence in polarized light between crossed polaroids. The birefringence appears in green polarization color. The relative sign of birefringence is positive related to the long axis of the amyloid fibers indicating an axiparallel binding of the congo red molecules to the amyloid fibrils. In Romhányi's technique, the birefringences of amyloid and collagen show opposite signs: the amyloid has positive and the collagen fibers have negative signs. It has been shown by Romhányi[72] that after a mild oxidation with potassium permanganate (freshly prepared 0.25% potassium permanganate in 0.15% aqueous H_2SO_4 for 3 min) secondary amyloid deposits can be dissolved by trypsin digestion (2 mg enzyme in 1 ml 0.15 mol/l phosphate buffer, pH 8.2, 37°C, 6 to 18 h). The dissolution can be followed by the gradual disappearance of the birefringence induced by Congo red staining. Primary amyloid deposits cannot be dissolved by the above treatment.

3. Toluidine Blue[73]

Procedure: The well-deparaffinized sections are brought to water or 50% ethanol through graded alcohols. Staining is carried out in 1% toluidine blue in 50% isopropanol for 30 min at 37°C. After staining, the sections are carefully blotted with filter paper, immersed for 1 min in absolute isopropanol, blotted again, cleared in two changes of xylene, and mounted in Canada balsam.

Comments: This is the so-called "standard toluidine blue" (STB) method. Amyloid deposits are stained orthochromatically (in blue color), and are birefringent exhibiting red or orange-red polarization colors in polarized light between crossed polaroids. The sign of birefringence is positive. Anomalous polarization colors are purple (after additive compensation) and greenish yellow (after subtractive compensation). Basophilic structures such as nuclei, mucin, etc., are stained mostly orthochromatically. Collagen is pale blue, and its birefringence appears silvery blue. Structures rich in sulfated GAGs are stained metachromatically, and they may exhibit optical anisotropy apearing in yellowish green polarization color (e.g., cartilage matrix). It has been shown in Chapter 10 that amyloid deposits are othochromatic and birefringent after staining with a 0.1% toluidine blue at pH 2.5 to 3.0 for 5 to 10 min using a poststaining stabilization and gum arabic as mounting medium as described in Section C.1 of this Chapter. The optical characteristics of amyloid observed after this reaction are similar to those obtained after STB procedure.

V. ENZYMATIC EXTRACTIONS

The enzymatic digestions are useful methods to control the specificity of different topo-optical reactions. They are carried out prior to the stainings, and it is important to note that three parallel sections are used for such experiments. One section is stained without any pretreatment, one section is treated only with the solvent of the enzyme under the same conditions used in digestion and then stained, and one section is digested and stained. The digestions are carried out in a moist chamber and the sections are covered by freshly prepared enzyme solutions. The efficacy of the enzymatic extraction can be evaluated by comparing digested sections to those treated with thc solvent.

Digestion with amylase[74] — 0.5% alfa-amylase is dissolved in 0.1 mol/l acetate buffer at pH 6, and the sections are incubated at 37°C for 30 to 120 min, occasionally for 10 to 12 h. After this digestion, ABT or KOH + ABT reactions are carried out. Amylase removes glycogen from the cells.

Digestion with neuraminidase [75,76] — The sections are treated with 2 U neuraminidase in 1 ml 0.05 mol/l acetate buffer containing 0.04 mol/l $CaCl_2$ at pH 5.5 for 1 to 2 h at 37°C. Alternative technique:[76] 1:1 (w/w) 500 U/ml enzyme and 0.1 mol/l acetate buffer containing 1% NaCl and 0.1% $CaCl_2$, pH 5.5, 16 to 24 h, 39 to 41°C. The enzyme digests the sialic acid residues of the different glycoproteins of the cell membrane, mucins, etc. After this digestion, a staining is carried out using a cationic dye at pH 4.5 to 7.2.

Digestion with testicular hyaluronidase — 0.5 to 1 mg enzyme in 1 ml 0.1 mol/l phosphate buffer, pH 6 to 6.5, 37°C, 18 to 24 h. The enzyme depolymerizes chondroitin sulfates and hyaluronic acid. After this digestion, staining with a cationic dye is carried out at pH 2.5 (for chondroitin sulfates) and at pH 5 (for hyaluronic acid + chondroitin sulfates).

Digestion with chondroitinases[77-79] — 0.25 to 2.5 U enzyme (chondroitinase AC, ABC, chondroitin 4- or 6-sulfatase) in 1 ml 0.1 to 0.5 mol/l *Tris*-HCl buffer at pH 7.3 (for chondroitinase AC) or at pH 8.0 (for chondroitinase ABC), 37°C, 12 to 24 h. The chondroitinase AC depolymerizes chondroitin 4- and 6-sulfate, the chondroitinase ABC digests chondroitin 4- and 6-sulfate and dermatan sulfate. Stainings with cationic dyes preferentially at low pH (at pH 2.5 or lower) are used after these digestions.

Digestion with heparitinase[75,79] — 0.5-5 U enzyme in 1 ml 0.5 mol/l *Tris*-HCl buffer containing 0.03 mol/l calcium acetate, pH 7.0, 37°, 2 to 12 h. Heparan sulfate is degraded by this enzyme. The recommended staining procedure is the same as mentioned for chondroitinase digestion.

Digestion with keratanase[80] — 0.1 U enzyme is dissolved in 1 ml 0.1 mol/l Tris-HCl buffer at pH 8.0 containing 30 mmol/l sodium acetate and proteinase inhibitors. The digestion is carried out at 37°C for 90 min. The enzyme digests keratan sulfates. Basophilic staining reactions are carried out at different pH values after this digestion.

Digestion with Streptomyces hyaluronidase[81-84] — 100 to 150 TRU enzyme in 1 ml 0.1 mol/l phosphate buffer, pH 5, 39 to 41°C, 2 to 4 h. This enzyme selectively depolymerizes hyaluronic acid. Basophilic staining reactions are carried out at different pH values after the digestion.

Collagenase extraction[76] — The sections are digested with collagenase (1 mg in 1 ml 0.9% NaCl) at 50°C for 45 to 60 min. Collagen-specific topo-optical reactions are carried out after this digestion.

VI. BLOCKING REACTIONS

Acetylation[74] — Sections are deparaffinized, then treated in absolute ethanol and acetic anhydride. Acetylation is carried out at room temperature for 48 h using a 2:3 mixture of acetic anhydride and anhydrous pyridine. Acetylation blocks the $-OH$ and $-NH_2$ groups. ABT reaction of carbohydrates is abolished by antecedent acetylation. It has been shown, furthermore, that the intensity of topo-optical reactions of GAGs with cationic dyes is increased if they are carried out after acetylation.[85] This is due to the blocking of tissue $-NH_2$ groups which — after dissociation to $-NH_3^+$ — may interact with the dye-binding anionic groups of different tissue constituents. Furthermore, the repulsive forces of the blocked $-NH_3^+$ groups against the cationic dyes decrease. Acetylation induces a topo-optical reaction of the collagen fibers without any additional stainings.[19] The sign of birefringence of the collagen is inverted from positive to negative because of the axiperpendicular orientation of the acetyl groups bound to the collagen fibers.

Methylation[74] — Dewaxed sections are brought to absolute ethanol and then to methanol. Methylation is carried out at 37°C for 12 to 24 h in the following solution: 1 ml concentrated HCl, 99 ml absolute methanol. During this procedure, the sulfate groups are removed and substituted by $-OH$ groups, and $-COOH$ groups are reversibly methylated.[86-88] If methylated sections are stained with cationic dyes — provided that the staining mechanism is electrostatic in nature — the staining is completely abolished where sulfates and carboxyls are responsible for the dye-uptake. The intensity of picrosirius red F3B staining for collagen is considerably increased after methylation, because the anionic groups — which repulse the anionic sirius red molecules — are blocked and removed. If methylated sections are subjected to a mild alkaline hydrolysis (saponification in 0.5% KOH in 70% ethanol, 30 min), the carboxyl groups can be restored and selectively stained.

Carboxymethylation[89] — Dewaxed sections are brought to absolute ethanol and then to methanol. Carboxymethylation is carried out at 37°C for 12 to 18 h in the following freshly prepared solution: 12.5% CH_3I in absolute methanol saturated with sodium carbonate. If carboxymethylated sections are stained with a cationic dye at higher pH values (pH 5 to 7), a selective staining of polysulfates (sulfated GAGs) can be obtained in the extracellular matrix. Phosphate groups are only partially blocked or not blocked at all, so a sulfate-selective dye-uptake cannot be expected intracellularly (owing to the presence of nucleic acids) and in structures rich in phosphoproteins (e.g., in dentine). Carboxymethylation combined with a basophilic staining reaction (e.g., with a toludine blue staining at pH 5) provides a reliable and simple method for selective histochemical demonstration of sulfated GAGs.

VII. QUANTIFICATION

The quantification of the topo-optical reactions is not yet solved. Earlier, the so-called anisotropic index[36] or quotient[6] calculation was used to compare the intensity of birefringence induced by topo-optical reactions in different structures. It has been shown, however, in our laboratory that the anisotropic index calculated after different topo-optical reactions did not correspond either to the submicroscopic orientation pattern or to the quantity of the constituent stained in a section.[90] Therefore, we propose the measurement of retardation values after different topo-optical reactions despite the fact that retardation values are also dependent upon the thickness of the structure. This thickness, however, is not necessarily equal to the thickness of the section. If we measure, for example, the retardation value of birefringence after a collagen-specific topo-optical reaction in hyaline cartilage matrix, we cannot determine the sum of the thickness of the collagen fibers which transmit the light beam. In a tendon or in other dense connective tissues, on the other hand, the section thickness is more or less equal to the sum thickness of the collagen fibers cut. In this case the measurement of section thickness would be useful for calculation of the birefringence which is independent of the thickness and which is proportional with the degree of orientation of the structure. Just to remind the reader, the retardation value (Γ) is described by the following equation:

$$\Gamma = d\Delta \qquad (1)$$

where d= the thickness of the birefringent structure; Δ = birefringence ($n_e - n_o$); n_e = extraordinary, and n_o = ordinary refractive index of the structure.
It follows that Δ is independent of the thickness:

$$\Gamma/d = \Delta \qquad (2)$$

Except in come cases, unfortunately, the real d cannot be determined, so, the measurement of retardation values remains the only possible way to quantify of the topo-optical reactions at this moment. To minimize the error deriving from the changes of d, the number of measurements of retardation values must be sufficiently high. We calculate the coefficient of variation (CV) during the measurements by a computer program, and if the CV falls below 5%, the minimum number of measurements necessary to have the required accuracy is established.

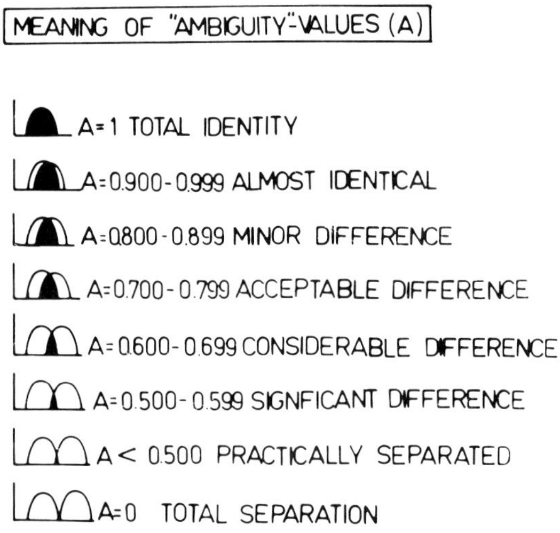

FIGURE 1. A simplified scheme of the meanings of the so-called "ambiguity" (A) values obtained by the ambiguity-probe when comparing two histograms. If A-values are smaller than 0.7, the difference between the histograms is significant. On the left side of the diagram the distributions of two data sets are schematically shown. If A = 1, the two histograms are identical. The closer the A value gets to zero, the more the two histograms differ from each other. If A = 0, the histograms are totally separated from each other.

We measure small tissue areas in the central part of the field using 400 to 1000 × microscopic magnifications. The retardation values are plotted in histogram forms and the histograms are compared mathematically. For this purpose, we prefer the so-called ambiguity-test[91-93] (Figure 1). Of course, other probes can also be used.

FINAL ACKNOWLEDGMENTS

After finishing the last Chapter of this volume, the author would like to take the opportunity to express his sincere thanks to friends and colleagues who have helped in this work with their contributions to the experiments, with ideas, hard criticism, and stimulating discussions. The names of my co-workers involved in the experimental works have been mentioned at the end of the relevant Chapters. I am especially indebted to my former and present graduate student co-workers for their skillful and enthusiastic help and access to some unpublished data.

I will forever keep the memory of my first director, Professor Stephan Krompecher in my heart. It was he who first focused my attention on matrix reserach. I will never forget the days and hours spent together with Professor George Romhányi, today Emeritus Professor of Pathology, University Medical School, Pécs, who convincingly showed the beauty of the microstructures explored by his favorite topo-optical reactions, who walked the hard road leading to an original observation, and who understood the significance of the happy serendipities in research. I acknowledge the countinous encouragment and helpful criticism of Professor George Székely, Director of our department, who — although being a neurobiologist — also appreciates the proteoglycan molecules and collagen fibers.

I am especially gratful to Ms. Ilona Oláh for her excellent and inestimable technical assistance in the experiments and also in editing the manuscript. It is a pleasure to express my gratitude to Ms. Ágnes M. Durkó for the photographic prints, and to Ms. Edit Ágh for the artwork. I wish to thank Mr. Lloyd Sparks for his indispensable help in editing the English manuscript. Thanks are due to Ms. Mária Kozma (Department of Clinical Chemistry) for her outstanding secretarial help. I am indebted to Professor Imre Zs.-Nagy (F. Verzár International Laboratory for Experimental Gerontology, Hungarian Section, Debrecen) for providing their IBM Displaywriter System.

The greater part of this research was supported by grants of the Hungarian Ministry of Health and Social Affairs.

I express my gratitude to my wife and co-worker, Dr. Róza Ádány for her help, criticism, and stimulation in science, and patience in private life.

I thank the publisher, CRC Press, for its interest in this work and its employees for their professional assistance in the editing of this Volume.

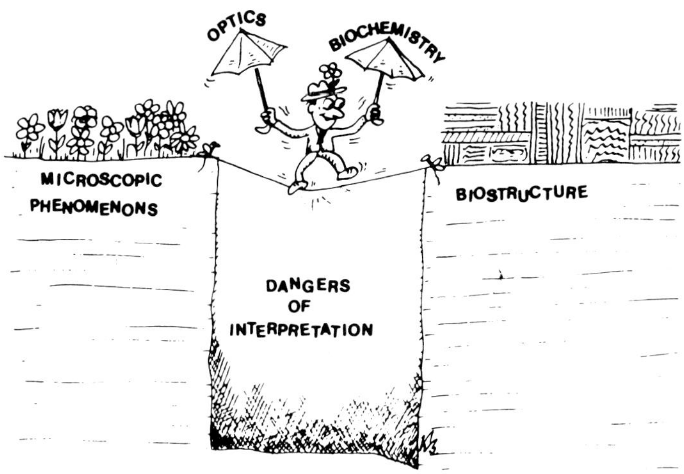

The two caricatures shown are drawn and recommended by me for those who are working with polarization microscopical techniques. The original copies decorate our laboratory just to remind us that even morphologists analyze their artifacts. There is, however, a narrow bridge between microscopical, sometimes attractive, artifacts and the reality of the biostructures. Our task is to find the trail to this bridge and to cross it.

REFERENCES

1. **Módis, L.**, Topo-Optical Investigation of Mucopolysaccharides (Acid Glycosaminoglycans), *Handbuch der Histochemie*, Vol. 2, Part 4, Graumann, W. and Neumann, K., Eds., G. Fischer Verlag, Stuttgart, 1974.
2. **Nedzel, G. A.**, Intranuclear birefringent inclusions, an artifact occuring in paraffin sections, *Q. J. Micr. Sci.*, 92, 343, 1951.
3. **Hamperl, H.**, Remnants of paraffin (''paraffin-inclusions'') in histological sections, *Virchows Arch. (A)*, 334, 79, 1961.
4. **Vlachos, J. D.**, Birefringence and paraffinophilia of cell nuclei, *Stain Technol.*, 43, 89, 1968.
5. **Frey-Wyssling, A.**, Ultrastructure research in biology before the electron microscope, *J. Microsc.*, 100, 21, 1974.
6. **Scheuner, G. and Hutschenreiter, J.**, *Polarisationsmikroskopie in der Histophysik*, VEB G. Thieme, Leipzig, 1972.
7. **Vidal, B. C.**, The part played by the mucopolysaccharides in the form birefringence of collagen, *Protoplasma*, 59, 472, 1964.
8. **Schmidt, W. J. and Keil, A.**, *Polarizing Microscopy in Dental Tissues. Theory, Methods and Results*, Pergamon Press, Oxford, 1971.
9. **Vidal, B. C.**, The part played by proteoglycans and structural glycoproteins in the molecular orientation of collagen bundles, *Cell. Mol. Biol.*, 26, 415, 1980.
10. **Vidal, B. C. and Vilarta, R.**, Articular cartilage: collagen type II-proteoglycan interactions. Availability of reactive groups. Variation in birefringence and differences as compared to collagen I, *Acta Histochem. (Jena)*, 83, 189, 1988.
11. **Mello, M. L. S., Godo, C., Vidal, B. C., and Abujadi, J. M.**, Changes in the macromolecular orientation of collagen fibers during the process of tendon repair in the rat, *Ann. Histochim.*, 20, 145, 1975.
12. **Whittaker, P., Boughner, D. R., Perkins, D. G., and Canham, P. B.**, Quantitative structural analysis of collagen in chordae tendineae and its relation to floppy mitral valves and proteoglycan infiltration, *Br. Heart J.*, 57, 264, 1987.
13. **Whittaker, P., Schwab, M. E., and Canham, P. B.**, The molecular organization of collagen in saccular aneurysms assessed by polarized light microscopy, *Connect. Tissue Res.*, 17, 43, 1988.
14. **Yarker, Y. E., Aspden, R. M., and Hukins, D. W. L.**, Birefringence of articular cartilage and the distribution of collagen fibril orientations, *Connect. Tissue Res.*, 11, 207, 1983.
15. **Ebner, V.**, Über eine optische Reaktion der Bindesubstanzen auf Phenole, *Sitz. Ber. Akad. Wiss. Wien., math. nat. Kl.*, 103, 162, 1894.
16. **Hutschenreiter, J. and Scheuner, G.**, Zur polarisationsoptischen Erfassung funktioneller Gruppen des Kollagens durch die Phenolreaktion, *Verh. Anat. Ges.*, 71, 1443, 1977.
17. **Neumark, T.**, Phenol reaction of collagen fibers and acid mucopolysaccharide content of connective tissue, *Acta Morphol. Acad. Sci. Hung.*, 14, 59, 1966.
18. **Romhányi, G., Deák, G., and Bukovinszky, A.**, Collagen-specific topo-optical staining reaction with congo red and its ultrastructural interpretation, *Acta Morphol. Acad. Sci. Hung.*, 18, 261, 1970.
19. **Romhányi, G., Bukovinszky, A., and Deák, G.**, On the ultrastructure and proteolytic susceptibility of acetylcollagen as studied by topo-optical reactions, *Acta Morphol. Acad. Sci. Hung.*, 20, 49, 1972.
20. **Vidal, B. C.**, Dichroism in collagen bundles stained with xylidine ponceau 2R, *Ann. Histochim.*, 15, 289, 1970.
21. **Romhányi, G., Bukovinszky, A., and Deák, G.**, Sulfation as a collagen-specific reaction. The ultrastructure of sulfate collagen, basement membranes and reticulin fibers as shown by topo-optical staining reactions, *Histochemie*, 36, 125, 1973.
22. **Constantine, V. S. and Mowry, R. W.**, The selective staining of human dermal collagen. II. The use of picrosirius red F3BA with polarization microscopy, *J. Invest. Dermatol.*, 50, 419, 1968.
23. **Junqueira, L. C. U., Bignolas, G., and Brentani, R. R.**, Picrosirius staining plus polarization microscopy, a specific method for collagen detection in tissue sections, *Histochem. J.*, 11, 447, 1979.
24. **Módis, L., Ádány, R., and Lakatos, I.**, Polarisationsoptische Analyse der menschlichen embryonalen Knorpelmatrix, *Acta Histochem.* (Jena), Suppl. 26, 305, 1982.
25. **Humason, G. L. and Lushbaugh, C. C.**, Sirius supra blue FGL-CF; superior to aniline blue in the combined elastin, reticulum and collagen stain, *Stain Technol.*, 44, 105, 1969.
26. **Rao, R. N., Fallman, P. M., Falls, D. G., and Meloan, S. N.**, A comparative study of PAS-phosphotungstic acid-diamine supra blue FGL and immunological reactions for type I collagen, *Histochemistry*, 91, 283, 1989.
27. **Piérard, G. E.**, Sirius red polarization method is useful to visualize the organization of connective tissues but not the molecular composition of their fibrous polymers, *Matrix*, 9, 68, 1989.
28. **Romhányi, G.**, Über die submikroskopischen Strukturen der elastischen Fasern, *Acta Morphol. Acad. Sci. Hung.*, 5, 311, 1955.

29. **Romhányi, G.**, Submicroscopic structure of elastic fibres as observed in the polarization microscope, *Nature,* 182, 930, 1958.
30. **Romhányi, G.**, On the submicroscopic structure of the elastic fibers of the bovine ligamentum nuchae as revealed by the polarization microscope, *Acta Morphol. Acad. Sci. Hung.*, 13, 397, 1964.
31. **Romhányi, G.**, Specific topo-optical reactions of connective tissue elements and their ultrastructural interpretation, *Connect. Tissue Res.*, 15, 13, 1986.
32. **Fischer, J.**, Ultrastructure of elastic fibers as shown by polarization optics after the topo-optical permanganate-bisulfute-toluidine blue (PBT) reaction, *Acta Histochem. (Jena),* 65, 87, 1979.
33. **Romhányi, G., Fischer, J., and Corless, J. M.**, Orientation of acidic polysaccharides and rhodopsin oligosaccharides in frog retinal rod outer segment, *Histochemistry,* 56, 65, 1978.
34. **Fischer, J.**, Optical studies on the molecular arrangement of RNA in tissues with a selective topo-optical reaction of RNA, *Histochemistry,* 59, 325, 1979.
35. **Malinin, G. I.**, Metachromatic staining of sodium bisulfite addition derivatives of glycogen, *J. Histochem. Cytochem,* 18, 834, 1970.
36. **Romhányi, G.**, Über die submikroskopische strukturelle Grundlage der metachromatischen Reaktion, *Acta Histochem. (Jena),* 15, 201, 1963.
37. **Jobst, K. and Kellermayer, M.**, Submicroscopic structure and dry weight of isolated thymus muclei following trypsin and salt treatment, *Acta Morphol. Acad. Sci. Hung.*, 15, 221, 1967.
38. **Kellermayer, M. and Jobst, K.**, Ion-dependent anisotropy of deoxyribonucleoprotein structures in tissue cultures, *Exp. Cell Res.*, 63, 204, 1970.
39. **Jobst, K. and Kellermayer, M.**, On the submicroscopic structure of chromosomes, *Acta Biol. Acad. Sci. Hung.*, 22, 25, 1971.
40. **Romhányi, G.**, Ultrastructure of biomembranes as shown by topo-optical reactions, *Acta Biol. Acad. Sci. Hung.*, 29, 311, 1978.
41. **Németh, Á.**, Topo-optical study of the pulmonary ciliary membrane surface, *Acta Morphol. Acad. Sci. Hung.*, 24, 247, 1976.
42. **Romhányi, G.**, Zur Frage der submikroskopischen Struktur des Amyloid, *Zentralbl. Allg. Pathol.*, 95, 130, 1956.
43. **Németh-Csóka, M.**, Untersuchungen über die Kollagenen Fasern. I. Über die submikroskopische Struktur der *in vitro* präzipitierten Kollagenen Fasern und die stabilisierende Rolle der saueren Mucopolysaccharide, *Acta Histochem. (Jena),* 8, 282, 1960.
44. **Németh-Csóka, M.**, Untersuchungen über die Kollagenfasern. II. Vergleichende chemische, polarisation- und elektronenmikroskopische Untersuchungen über die Kollagenfasern mit Chondroitinschwefelsäure als Grundsubstanz, *Acta Histochem. (Jena),* 12, 255, 1961.
45. **Scheuner, G. and Hantschel, H.**, Submikroskopische Untersuchungen an der Zonula ciliaris, gleichzeitig ein Beitrag zur polarisationsoptischen Bestimmung gerichtet angeordneter saurer Gruppen, *Histochemie,* 8, 9, 1967.
46. **Scheuner, G.**, Zum polarisationsoptisch-histochemischen Nachweis gerichtet angeordneter saurer Gruppen mit Akridinderivaten, *Acta Histochem. (Jena),* 31, 36, 1968.
47. **Gyeskó, A. and Sugár, J.**, Quantitative cytochemical investigations on NK/Ly mouse ascites tumor cells treated with degranol, *Virchows Arch. (Cell Pathol.),* 5, 60, 1970.
48. **Surján, L., Juhász, J., and Sebök, J.**, Lymphocyte activation and the increase of nuclear birefringence, *Acta Histochem. (Jena),* 64, 139, 1979.
49. **Wezeman, F. H. and Kuettner, K. E.**, Selective histochemical staining of cartilage matrix by rivanol, *Anat. Rec.*, 180, 481, 1974.
50. **Kawiak, J. and Kowalski, H.**, Evaluation of the method of rivanol precipitation of (^{35}S) sulphated mucopolysaccharides present in the material obtained from a small group of cells, *Experientia,* 20, 397, 1964.
51. **Németh-Csóka, M.**, A turbidometric method for the determination of chondroitin sulphuric acid by Rivanol, *Biochim. Biophys. Acta.* 50, 585, 1961.
52. **Farndale, R. W., Sayers, C. A., and Barrett, A. J.**, A direct spectrophotometric microassay for sulfated glycosaminoglycans in cartilage cultures, *Connect. Tissue Res.*, 9, 247, 1982.
53. **Templeton, D. M.**, The basis and applicability of the dimethylmethylene blue binding assay for sulfated glycosaminoglycans, *Connect. Tissue Res.*, 17, 23, 1988.
54. **Scheuner, G. and Flemming, G.**, Zum Einsatz von *N,N'*-Diaethyl-pseudoisocyaninchlorid für den polarisationsoptischen Nachweis von Polyanionen im histologischen Schnitt, *Acta Histochem. (Jena),* 64, 249, 1979.
55. **Makovitzky, J.**, Untersuchungen über die anisotrope 1:9-Dimethyl-methylenblau- und *N,N'*-Diaethylpseudoisocyaninchloridfarbung der Glykokalyx, *Histochemistry,* 59, 55, 1978.
56. **Scott, J. E. and Dorling, J.**, Differential staining of acid glycosaminoglycans (mucopolysaccharides) by alcian blue staining in salt solutions, *Histochemie,* 5, 221, 1965.

57. **Kent, G. N., Dodds, R. A., Klenerman, L., Watts, R. W. E., Bitensky, L., and Chayen, J.**, Changes in crystal size and orientation of acidic glycosaminoglycans at the fracture site in fractured necks of femur, *J. Bone Joint Surg. (Br.)*, 65, 189, 1983.
58. **Feris, B. D., Klenerman, L., Dodds, R. A., Bitensky, L., and Chayen, J.**, Altered organization of non-collagenous bone matrix in osteoporosis, *Bone*, 8, 285, 1987.
59. **Skerry, T. M., Bitensky, L., Chayen, J., and Lanyon, L. E.**, Loading-related reorientation of bone proteoglycan *in vivo*. Strain memory in bone tissue? *J. Orthop. Res.*, 6, 547, 1988.
60. **Dunham, J., Shackleton, D. R., Billingham, M. E. J., Bitensky, L., Chayen, J., and Muir, H. I.**, A reappraisal of the structure of normal canine articular cartilage, *J. Anat.*, 157, 89, 1988.
61. **Goldstein, D. J. and Horobin, R. W.**, Surface staining of cartilage by alcian blue, with reference to the role of microscopic dye aggregates in histological staining, *Histochem. J.*, 6, 175, 1974.
62. **Scott, J. E. and Harbinson, R. J.**, Periodate oxidation of acid polysaccharides. II. Rates of oxidation of uronic acids in polyuronides and acid mucopolysaccharides, *Histochemie*, 19, 155, 1969.
63. **Scott, J. E. and Dorling, J.**, Periodate oxidation of acid polysaccharides. III. A PAS method for chondroitin sulphates and other glycosamino-glycuronans, *Histochemie*, 19, 295, 1969.
64. **Kiviranta, I., Tammi, M., Jurvelin, J., Saamenen, A.-M., and Helminen, H. J.**, Demonstration of chondroitin sulphate and glycoproteins in articular cartilage matrix using periodic acid-Schiff (PAS) method, *Histochemistry*, 83, 303, 1985.
65. **Romhányi, G., Deák, G., and Fischer, J.**, Aldehyde bisulfite-toluidine blue (ABT) staining as a topo-optical reaction for demonstration of linear order of vicinal OH groups in biological structures, *Histochemistry*, 43, 333, 1975.
66. **Culling, C. F. A., Reid, P. E., Burton, J. D., and Dunn, W. L.**, A histochemical method of differentiating lower gastrointestinal tract mucin from other mucins in primary or metastatic tumors, *J. Clin. Pathol.*, 28, 656, 1975.
67. **Fischer, J.**, Demonstration of microorganisms in tissues by the ABT and KOH-ABT topo-optical reactions, *Acta Morphol. Acad. Sci. Hung.*, 24, 203, 1976.
68. **Fischer, J. and Emödy, L.**, Molecular order of carbohydrate components in cell walls of bacteria, fungi, and algae according to the topo-optical reaction of the vicinal OH groups, *Acta Microbiol. Acad. Sci. Hung.*, 23, 97, 1976.
69. **Cooper, J. H.**, Selective amyloid staining as a function of amyloid composition and structure. Histochemical analysis of the alkaline congo red, standardized toluidine blue, and iodine methods, *Lab. Invest.*, 31, 232, 1974.
70. **Puchtler, H. Sweat, F., and Levene, M.**, On the binding of congo red by amyloid, *J. Histochem. Cytochem.*, 10, 355, 1962.
71. **Romhányi, G.**, Selective differentiation between amyloid and connective tissue structures based on the collagen specific topo-optical staining reaction with congo red, *Virchows Arch. (A)*, 354, 209, 1971.
72. **Romhányi, G.**, Differences in ultrastructural organization of amyloid as revealed by sensitivity or resistance to induced proteolysis, *Virchows Arch. (A)*, 357, 29, 1972.
73. **Wolman, M.**, Amyloid, its nature and molecular structure. Comparison of a new toluidine blue polarized light method with traditional procedures, *Lab. Invest.*, 25, 104, 1971.
74. **Lillie, R. D. and Fullmer, H. M.**, *Histopathologic Technique and Practical Histochemistry*, 4th. ed., McGraw-Hill, A Blakiston Publication, New York, 1976, 287.
75. **Avasthi, P. S. and Koshy, V.**, The anionic matrix at the rat glomerular endothelial surface, *Anat. Rec.*, 220, 258, 1988.
76. **Pearse, A. G. E.**, *Histochemistry. Theoretical and Applied*, 4th ed., Vol. 1, Churchill-Livingstone, Edinburgh, 1980.
77. **Yamada, K. and Hoshino, M.**, Digestion with chondroitinases of acid mucopolysaccharides in rabbit cartilages as revealed by electron microscopy, *Histochem. J.*, 5, 195, 1973.
78. **Reeves, W. H., Kanwar, Y. S., and Farquhar, M. G.**, Assembly of the glomerular filtration surface. Differentiation of anionic sites in glomerular capillaries of newborn rat kidney, *J. Cell Biol.*, 85, 735, 1980.
79. **Kimura, A., Kawaguchi, T., Ono, T., Sakuma, A., Yokoya, Y., Kochi, H., and Nakamura, K.**, Cell surface heparan sulphate and adhesive property of sublines of rat ascites hepatoma AH7974, *J. Cell Sci.*, 90, 683, 1988.
80. **Yamada, K., Fujita, Y., and Shimizu, S.**, The effect of digestion with keratanase *(Pseudomonas sp.)* on certain histochemical reactions for glycosaminoglycans in cartilaginous and corneal tissues, *Histochem. J.*, 14, 897, 1982.
81. **Yamada, K.**, Effects of novel (Streptomyces) hyaluronidase digestion upon some mucosaccharide stainings of the cartilages and aortas in the rabbit and rat, *Histochemie*, 27, 277, 1971.
82. **Yamada, K.**, The effect of digestion with streptomyces hyaluronidase upon certain histochemical reactions of hyaluronic acid-containing tissues, *J. Histochem. Cytochem.*, 21, 794, 1973.
83. **Yamada, K. and Hirano, K.**, The histochemistry of hyaluronic acid-containing mucosubstances, *J. Histochem. Cytochem.*, 21, 469, 1973.

84. **Derby, M. A. and Pintar, J. E.,** The histochemical specificity of Streptomyces hyaluronidase and chondroitinase ABC, *Histochem. J.*, 10, 529, 1978.
85. **Ádány, R., Módis, L., and Zsindely, T.,** Sensitization of topo-optical reactions in showing tissue glycosaminoglycans of low local concentration, *Acta Histochem. (Jena)*, 67, 247, 1980.
86. **Kantor, T. G. and Schubert, M.,** A method for the desulfation of chondroitin sulfate, *J. Am. Chem. Soc.*, 79, 152, 1957.
87. **Spicer, S. S. and Lillie, R. D.,** Saponification as a means of selectively reversing the methylation blockade of tissue basophilia, *J. Histochem. Cytochem.*, 7, 123, 1959.
88. **Scudder, P., Tang, P. W., Hounsell, E. F., Lawson, A. M., Mehmet, H., and Feizi, T.,** Isolation and characterization of sulphated oligosaccharides released from bovine corneal keratan sulphate by the action of endo-β-galactosidase, *Eur. J. Biochem.*, 157, 365, 1986.
89. **Terner, J. Y.,** Histochemical alkylation: a study of methyliodide and its effect on tissues, *J. Histochem. Cytochem.*, 12, 504, 1964.
90. **Ádány, R., Módis, L., and Balogh, G.,** A critical re-evaluation of the anisotropic index used in polarization microscopy: a study on epiphyseal cartilage, *Acta Histochem. (Jena)*, 65, 243, 1979.
91. **Genchi, H. and Mori, K.,** Evaluation and feature extraction on automatic pattern recognition system. Denki-Tsushin Gakkai, Part I, 196 (in Japanese). Quoted in Tanaka, N., Ikeda, H., Ueno, T., Watanabe, S., Imasato, Y., and Kashida, R., Fundamental study of an automated cytoscreening system utilizing the pattern recognition system, in *The Automation of Uterine Cancer Cytology*, Wied, G. L., Bahr, G. F., and Bartels, P. H., Eds., Tutorials of Cytology, Chicago, 1976, 223.
92. **Bartels, P. H.,** Numerical evaluation of cytologic data. III. Selection of features for discrimination, *Anal. Quant. Cytol.*, 1, 153, 1979.
93. **Bodó, M., Szentirmay, Z., Sugár, J., Bak, M., and Svastits, E.,** Role of biometric data in evaluation of aspiration cytology in mammary diseases (in Hungarian), *Orvostudomány (Budapest)*, 32, 65, 1981.

INDEX

D

E

F

G

H

I

J

K

L

M

N

O